THE SUSTAINABLE URBAN DEVELOPMENT READER

The Sustainable Urban Development Reader brings together classic readings from a wide variety of sources to investigate how our cities and towns can become more sustainable.

Thirty-eight selections span issues such as land-use planning, urban design, transportation, ecological restoration, economic development, resource use, and equity planning. Part introductions outline the major themes, while the editors' introductions to the individual writings explain their interest and significance to wider debates. Additional parts present twenty-four case studies of real-world sustainable urban planning examples, sustainability planning exercises, and further reading.

Providing background in theory, practical application, and vision, in a clear, accessible format, *The Sustainable Urban Development Reader* is an essential resource for students, professionals, and indeed anyone interested in the future of urban environments.

Stephen M. Wheeler is Assistant Professor of Community and Regional Planning and Design at the University of New Mexico, and **Timothy Beatley** is Teresa Heinz Professor of Sustainable Communities, in the School of Architecture at the University of Virginia.

THE ROUTLEDGE URBAN READER SERIES

Series editors

Richard T. LeGates

Professor of Urban Studies, San Francisco State University

Frederic Stout

Lecturer in Urban Studies, Stanford University

The Routledge Urban Reader Series responds to the need for comprehensive coverage of the classic and essential texts that form the basis of intellectual work in the various academic disciplines and professional fields concerned with cities.

The readers focus on the key topics encountered by undergraduates, graduates and scholars in urban studies and allied fields. They discuss the contributions of major theoreticians and practitioners and other individuals, groups, and organizations that study the city or practise in a field that directly affects the city.

As well as drawing together the best of classic and contemporary writings on the city, each reader features extensive general, section and selection introductions prepared by the volume editors to place the selections in context, illustrate relations among topics, provide information on the author and point readers towards additional related bibliographic material.

Each reader will contain:

- Approximately thirty-six *selections* divided into approximately six sections. Almost all of the selections will be previously published works that have appeared as journal articles or portions of books.
- A *general introduction* describing the nature and purpose of the reader.
- Two- to three-page *section introductions* for each section of the reader to place the readings in context.
- A one-page *selection introduction* for each selection describing the author, the intellectual background of the selection, competing views of the subject matter of the selection and bibliographic references to other readings by the same author and other readings related to the topic.
- A plate section with twelve to fifteen plates and illustrations at the beginning of each section.
- An index.

The types of readers and forthcoming titles are as follows:

THE CITY READER

The City Reader: third edition – an interdisciplinary urban reader aimed at urban studies, urban planning, urban geography and urban sociology courses – will be the *anchor urban reader*. Routledge published a first edition of *The City Reader* in 1996 and a second edition in 2000. *The City Reader* has become one of the most widely used anthologies in urban studies, urban geography, urban sociology and urban planning courses in the world.

URBAN DISCIPLINARY READERS

The series will contain *urban disciplinary readers* organized around social science disciplines. The urban disciplinary readers will include both classic writings and recent, cutting-edge contributions to the respective disciplines. They will be lively, high-quality, competitively priced readers which faculty can adopt as course texts and which will also appeal to a wider audience.

TOPICAL URBAN ANTHOLOGIES

The urban series will also include *topical urban readers* intended both as primary and supplemental course texts and for the trade and professional market.

INTERDISCIPLINARY ANCHOR TITLE

The City Reader: third edition
Richard T. LeGates and Frederic Stout (eds)

URBAN DISCIPLINARY READERS

The Urban Geography Reader
Nick Fyfe and Judith Kenny (eds)

The Urban Sociology Reader
Jan Lin and Christopher Mele (eds)

The Urban Politics Reader
Elizabeth Strom and John Mollenkopf (eds)

The Urban and Regional Planning Reader
Eugenie Birch (ed.)

TOPICAL URBAN READERS

The City Cultures Reader: second edition
Malcolm Miles, Tim Hall with Iain Borden (eds)

The Cybercities Reader
Stephen Graham (ed.)

The Sustainable Urban Development Reader
Stephen M. Wheeler and Timothy Beatley (eds)

The Global Cities Reader
Neil Brenner and Roger Keil (eds)

For further information on The Routledge Urban Reader Series
please visit our website:

www.geographyareana.com/geographyareana/urbanreaderseries

or contact:

Andrew Mould
Routledge
11 New Fetter Lane
London EC4P 4EE
England
andrew.mould@routledge.co.uk

Richard T. LeGates
Urban Studies Program
San Francisco State University
1600 Holloway Avenue
San Francisco, California 94132
(415) 338-2875
dlegates@sfsu.edu

Frederic Stout
Urban Studies Program
Stanford University
Stanford, California 94305-6050
(650) 725-6321
fstout@stanford.edu

The Sustainable Urban Development Reader

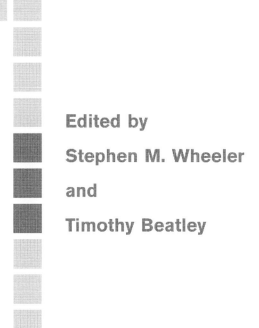

Edited by

Stephen M. Wheeler

and

Timothy Beatley

Routledge
Taylor & Francis Group

LONDON AND NEW YORK

First published 2004
by Routledge
11 New Fetter Lane, London EC4P 4EE

Simultaneously published in the USA and Canada
by Routledge
29 West 35th Street, New York, NY 10001

Routledge is an imprint of the Taylor & Francis Group

© 2004 Selection and editorial matter Stephen M. Wheeler and Timothy Beatley

Typeset in Amasis MT Lt and Akzidenz Grotesk by Graphicraft Limited, Hong Kong
Printed and bound in Great Britain by Bell & Bain Ltd, Glasgow

British Library Cataloguing in Publication Data
A catalogue record for this book is available from the British Library

Library of Congress Cataloging in Publication Data
The sustainable urban development reader / edited by Stephen M. Wheeler
and Timothy Beatley.
    p. cm. — (The Routledge urban reader series)
Includes bibliographical references and index.
1. City planning.   2. Community development, Urban.   3. Sustainable development.
I. Wheeler, Stephen (Stephen Maxwell)   II. Beatley, Timothy, 1957–   III. Series.
HT166.S9135 2004
307.1′216—dc22   2003015200

ISBN 0-415-31186-1 (hbk)
ISBN 0-415-31187-X (pbk)

To Mimi, and to the late Dave Brower, whose example and encouragement have been invaluable (SW)

To Anneke and Carolena (TB)

Contents

Acknowledgments   xiv

Introduction   1

PART 1 ORIGINS OF THE SUSTAINABILITY CONCEPT   5

Introduction   7

"The Three Magnets"   11
Ebenezer Howard

"Cities and the Crisis of Civilization"   15
Lewis Mumford

"The Land Ethic"   20
Aldo Leopold

"Orthodox Planning and The North End"   30
Jane Jacobs

"Plight and Prospect"   35
Ian L. McHarg

"The Development of Underdevelopment"   38
Andre Gunder Frank

"Perspectives, Problems, and Models"   42
Donella H. Meadows, Dennis L. Meadows, Jörgen Randers,
and William W. Behrens III

"The Steady-State Economy"   47
Herman E. Daly

"Towards Sustainable Development"   53
World Commission on Environment and Development
(The Brundtland Commission)

"The Rio Declaration on Environment and Development", "Introduction
to Chapter 7" from *Agenda 21* (1992), and the "Istanbul Declaration
on Human Settlements"   58
United Nations

PART 2 DIMENSIONS OF URBAN SUSTAINABILITY   **67**

Introduction   69

Land use and urban design   **71**

"The Next American Metropolis"   73
Peter Calthorpe

"Outdoor Space and Outdoor Activities"   81
Jan Gehl

Transportation   **87**

"Transit and the Metropolis: Finding Harmony"   89
Robert Cervero

"Traffic Calming"   97
Peter Newman and Jeffrey Kenworthy

"Bicycling Renaissance in North America?"   104
John Pucher, Charles Komanoff, and Paul Shimek

Urban ecology and restoration   **111**

"City and Nature"   113
Anne Whiston Spirn

"Land Development and Endangered Species: Emerging Conflicts"   116
Timothy Beatley

"What Is Restoration?"   120
Ann L. Riley

Energy and materials use   **123**

"The Metabolism of Cities"   125
Herbert Girardet

"Waste as a Resource"   133
John Tillman Lyle

Environmental justice and social equity   **141**

"People-of-Color Environmentalism"   143
Robert Bullard

"Domesticating Urban Space"   150
Dolores Hayden

Economic development   **157**

"The Economic System and Natural Environments"   159
David Pearce and Edward B. Barbier

"Natural Capitalism"   162
Paul Hawken

"Import Replacement"   171
Michael Shuman

Green architecture and building   **179**

"Design, Ecology, Ethics and the Making of Things"   181
William McDonough

"Principles of Green Architecture"   188
Brenda and Robert Vale

"Sustainability and Building Codes"   193
David Eisenberg and Peter Yost

PART 3 TOOLS FOR SUSTAINABILITY PLANNING   **199**

Introduction   201

"Urban Sustainability Reporting"   203
Virginia W. Maclaren

"What *Is* an Ecological Footprint?"   211
Mathis Wackernagel and William Rees

"Seeing Change"   220
Allan B. Jacobs

"A Progressive Politics of Meaning"   225
Michael Lerner

PART 4 SUSTAINABLE URBAN DEVELOPMENT INTERNATIONALLY 233

Introduction 235

"Urban Planning in Curitiba" 237
Jonas Rabinovitch and Josef Leitman

"Planning for Sustainability in European Cities: A Review of
Practice in Leading Cities" 249
Timothy Beatley

"Collective Action Toward a Sustainable City: Citizens' Movements
and Environmental Politics in Taipei" 259
Hsin-Huang Michael Hsiao and Hwa-Jen Liu

PART 5 VISIONS OF SUSTAINABLE COMMUNITY 275

Introduction 277

"The Town–Country Magnet" 279
Ebenezer Howard

"The Streets of Ecotopia's Capital" and "Car-Less Living
in Ecotopia's New Towns" 282
Ernest Callenbach

"Description of Abbenay" 288
Ursula K. Le Guin

PART 6 CASE STUDIES OF URBAN SUSTAINABILITY 293

Introduction 295

Urban sustainability at the building and site scale 297

Commerzbank Headquarters, Frankfurt 299
Menara Mesiniaga bio-climatic skyscraper, Kuala Lumpur, Malaysia 299
Adelaide EcoVillage (Christie Walk) 300
Condé Nast building (4 Times Square), New York 300

Urban sustainability at the neighborhood or district scale 301

Kronsberg Ecological District, Hannover, Germany 303
Beddington Zero Energy Development (BedZED), London 303
Greenwich Millennium Village, London 304
Nieuwland (solar suburb), Amersfoort, The Netherlands 304
Village Homes, Davis, California 304
Los Angeles Eco-Village 305

Civano and Armory Park, Tucson, Arizona 305
EcoCity Cleveland and Cleveland EcoVillage 306

Urban sustainability at the city and regional scale 307

Vancouver, British Columbia 309
Bogotá, Colombia 309
Gaviotas, Colombia 310
Auroville, India 310
IBA Emscher Park, Germany 311
London, England 312
Chicago, Illinois 312
Austin, Texas 313
Portland, Oregon 313
Burlington, Vermont 314
San José, California 315
Santa Monica, California 316

Plate Section

PART 7 SUSTAINABILITY PLANNING EXERCISES 317

Introduction 319

Cognitive Mapping Exercise 320
Future Visions Exercise 320
Definitions of Sustainable Development 321
Analyzing the Three Es in an Urban Planning Debate 322
Sustainability Indicators Exercise 324
Personal Ecological Footprints/Household Sustainability Audit 324
Firsthand Analysis of Urban Environments 325
Regional Vision Exercise 326
Economic Development Exercise 327
Creek Mapping Exercise 329
Neighborhood Planning Exercise 330
An Ecological Site Plan 331
International Development Exercise 332

Further reading 335
Illustration credits 338
Copyright information 340
Index 343

Acknowledgments

We would like to thank the many people who have made this volume possible, above all the contributors, whose work continues to inspire us.

Series editor Richard T. LeGates first approached us with the idea of doing such a reader, and has been a wonderful source of support and guidance during the process. Routledge's *The City Reader*, edited by LeGates and Frederic Stout, has been an excellent model and high standard for us to follow.

David Orr, Marcia McNally, Wicak Sarosa, Keiro Hattori, Kang-Li Wu, Herbert Girardet, Mark Roseland, Richard LeGates, and four anonymous reviewers supplied very helpful comments on the content and structure of this book. Thanks to all. Andrew Mould at Routledge has played a central role in making this book possible, while editorial assistant Melanie Attridge, production editor Nicola Cooper, and copy editor Ann King have has skillfully guided the manuscript through production.

Over the years our students have been a great source of feedback on particular readings, and have challenged us to find material that does a good job of communicating sustainability concepts to those not yet familiar with the field. In addition, Stephen Wheeler would like to thank his wife, Mimi, for her constant love and support as well as astute editorial comments during the process of preparing this book. Tim Beatley would like to thank, as always, his wife Anneke for her patience and love, and his daughter Carolena for her great energy and spirit (that keep him going).

INTRODUCTION

What will our cities and suburban landscapes be like in fifty years' time? In a hundred? How can we plan and develop communities that will meet long-term human and environmental needs? The concept of sustainable urban development provides a way for citizens, planners, and policymakers worldwide to explore such questions. Increasingly, sustainability is becoming a goal of official city plans, and is also informing architecture, landscape architecture, environmental planning, and many other disciplines.

This book aims to provide readers with a wide, thought-provoking selection of writings on this timely subject. We present thirty-six classic readings related to sustainable community planning, drawn from books, academic journals, and general-interest magazines. Extensive introductions put each reading in context, and twenty-four case studies of sustainable urban development initiatives help illustrate the range of projects now underway. Since many of us learn most "by doing," a final section of exercises related to sustainability planning helps individuals, students, or community groups work out their own detailed understandings of sustainable community planning.

Although many of these writings are from North America, we have included pieces from Europe, Asia, Latin America, and elsewhere, and consider many of the urban sustainability challenges addressed here to be universal. Cities and towns worldwide are facing similar problems of growing automobile use, suburban sprawl, pollution, profligate use of natural resources, rising inequities, and loss of indigenous landscapes and ecosystems. Communities in most parts of the world are also now confronted by a global economic system that frequently undercuts local traditions, businesses, community, environment, and sense of place. Although the context of urban development varies considerably from country to country, many sustainability strategies will be the same (e.g. seeking to co-ordinate transportation and land use, restore urban ecosystems, or design the public realm so as to be friendly for women, children, and the elderly). Moreover, every society these days can learn from innovations in other places. Hence we have tried to keep our perspective as global as possible.

Part 1 highlights classic historic writings that have paved the way for more recent discussions of urban sustainability. Writers such as Ebenezer Howard, Lewis Mumford, and Aldo Leopold raised questions in the early twentieth century about the nature of the industrial city and the fundamental relationship between human development and the natural world. Later writers such as Jane Jacobs, Ian McHarg, Herman Daly, Andre Gunder Frank, and the *Limits to Growth* team spurred a re-evaluation of unsustainable development practices during that crucial period of the 1960s and 1970s when many of the ecological and social implications of global development were first widely understood. Subsequent United Nations conferences and commissions – especially the mid-1980s Brundtland Commission and the 1992 Earth Summit – also helped to draw attention to the need for a new development paradigm, issuing declarations such as *Agenda 21* that were influential in stimulating local planning initiatives in many parts of the world. These and other influences have laid the groundwork for more recent sustainability planning. Understanding particular historical themes – which continue to be echoed today – is important in order to understand how cities and towns can become more sustainable in the future.

After this look at the origins of the sustainability concept, we survey classic writings in a number of issue areas important to urban development (Part 2). Through selections from the American New Urbanist planner Peter Calthorpe and the Danish designer Jan Gehl, the chapter on land use and urban design addresses questions such as: How can more compact and walkable neighborhoods be designed? What forms and densities of housing are appropriate for transit-oriented development? and How can streets and public spaces work better for people? Our introduction to this chapter, as with introductions throughout the book, points interested readers toward additional writings and resources, in this case especially those connected with recent movements such as the New Urbanism and Smart Growth.

Transportation systems are fundamental in shaping the land use and physical form of urban areas, as well as determining much about the livability of our communities. Our chapter on this subject starts with a selection by Robert Cervero, who describes ways that urban regions might make progress towards his vision of the "transit metropolis." Cervero also outlines the huge variety of public transportation modes that might play a role in reducing automobile use. Australian planners Peter Newman and Jeffrey Kenworthy then describe the history of the international movement known as "traffic calming," and explore the techniques, goals, and results of this approach. John Pucher, Charles Komanoff, and Paul Shimek conclude the chapter with a discussion of ways to make cities and towns more bicycle-friendly, especially in North America where the design of communities often works against walking and cycling.

Our discussion of urban ecology and greening begins with a selection from Ann Whiston Spirn's classic 1984 book *The Granite Garden*, exploring the profound interrelationships between nature and the city. Spirn was one of the first to treat cities as a natural landscape, and helped pioneer a new understanding of how natural forces shape even the most urban of settings. In a selection from his book *Habitat Conservation Planning*, Tim Beatley then discusses the growing worldwide concern about biodiversity, focusing on habitat conservation plans as a key mechanism to preserve ecosystems in or near urban areas. Finally, stream restoration pioneer Ann Riley investigates the concept of restoration as it applies to urban watershed features. Whereas "conservation" was the watchword of previous generations of environmentalists, "restoration" has become the mantra of many recent activists, and offers exciting possibilities for constructive, hands-on action in cities and towns everywhere.

One of the most unsustainable dimensions of current urban development has to do with energy and materials use, and the waste, pollution, and greenhouse gas emissions that usually result from this consumption. In our chapter on resource use, we include a selection from British sustainability pioneer Herbert Girardet analyzing the flow of raw materials through the urban system. To dramatize his points, Girardet calculates the metabolism of Greater London in terms of energy and resources consumed. The late regenerative design pioneer John Tillman Lyle then examines the unglamorous but fascinating topic of urban waste, and analyzes potential methods of ecological sewage treatment. Lyle profiles the ecological sewage treatment marsh in the city of Arcata, California, which processes waste for 15,000 people while also creating important wildlife habitat and an attractive and well-used recreational trail system.

The rising tide of inequity in many societies – in which some groups within society prosper while others suffer – is profoundly rooted in current patterns of urban development. Two selections serve to highlight some urban dimensions of equity issues. In an excerpt from his classic 1990 book *Dumping in Dixie*, African-American sociologist Robert Bullard describes the emergence and nature of the environmental justice movement. Bullard calls for a new environmentalism that takes into account equity impacts on particular urban communities, and that fights the institutional forces perpetuating environmental injustice. In a selection from her 1984 book *Redesigning the American Dream*, Dolores Hayden, one of the foremost feminist critics of urban design and planning, examines how women have been excluded from or made to feel uncomfortable within urban environments. Hayden calls for "small, commonsense improvements in urban design" as well as larger changes in the ways society views gender roles, nurturing, and the split between private and public life.

One of the most fundamental challenges to sustainable urban development is the need to redirect our economic engine into paths that are restorative rather than exploitative; for example, that are not reliant on long-distance trade, ever-growing consumption of material products, and replacement of local businesses by local branches of multinational corporations. In our chapter on economic development,

British economists David Pearce and Edward B. Barbier first of all describe the basic failure of current market economics to take into account many aspects of the world around us, especially natural environments. In the following selection California businessman Paul Hawken outlines his concept of "natural capitalism," in which the energies of capitalist markets are harnessed for constructive rather than destructive purposes. Michael Shuman then follows by describing a vision of community self-reliance and import replacement, a path that runs counter to the current emphasis on global free trade, but that can potentially offer many environmental and social benefits for local communities.

The buildings in which we live and work are one of the most basic features of urban environments, and so we conclude our survey of sustainable city dimensions with a discussion of green architecture and building practices. In a sermon given at the Cathedral of St. John the Divine in New York City in 1993, ecological architect William McDonough eloquently describes his philosophy of placing building design within the context of all of nature. McDonough includes the needs of human users and surrounding communities within this context, and looks to vernacular local traditions for clues as to how to design for particular climates and cultures. Long-time British pioneers Brenda and Robert Vale then outline what they see as basic principles of a "green" approach to architecture, likewise stressing that the wisdom of historic cultures can be a powerful guide to improve the sustainability of modern building practices. In a different vein, David Eisenberg and Peter Yost close the chapter by analyzing how modern building codes constrain green building practices and how these regulations might be revised so as to make ecological design more possible. These authors succinctly describe the nature, emergence, and limitations of building codes, and urge the environmental design community to become more involved in revising these basic frameworks within which urban construction takes place.

Although many traditional urban planning techniques, such as the preparation of general plans and zoning codes, may be adapted to promote sustainability goals, certain new or rediscovered planning tools can be particularly useful for sustainable urban development. Part 3, "Tools for sustainability planning", investigates the subject of sustainability indicators through Virginia Maclaren's analysis of these in the *Journal of the American Planning Association*. We next delve into the topic of ecological footprint analysis through a piece by two of the originators of this concept, William Rees and Mathis Wackernagel, from their 1996 book *Our Ecological Footprint*. In a selection from his classic 1985 volume *Looking at Cities*, Allan Jacobs provides a helpful guide to that most basic and essential of urban analysis techniques, the process of simply observing the city. Too often ignored by planners holed up in rooms full of computers, skillful firsthand analysis is essential in order to analyze how urban places function and to see their current handicaps and future possibilities. Finally, Michael Lerner provides an inspiring description of a "politics of meaning" which might become one of the most fundamental tools for changing the status quo. Lerner calls upon professionals and politicians to re-evaluate the spiritual foundation of their work, and to develop a new commitment to meaningful collective challenges such as the task of creating more sustainable and livable communities.

Part 4 examines sustainable urban development efforts internationally, with the aim of giving readers a taste of the wide variety of opportunities and challenges facing cities in different parts of the world. One of the most celebrated examples of innovative urban planning is Curitiba, a city of 1.6 million in southern Brazil. Described here in an article by planners Jonas Rabinovitch and Josef Leitman reprinted from *Scientific American*, Curitiba has reshaped its physical form and transportation network over more than four decades, and has also been on the cutting edge of creative social planning. Tim Beatley then updates us on recent European sustainable community initiatives, profiling projects and areas of current work in The Netherlands, Germany, England, Denmark, and other nations. Looking at a different hemisphere, Hsin-Huang Michael Hsiao and Hwa-Jen Liu provide an excellent analysis of the dynamics of urban environmentalism in Taipei. They highlight in particular the tensions between the environmental interests of relatively well-off urban classes and the basic survival needs of the truly poor who crowd many cities in the developing world.

In Part 5 we turn to the subject of vision, through utopian descriptions of more sustainable communities. Such visions have been important historically in calling people's attention to the need for change and to specific alternative philosophies. These visions start with Ebenezer Howard's classic turn-of-the-century Garden City ideal, which laid the groundwork for many actual twentieth-century planned communities. We then move to Ernest Callenbach's *Ecotopia*, which inspired environmental activists in the 1970s and, like all ecological utopias, has yet to be realized. Our vision section concludes with Ursula K. Le Guin's science fiction world of Annares in *The Dispossessed*, specifically her description of the city of Abbaney. Le Guin portrays an even more radical social transformation far into the future, when a secessionist movement seeks to build a society based on cooperation, equity, and modest consumption, in contrast to its corrupt capitalist home world.

Part 6 of the Reader provides twenty-four case studies of innovative sustainable urban development practices. These take place on several different scales: that of the individual building or site, that of the neighborhood or district, and that of the city or region. In Part 7 we conclude with a selection of exercises that readers may find interesting to complete either individually, in groups, or through classes. These exercises have been developed by one of us (Wheeler) in conjunction with courses at the University of California at Berkeley, but may be adapted for many other types of group or situation.

In this Reader we have sought selections that are classic, readable, diverse, and if possible relatively specific in their recommendations. We hope that readers of many types will find this book useful – students, academics, planning professionals, and architects certainly, but also environmental activists, community leaders, and urban and suburban residents of all sorts. Each of us, after all, is confronted on a daily basis with the problems resulting from current modes of community development. Whether the issue is traffic congestion, lack of parks and open space, unfriendly streets and public spaces, poor schools, lack of decent-paying, meaningful work, or the frequent absence of community, finding solutions often depends on an understanding of the urban systems around us – both ways they have arisen in the past, and ways they can be improved in the future. This book aims to provide a foundation for that understanding.

PART ONE

Origins of the sustainability concept

INTRODUCTION TO PART ONE

The term "sustainable development" appears to have been used for the first time in the 1972 book *Limits to Growth*, in a passage excerpted later in this section, and has been applied widely to urban planning and architecture only since the early 1990s. However, concerns about the unsustainability of modern urban development patterns have a much longer history. To put our future efforts into perspective, it is important to be aware of this history and of the various themes that have shaped debates about sustainability.

At least as far back as the early nineteenth century, many commentators in Britain, continental Europe, and the United States were worried about the rapid growth of industrial cities. Urban expansion was a subtext underlying Henry David Thoreau's writings at Walden Pond in the 1840s. Thoreau's retreat was in large part a flight from the pace and pressures of urban life, but, new rail lines emanating from Boston plus a growing suburban population were encroaching on the pond itself. In 1840s England, the deplorable social conditions of the working class in Manchester and the increasing spatial segregation between the suburban estates of wealthy mill owners and the urban tenements of their workers helped motivate the writings of Frederick Engels. At the same time Romantic poets such as Keats, Wordsworth, and Shelley extolled the virtues of nature in a reaction against industrial society, while novelists ranging from Charles Dickens beginning in the 1830s to D.H. Lawrence in the 1920s described the horrors of industrial cities and towns, and the efforts of individuals to surmount or come to terms with these circumstances.

Cities of a million or more residents were virtually unknown before 1800, when London achieved this level of population (ancient Rome may have been home to a similar number). But coal-fired industrial factories drew workers from the countryside, while factors such as overpopulation, the privatization of formerly commonly held rural land, and increasingly centralized rural landownership pushed country dwellers away from their traditional communities. For the first time in the middle and late nineteenth century large numbers of people lived in crowded urban environments far from the countryside, and new technological advances such as the streetcar, the railroad, macadam road paving, modern plumbing, and electric lights helped distance people from the natural world. These forerunners of late twentieth-century megacities experienced enormous problems related to public health, sanitation, residential overcrowding, and nonexistent infrastructure. Deforestation of the countryside and pollution of air and water also reached new, and in some cases still unmatched, heights. Not surprisingly, many observers felt that the balance between human beings and the natural world had been tipped too far in one direction.

Late nineteenth- and early twentieth-century social reformers sought to draw attention to the deterioration of urban conditions and the need for alternative living environments. One of the most influential of these writers was Ebenezer Howard, a court stenographer whose slim book on "garden cities," first published in 1898, inspired generations of urban planners and designers. Howard, Scottish visionary Patrick Geddes, and their American follower Lewis Mumford saw the extreme overcrowding of early industrial cities – with its accompanying problems of sanitation, services, pollution, and public health – as the main problem to be addressed. In response, they called for a new balance between city and country in which populations were decentralized into carefully planned new communities in the countryside. While this idea had many merits, these authors wrote before automobile use became widespread or its implications understood, and before the huge wave of twentieth-century suburbanization turned Howard's "garden city" idea into much-simplified "garden suburbs" and created a whole new set of development problems in the process. Yet these early writers did much to focus public attention on the unsustainability of urban development trends at that time, the inability of private sector forces to deal with these problems, and the need for thoughtful planning of better alternatives.

The professions of landscape architecture and city planning emerged largely in reaction to the rapid nineteenth-century expansion of industrial cities. The former focused in large part on providing picturesque parks and living environments to urban residents, and in the process helped lay the aesthetic groundwork for twentieth-century suburbia. The latter sought to ensure the forms of infrastructure, housing, land use, and transportation that were viewed as necessary for orderly urban growth. However, in their response to nineteenth-century problems, both professions inadvertently established the conditions for another set of sustainability problems in the twentieth century, namely those related to low-density suburban sprawl.

So one main planning theme that emerged in the nineteenth century – and that sustainability-oriented writers have returned to ever since – was the balance between city and nature. Another theme, much less acted upon, had to do with the challenge of promoting equity. From oppressive working conditions within factories to the mile after mile of dreary tenements and working-class suburbs that were constructed in the early industrial era, the environment in which working individuals lived was often grim, unhealthy, and unjust. Politicians eventually enacted some reforms (for example, in the form of housing codes ensuring adequate light and air), but many other inequities continued. In the post-World War II period, and especially in the 1960s and 1970s, a growing number of writers criticized the rifts between rich and poor, between different ethnic or racial groups, and between how men and women are affected by urban environments. Many critics began to realize that many twentieth-century development practices, both within cities and worldwide, were worsening inequities rather than improving them. The need for a more equitable society became a second cornerstone of sustainable development, one that is more difficult to address than environmental clean-up because it challenges so directly the structure of wealth and power within nations.

A third recurrent theme has to do with the notion of economic growth and the limitations of economics in regulating human and natural systems. In the mid-nineteenth century, British economist and social critic John Stuart Mill first raised the notion that a steady-state economy might be desirable, as opposed to one based on endlessly growing production and consumption. This theme was taken up a century later by Kenneth Boulding, E.F. Shumacher, and Herman Daly, all economists who considered whether existing concepts of economic development were compatible with the notion of a limited planet. Daly in particular further developed the concept of steady-state economics, and Part 1 includes one of his classic essays on the topic. The chapter on economic development in Part 2 explores other economic implications of sustainable urban development.

Various debates around these three concerns – environment, equity, and economy, frequently referred to as the "three e's" of sustainable development – have thus been gestating for a century or more. Sustainability advocates have sought ways to maximize all three valuesets at once, rather than playing them off against one another as more traditional development strategies have often done. But the process is not easy, and is likely to require a closer look at each of these areas.

The concept of "sustained" development itself emerged most directly from the field of natural resource management. In the late nineteenth century Germany faced severe problems with overcutting of forests, and developed sustained-yield forest management techniques to compensate. Americans such as Gifford Pinchot learned these approaches at continental forestry schools and imported them to the USA, which despite its vast natural resource holdings was beginning to confront the notion of limits as well. The concept of managing ecosystems for sustained resource yield was quickly applied to wildlife species and fisheries as well to forestry. This "conservationist" perspective pioneered in the late nineteenth century – based on a view of humans as apart from nature and managing natural resources for their own use – is frequently contrasted with the "preservationist" perspective advocated by Sierra Club founder John Muir and others at the same time. In the latter view, nature has intrinsic value and should be protected for its own sake. Both perspectives have played a role in recent sustainable development discourses. Aldo Leopold, whose pivotal essay on "the land ethic" is included in Part 1, was central between these two camps. Although his career began within the conservationist tradition, in later life he came to see humans as part of a larger organic whole, and so helped lay the groundwork for more radical environmental movements such as "deep ecology." His assertion of a profound human responsibility to care for and heal natural systems is an important philosophy behind many sustainable community initiatives.

Selections from the pivotal reports of the *Limits to Growth* research team, the Brundtland Commission, and the 1992 Rio Earth Summit round out this section. These classic documents have all been extremely influential internationally in fueling sustainable development activity, though each is open to criticism on various grounds. Unfortunately there is no single, universally acknowledged manifesto that by itself sets out a sustainable *urban* development agenda. The 1996 United Nations Habitat II Conference, the "City Summit," sought to produce such a document, but the so-called Habitat Declaration has not attracted a wide following. However, other declarations of sustainable development principles have been put forth by architects, urban designers, and activists, and consensus is indeed emerging on many directions for sustainable urban development.

O
N
E

"The Three Magnets"

from the "Author's Introduction" to
Garden Cities of To-morrow (1898)

Ebenezer Howard

Editors' Introduction

Perhaps the single most influential and visionary book in the history of urban planning has been Ebenezer Howard's 1898 volume originally entitled simply *To-morrow*, and four years later reissued as *Garden Cities of To-morrow*. In a more detailed fashion than had ever been attempted before, Howard outlined a strategy for addressing the problems of the industrial city, one that attempts to balance city and country in what we might view today as a sustainable fashion.

To be sure, previous visionaries had suggested or even built new towns outside of cities. Scottish mill owner Robert Owen, for example, had constructed the town of New Lanark in 1800–10 near Glasgow for his workers, British soap manufacturer William Lever had built Port Sunlight in 1888 near Liverpool to house *his* workers, chocolate manufacturer George Cadbury had created Bournville near Birmingham in the 1880s, and American railroad magnate George Pullman had developed the town of Pullman outside Chicago for his employees at about the same time. Meanwhile, French philosophers Pierre-Joseph Proudhon and Charles Fourier as well as the Russian Peter Kropotkin had suggested principles for utopian new communities. Catalán engineer Ildefons Cerdá had laid out a large new extension to Barcelona in 1859, and had authored a pioneering book of urban planning philosophy in 1863, calling for a holistic, integrated approach to urbanization. But Howard's vision of systematically deconcentrating the population of an industrial city such as London into a ring of carefully organized garden cities surrounded by countryside and connected by railroads spoke powerfully to the needs of the time.

Later urbanists including Raymond Unwin, John Nolen, Lewis Mumford, Patrick Abercrombie, Ian McHarg, and Peter Calthorpe would seek different implementations of this basic idea. Two English garden cities were actually built in the early twentieth century, namely Letchworth and Welwyn, and the concept inspired the British New Town program that constructed eleven satellite cities around London between the 1940s and the 1960s. Swedish new towns such as Vällingby and Farsta, Dutch new towns such as Houten, and German new communities near Frankfurt have been built following many of the same principles. For its part, the US government sponsored three garden cities in the 1930s – Greenbelt, Maryland, Greenhills, Ohio, and Greendale, Wisconsin – while private developers built a handful of new towns along the garden city model, including Radburn, New Jersey (of which only one neighborhood was completed), Baldwin Hills Village in Los Angeles, and (much later) Reston, Virginia.

A court stenographer by profession, Howard exemplifies how some of the most revolutionary ideas in city planning have come from concerned citizens rather than from professional planners or architects. Jane Jacobs and Lewis Mumford – both writers rather than planners – also fall into this category. Howard's style was cautious, pragmatic, and designed to appear reasonable to the average citizen. He quoted extensively from

Personal investment

leading authorities of the day, provided conceptual graphics, and included financial information attempting to show how garden cities could be developed economically. His "three magnets" diagram (Figure 1) was a simple but effective metaphor to get readers to see that a new concept of urban development was needed, one that balanced city and country.

Howard's search for a balance between city and country life is still central to the task of creating more sustainable communities, but the emphasis has shifted. Instead of the extremely dense nineteenth-century city with a frequent shortage of decent housing, clean water, and basic sanitation, we now have relatively low-density, automobile-dependent suburbs with a much higher quality of housing and infrastructure, but with many other problems. So the question now as at the turn of the nineteenth century remains how to rethink this balance, perhaps creating new forms of garden city that avoid the problems of both overcrowded industrial cities and of low-density suburban sprawl.

There is, however, a question in regard to which no one can scarcely find any difference of opinion. It is wellnigh universally agreed by men of all parties, not only in England, but all over Europe and America and our colonies, that it is deeply to be deplored that people should continue to stream into the already over-crowded cities, and should thus further deplete the country districts.

Lord Rosebery, speaking some years ago as Chairman of the London County Council, dwelt with very special emphasis on this point:

> There is no thought of pride associated in my mind with the idea of London. I am always haunted by the awfulness of London: by the great appalling fact of these millions cast down, as it would appear by hazard, on the banks of this noble stream, working each in their own grove and their own cell, without regard or knowledge of each other, without heeding each other, without having the slightest idea how the other lives – the heedless casualty of unnumbered thousands of men. Sixty years ago a great Englishman, Cobbett, called it a wen. If it was a wen then, what is it now? A tumour, an elephantiasis sucking into its gorged system half the life and the blood and the bone of the rural districts. (March 1891)

Sir John Gorst points out the evil, and suggests the remedy:

> If they wanted a permanent remedy of the evil they must remove the cause; they must back the tide, and stop the migration of the people into the towns, and get the people back to the land. The interest and the safety of the towns themselves were involved in the solution of the problem. (*Daily Chronicle*, 6 November 1891)

Dean Farrar says:

> We are becoming a land of great cities. Villages are stationary or receding; cities are enormously increasing. And if it be true that great cities tend more and more to become the graves of the physique of our race, can we wonder at it when we see the houses so foul, so squalid, so ill-drained, so vitiated by neglect and dirt?

> . . . All, then, are agreed on the pressing nature of this problem, all are bent on its solution, and though it would doubtless be quite Utopian to expect a similar agreement as to the value of any remedy that may be proposed, it is at least of immense importance that, on a subject thus universally regarded as of supreme importance, we have such a consensus of opinion at the outset. . . .

Whatever may have been the causes which have operated in the past, and are operating now, to draw the people into the cities, those causes may all be summed up as 'attractions'; and it is obvious, therefore, that no remedy can possibly be effective which will not present to the people, or at least to considerable portions of them, greater 'attractions' that our cities now possess, so that the force of the old 'attractions' shall be overcome by the force of new 'attractions' which are to be created. Each city may be regarded as a magnet,

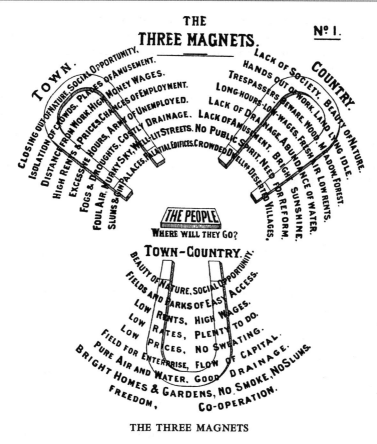

Figure 1. Ebenezer Howard's diagram of the "Three Magnets" (1898).

each person as a needle; and, so viewed, it is at once seen that nothing short of the discovery of a method for constructing magnets of yet greater power than our cities possess can be effective for redistributing the population in a spontaneous and healthy manner.

So presented, the problem may appear at first sight to be difficult, if not impossible, of solution. 'What', some may be disposed to ask, 'can possibly be done to make the country more attractive to a workaday people than the town – to make wages, or at least the standard of physical comfort, higher in the country than in the town; to secure in the country equal possibilities of social intercourse, and to make the prospects of advancement for the average man or woman equal, not to say superior, to those enjoyed in our large cities?' . . .

There are in reality not only, as is so constantly assumed, two alternatives – town life and country life – but a third alternative, in which all the advantages of the most energetic and active town life, with all the beauty and delight of the country, may be secured in perfect combination; and the certainty of being able to live this life will be the magnet which will produce the effect for which we are all striving – the spontaneous movement of the people from our crowded cities to the bosom of our kindly mother earth, at once the source of life, of happiness, of wealth, and of power. The town and the country may, therefore, be regarded as two magnets, each striving to draw the people to itself – a rivalry which a new form of life, partaking of the nature of both, comes to take part in. . . .

Neither the Town magnet nor the Country magnet represents the full plan and purpose of nature. Human society and the beauty of nature are meant to be enjoyed together. The two magnets must be made one. As man and woman by their varied gifts and faculties supplement each other, so should town and country. The town is the symbol of society – of mutual help and friendly co-operation, of fatherhood, motherhood, brother-hood, sisterhood, of wide relations between man and man – of broad, expanding sympathies – of science, art, culture, religion. And the country! The country is the symbol of God's love and care for man. All that we are and all that we have comes from it. Our bodies are formed of it; to it they return. We are fed by it, clothed by it, and by it are we warmed and sheltered. On its bosom we rest. Its beauty is the inspiration of art, of music, of poetry. Its forces propel all the wheels of industry. It is the source of all health, all wealth, all knowledge. But its fullness of joy and wisdom has not revealed itself to man. Nor can it ever, so long as this unholy, unnatural separation of society and nature endures. Town and country *must be married*, and out of this joyous union will spring a new hope, a new life, a new civilization. It is the purpose of this work to show how a first step can be taken in this direction by the con-struction of a Town–country magnet; and I hope to convince the reader that this is practicable, here and now, and that on principles which are the very soundest, whether viewed from the ethical or the economic standpoint.

I will undertake, then, to show how in 'Town–country' equal, nay better, opportunities of social intercourse may be enjoyed than are enjoyed in any crowded city, while yet the beauties of nature may encompass and enfold each dweller therein; how higher wages are compatible with reduced rents and rates; how abundant opportunities for employ-ment and bright prospects of advancement may be secured for all; how capital may be attracted and wealth created; how the most admirable sanitary conditions may be ensured; how beautiful homes and gardens may be seen on every hand; how the bounds of freedom may be widened, and yet all the best results of concert and co-operation gathered in by a happy people.

The construction of such a magnet, could it be effected, followed, as it would be, by the con-struction of many more, would certainly afford a solution of the burning question set before us by Sir John Gorst, 'how to back the tide of migration of the people into the towns, and how to get them back upon the land.'

* The incept of the suburban ideal.

"Cities and the Crisis of Civilization"

from the "Introduction" to *The Culture of Cities* (1938)

Lewis Mumford

Editors' Introduction

Howard's concern with the rapid nineteenth-century growth of industrial cities was taken up by many others in the United Kingdom, continental Europe, and the USA. British architect Raymond Unwin, for example, in 1909 lamented the dramatic expansion of urban areas in terms we might use today:

> The last century has been remarkable, not only in this country but in some others, for an exceedingly rapid and extensive growth of towns. In England this growth has produced most serious results. For many years social reformers have been protesting against the evils which have arisen owning to this rapid and disorderly increase in the size of towns and their populations. Miles and miles of ground, which people not yet elderly can remember as open green fields, are now covered with dense masses of buildings packed together in rows along streets which have been laid out in a perfectly haphazard manner, without any consideration for the common interests of the people.[1]

The solution of Unwin and his associate Barry Parker was a better-designed garden suburb, emphasizing the aesthetic, place-making themes pioneered a decade earlier by German architect Camillo Sitte. Many public officials took an even more pragmatic approach focused on regulation rather than physical design. Beginning in Germany in the 1890s and continuing in Britain and the USA during the early decades of the twentieth century, they enacted zoning codes designed to control densities and enforce segregation of housing, shops, and workplaces – presumably protecting neighborhood quality and property values. But visionaries such as Howard and Geddes sought a broader rethinking of development principles on a metropolitan scale.

Lewis Mumford was in the latter camp, a brilliant humanist critic of architecture and society. During his long life he played a central role as American popularizer of garden city ideas. Like Howard and Geddes, Mumford and his colleagues in the Regional Plan Association of America (RPAA) sought to respond to the problems of the overcrowded industrial city by advocating the decentralization of population so as to achieve a better balance of city and countryside. In his many books, Mumford displayed an unparalleled ability to weave together an encyclopedic knowledge of history with an eloquent rhetorical style and a passionate concern for human culture and welfare. Although its prose style now seems from another era, books such as *The Culture of Cities* (New York: Harcourt, Brace, 1938), from which this passage is taken, inspired generations of later urbanists. One of these was MIT urban design professor Kevin Lynch, who would continue Mumford's emphasis on developing a normative urban planning agenda in books such as *Good City Form* (Cambridge, MA: MIT Press, 1981).

A writer rather than a professional architect or planner, Mumford nevertheless proved one of the single greatest influences on American planning in the twentieth century. An overview of his work is given by Mark Luccarelli in his book *Lewis Mumford and the Ecological Region: The Politics of Planning* (New York: The Guilford Press, 1995); a leading biography is Donald L. Miller's *Lewis Mumford: A Life* (New York: Grove Press, 1989). Living through two world wars, a cold war, and the devastation of urban landscapes by the automobile and urban renewal, Mumford had a deep fear that mechanistic, warlike forces would subvert the humane values and small-scale relationships he saw as characterizing pre-industrial cities. "Ours is an age in which the increasingly automatic processes of production and urban expansion have displaced the human goals they are supposed to serve," he wrote in 1961.[2] Taking a sweeping view of history, Mumford used terms such as "paleotechnic," "neotechnic," and "biotechnic" to denote different eras of human activity and mindsets. Optimistically, he saw the age of nineteenth-century coal-based industrialization giving way to a cleaner, less exploitative neotechnic era, based on electricity as a power source, and eventually to a more restorative biotechnic era, based on biological science and a more organic philosophy. Within this evolution, he saw cities playing a central role in nurturing human culture. Although often not specific on details, Mumford was clear about the general change of direction needed, which was toward "the development of a more organic world picture, which shall do justice to all the dimensions of living organisms and human personalities."[3]

* to be included in the synthetic definition of "S.D."

The city, as one finds it in history, is the point of maximum concentration for the power and culture of a community. It is the place where the diffused rays of many separate beams of life fall into focus, with gains in both social effectiveness and significance. The city is the form and symbol of an integrated social relationship: it is the seat of the temple, the market, the hall of justice, the academy of learning. Here in the city the goods of civilization are multiplied and manifolded; here is where human experience is transformed into viable signs, symbols, patterns of conduct, systems of order. Here is where the issues of civilization are focused: here, too, ritual passes on occasion into the active drama of a fully differentiated and self-conscious society.

Cities are a product of the earth. They reflect the peasant's cunning in dominating the earth; technically they but carry his skill in turning the soil to productive uses, in enfolding his cattle for safety, in regulating the waters that moisten his fields, in providing storage bins and barns for his crops. Cities are emblems of that settled life which began with permanent agriculture: a life conducted with the aid of permanent shelters, permanent utilities like orchards, vineyards, and irrigation works, and permanent buildings for protection and storage.

Every phase of life in the countryside contributes to the existence of cities. What the shepherd, the woodman, and the miner know, becomes transformed and "etherealized" through the city into durable elements in the human heritage: the textiles and butter of one, the moats and dams and wooden pipes and lathes of another, the metals and jewels of the third, are finally converted into the instruments of urban living: underpinning the city's economic existence, contributing art and wisdom to its daily routine. Within the city the essence of each type of soil and labor and economic goal is concentrated: thus arise greater possibilities for interchange and for new combinations not given in the isolation of their original habitats.

Cities are a product of time. They are the molds in which men's lifetimes have cooled and congealed, giving lasting shape, by way of art, to moments that would otherwise vanish with the living and leave no means of renewal or wider participation behind them. In the city, time becomes visible: buildings and monuments and public ways, more open than the written record, more subject to the gaze of many men than the scattered artifacts of the countryside, leave an imprint upon the minds even of the ignorant or the indifferent. Through the material fact of preservation, time challenges time, time clashes with time: habits and values carry over beyond the living group, streaking with different strata of time the character of any single generation. Layer upon layer, past times preserve themselves in the city until life itself is finally threatened with suffocation: then, in sheer defense, modern man invents the museum.

By the diversity of its time-structures, the city in part escapes the tyranny of a single present, and the monotony of a future that consists in repeating only a single beat heard in the past. Through its complex orchestration of time and space, no less than through the social division of labor, life in the city takes on the character of a symphony: specialized human aptitudes, specialized instruments, give rise to sonorous results which, neither in volume nor in quality, could be achieved by any single piece.

Cities arise out of man's social needs and multiply both their modes and their methods of expression. In the city remote forces and influences intermingle with the local: their conflicts are no less significant than their harmonies. And here, through the concentration of the means of intercourse in the market and the meeting place, alternative modes of living present themselves: the deeply rutted ways of the village cease to be coercive and the ancestral goals cease to be all-sufficient: strange men and women, strange interests, and stranger gods loosen the traditional ties of blood and neighborhood. . . .

The city is a fact in nature, like a cave, a run of mackerel or an ant-heap. But it is also a conscious work of art, and it holds within its communal framework many simpler and more personal forms of art. Mind *takes form* in the city; and in turn, urban forms condition mind. For space, no less than time, is artfully reorganized in cities: in boundary lines and silhouettes, in the fixing of horizontal planes and vertical peaks, in utilizing or denying the natural site, the city records the attitude of a culture and an epoch to the fundamental facts of its existence. The dome and the spire, the open avenue and the closed court, tell the story, not merely of different physical accommodations, but of essentially different conceptions of man's destiny. The city is both a physical utility for collective living and a symbol of those collective purposes and unanimities that arise under such favoring circumstance. With language itself, it remains man's greatest work of art. . . .

Today a great many things stand in the way of grasping the role of the city and of transforming this basic means of communal existence. During the last few centuries the strenuous mechanical organization

of industry, and the setting up of tyrannous political states, have blinded most men to the importance of facts that do not easily fit into the general pattern of mechanical conquest, capitalistic forms of exploitation, and power politics. Habitually, people treat the realities of personality and association and city as abstractions, while they treat confused pragmatic abstractions such as money, credit, political sovereignty, as if they were concrete realities that had an existence independent of human conventions.

Looking back over the course of Western Civilization since the fifteenth century, it is fairly plain that mechanical integration and social disruption have gone on side by side. Our capacity for effective physical organization has enormously increased; but our ability to create a harmonious counterpoise to these external linkages by means of co-operative and civic associations on both a regional and a world-wide basis, like the Christian Church in the Middle Ages, has not kept pace with these mechanical triumphs. By one of those mischievous turns, from which history is rarely free, it was precisely during this period of flowing physical energies, social disintegration, and bewildered political experiment that the populations of the world as a whole began mightily to increase, and the cities of the Western World began to grow at an inordinate rate. Forms of social life that the wisest no longer understood, the more ignorant were prepared to build. Or rather: the ignorant were completely unprepared, but that did not prevent the building.

The result was not a temporary confusion and an occasional lapse in efficiency. What followed was a crystallization of chaos: disorder hardened uncouthly in metropolitan slum and industrial factory districts; and the exodus into the dormitory suburbs and factory spores that surrounded the growing cities merely widened the area of social derangement. The mechanized physical shell took precedence in every growing town over the civic nucleus: men became dissociated as citizens in the very process of coming together in imposing economic organizations. Even industry, which was supposedly served by this planless building and random physical organization, lost seriously in efficiency: it failed to produce a new urban form that served directly its complicated processes. As for the growing urban populations, they lacked the

Community, community, community

most elementary facilities for urban living, even sunlight and fresh air, to say nothing of the means to a more vivid social life. The new cities grew up without the benefit of coherent social knowledge or orderly social effort: they lacked the useful urban folkways of the Middle Ages or the confident esthetic command of the Baroque period: indeed, a seventeenth-century Dutch peasant, in his little village, knew more about the art of living in communities than a nineteenth-century municipal councilor in London or Berlin. Statesmen who did not hesitate to weld together a diversity of regional interests into national states, or who wove together an empire that girdled the planet, failed to produce even a rough draft of a decent neighborhood.

In every department, form disintegrated: except in its heritage from the past, the city vanished as an embodiment of collective art and technics. And where, as in North America, the loss was not alleviated by the continued presence of great monuments from the past and persistent habits of social living, the result was a raw, dissolute environment, and a narrow, constricted, and baffled social life. Even in Germany and the Low Countries, where the traditions of urban life had lingered on from the Middle Ages, the most colossal blunders were committed in the most ordinary tasks of urban planning and building. As the pace of urbanization increased, the circle of devastation widened.

Today we face not only the original social disruption. We likewise face the accumulated physical and social results of that disruption: ravaged landscapes, disorderly urban districts, pockets of disease, patches of blight, mile upon mile of standardized slums, worming into the outlying areas of big cities, and fusing with their ineffectual suburbs. In short: a general miscarriage and defeat of civilized effort. So far have our achievements fallen short of our needs that even a hundred years of persistent reform in England, the first country to suffer heavily from disurbanization, have only in the last decade begun to leave an imprint. True: here and there patches of good building and coherent social form exist: new nodes of integration can be detected, and since 1920 these patches have been spreading. But the main results of more than a century of misbuilding and malformation, dissociation and disorganization still hold. Whether the

observer focuses his gaze on the physical structure of communal living or upon the social processes that must be embodied and expressed, the report remains the same.

Today we begin to see that the improvement of cities is no matter for small one-sided reforms: the task of city design involves the vaster task of rebuilding our civilization. We must alter the parasitic and predatory modes of life that now play so large a part, and we must create region by region, continent by continent, an effective symbiosis, or co-operative living together. The problem is to co-ordinate, on the basis of more essential human values than the will-to-power and the will-to-profits, a host of social functions and processes that we have hitherto misused in the building of cities and polities, or of which we have never rationally taken advantage.

Unfortunately, the fashionable political philosophies of the past century are of but small help in defining this new task: they dealt with legal abstractions, like Individual and State, with cultural abstractions, like Humanity, the Nation, the Folk, or with bare economic abstractions like the Capitalist Class or the Proletariat – while life as it was lived in the concrete, in regions and cities and villages, in wheatland and cornland and vineland, in the mine, the quarry, the fishery, was conceived as but a shadow of the prevailing myths and arrogant fantasies of the ruling classes – or the often no less shadowy fantasies of those who challenged them.

Here and there one notes, of course, valiant exceptions both in theory and in action. Le Play and Reclus in France, W.H. Riehl in Germany, Kropotkin in Russia, Howard in England, Grundtvig in Denmark, Geddes in Scotland, began half a century ago to lay the ideological basis for a new order. The insights of these men may prove as important for the new biotechnic regime, based on the deliberate culture of life, as the formulations of Leonardo, Galileo, Newton, and Descartes were for the more limited mechanical order upon which the past triumphs of our machine civilization were founded. In the piecemeal improvement of cities, the work of sanitarians like Chadwick and Richardson, community designers like Olmsted, far-seeing architects like Parker and Wright, laid the concrete basis for a collective environment in which the needs of reproduction and nurture and psychological development and the social processes themselves would be adequately served.

Now the dominant urban environment of the past century has been mainly a narrow by-product of the machine ideology. And the greater part of it has already been made obsolete by the rapid advance of the biological arts and sciences, and by the steady penetration of sociological thought into every department. We have now reached a point where these fresh accumulations of historical insight and scientific knowledge are ready to flow over into social life, to mold anew the forms of cities, to assist in the transformation of both the instruments and the goals of our civilization. Profound changes, which will affect the distribution and increase of population, the efficiency of industry, and the quality of Western Culture, have already become visible. To form an accurate estimate of these new potentialities and to suggest their direction into channels of human welfare, is one of the major offices of the contemporary student of cities. Ultimately, such studies, forecasts, and imaginative projects must bear directly upon the life of every human being in our civilization.

What is the city? How has it functioned in the Western World since the tenth century, when the renewal of cities began, and in particular, what changes have come about in its physical and social composition during the last century? What factors have conditioned the size of cities, the extent of their growth, the type of order manifested in street plan and in building, their manner of nucleation, the composition of their economic and social classes, their physical manner of existence and their cultural style? By what political processes of federation or amalgamation, co-operative union or centralization, have cities existed; and what new units of administration does the present age suggest? Have we yet found an adequate urban form to harness all the complex technical and social forces in our civilization; and if a new order is discernible, what are its main outlines? What are the relations between city and region? And what steps are necessary in order to redefine and reconstruct the region itself, as a collective human habitation? What, in short, are the possibilities for creating form and order and design in our present civilization? . . .

Today our world faces a crisis: a crisis which, if its consequences are as grave as now seems, may not fully be resolved for another century. If the destructive forces in civilization gain ascendancy, our new urban culture will be stricken in every part. Our cities, blasted and deserted, will be cemeteries for the dead: cold lairs given over to less destructive beasts than man. But we may avert that fate: perhaps only in facing such a desperate challenge can the necessary creative forces be effectually welded together. Instead of clinging to the sardonic funeral towers of metropolitan finance, ours to march out to newly plowed fields, to create fresh patterns of political action, to alter for human purposes the perverse mechanisms of our economic regime, to conceive and to germinate fresh forms of human culture.

Instead of accepting the stale cult of death that the fascists have erected, as the proper crown for the servility and the brutality that are the pillars of their states, we must erect a cult of life: life in action, as the farmer or the mechanic knows it: life in expression, as the artist knows it: life as the lover feels it and the parent practices it: life as it is known to men of good will who meditate in the cloister, experiment in the laboratory, or plan intelligently in the factory or the government office.

Nothing is permanent: certainly not the frozen images of barbarous power with which fascism now confronts us. Those images may easily be smashed by an external shock, cracked as ignominiously as fallen Dagon, the massive idol of the heathen: or they may be melted, eventually, by the internal warmth of normal men and women. Nothing endures except life: the capacity for birth, growth, and daily renewal. As life becomes insurgent once more in our civilization, conquering the reckless thrust of barbarism, the culture of cities will be both instrument and goal.

Ninja

NOTES

1 Unwin, Raymond. 1909. *Town Planning in Practice: An Introduction to the Art of Designing Cities and Suburbs.* London: Ernest Benn, p. 2.

2 Mumford, Lewis. 1961. *The City in History: Its Origins, Its Transformations, and Its Prospects.* New York: Harcourt, Brace, and World, p. 570.

3 Ibid., p. 567.

"The Land Ethic"

from *A Sand County Almanac* (1949)

Aldo Leopold

Editors' Introduction

More than half a century after his death, Aldo Leopold (1887–1948) is seen as one of the seminal figures in the development of modern environmentalism. His career included service as a conservation ecologist with the US Forest Service in New Mexico and as a professor at the University of Wisconsin. Starting within the conservationist tradition, which emphasized managing natural resources for sustained yield, Leopold expanded his perspective toward an acknowledgment of the intrinsic value of ecosystems and a view of the world as an organic, evolving unity. This philosophy foreshadowed the worldview of deep ecologists and other radical environmentalists of the 1970s and after.

Leopold was an ethicist as well as an environmentalist. His later work included a profound questioning of anthropocentric and economic values, and was based on a belief in the necessity of moral evolution in order for societies to live within "their sustained carrying capacity." In his essay "The Land Ethic," he equates the historic spread of ethical notions of human rights with the more recent growth of an understanding that entire ecosystems – not just certain elements of these – have value. He explicitly refutes the possibility of bringing about environmentally sound practices through economics alone, and believes that the only means through which this can be done is a process of social and moral growth.

In addition to *A Sand County Almanac* (New York: Oxford University Press, 1949), in which "The Land Ethic" is contained, an excellent collection of Leopold's writings is provided by *The River of the Mother Of God and other Essays by Aldo Leopold* (Madison: University of Wisconsin Press, 1991), edited by Susan L. Flader and J. Baird Callicott. Biographies of Leopold include Marybeth Lorbiecke's *Aldo Leopold: A Fierce Green Fire* (Helena, MT: Falcon Press, 1996). More information on the development of environmental philosophies in the twentieth century and the range of these philosophies is contained in Carolyn Merchant's book *Radical Ecology: The Search for a Livable World* (New York: Routledge, 1993), Robert Gottlieb's excellent history of the movement *Forcing the Spring: The Transformation of the American Environmental Movement* (Washington, DC: Island Press, 1993), and Philip Shabecoff's books *A Fierce Green Fire: The American Environmental Movement* (New York: Hill and Wang, 1993) and *Earth Rising: American Environmentalism in the 21st Century* (Washington, DC: Island Press, 2000).

When God-like Odysseus returned from the wars in Troy, he hanged all on one rope a dozen slave-girls of his household whom he suspected of misbehavior during his absence.

This hanging involved no question of propriety. The girls were property. The disposal of property was then, as now, a matter of expediency, not of right and wrong.

Concepts of right and wrong were not lacking from Odysseus' Greece: witness the fidelity of his wife through the lonely years before at last his black-prowed galleys clove the wine-dark seas for home. The ethical structure of that day covered wives, but had not yet been extended to human chattels. During the three thousand years which have since elapsed, ethical criteria have been extended to many fields of conduct, with corresponding shrinkages in those judged by expediency only.

THE ETHICAL SEQUENCE

This extension of ethics, so far studied only by philosophers, is actually a process in ecological evolution. Its sequences may be described in ecological as well as in philosophical terms. An ethic, ecologically, is a limitation on freedom of action in the struggle for existence. An ethic, philosophically, is a differentiation of social from anti-social conduct. These are two definitions of one thing. The thing has its origin in the tendency of interdependent individuals or groups to evolve modes of co-operation. The ecologist calls these symbioses. Politics and economics are advanced symbioses in which the original free-for-all competition has been replaced, in part, by co-operative mechanisms with an ethical content.

The complexity of co-operative mechanisms has increased with population density, and with the efficiency of tools. It was simpler, for example, to define the anti-social uses of sticks and stones in the days of the mastodons than of bullets and billboards in the age of motors.

The first ethics dealt with the relation between individuals; the Mosaic Decalogue is an example. Later accretions dealt with the relation between the individual and society. The Golden Rule tries to integrate the individual to society; democracy to integrate social organization to the individual. There is as yet no ethic dealing with man's relation to land and to the animals and plants which grow upon it. Land, like Odysseus' slave-girls, is still property. The land-relation is still strictly economic, entailing privileges but not obligations.

The extension of ethics to this third element in human environment is, if I read the evidence correctly, an evolutionary possibility and an ecological necessity. It is the third step in a sequence.

The first two have already been taken. Individual thinkers since the days of Ezekiel and Isaiah have asserted that the despoliation of land is not only inexpedient but wrong. Society, however, has not yet affirmed their belief. I regard the present conservation movement as the embryo of such an affirmation.

An ethic may be regarded as a mode of guidance for meeting ecological situations so new or intricate, or involving such deferred reactions, that the path of social expediency is not discernible to the average individual. Animal instincts are modes of guidance for the individual in meeting such situations. Ethics are possibly a kind of community instinct in-the-making.

THE COMMUNITY CONCEPT

All ethics so far evolved rest upon a single premise: that the individual is a member of a community of interdependent parts. His instincts prompt him to compete for his place in the community, but his ethics prompt him also to co-operate (perhaps in order that there may be a place to compete for).

The land ethic simply enlarges the boundaries of the community to include soils, waters, plants, and animals, or collectively: the land.

This sounds simple: do we not already sing our love for and obligation to the land of the free and the home of the brave? Yes, but just what and whom do we love? Certainly not the soil, which we are sending helter-skelter downriver. Certainly not the waters, which we assume have no function except to turn turbines, float barges, and carry off sewage. Certainly not the plants, of which we exterminate whole communities without batting an eye. Certainly not the animals, of which we have already extirpated many of the largest and most beautiful species. A land ethic of course cannot prevent the alteration, management, and use of these 'resources,' but it does affirm their right to continued existence, and, at least in spots, their continued existence in a natural state.

In short, a land ethic changes the role of *Homo sapiens* from conqueror of the land-community to plain member and citizen of it. It implies respect for his fellow-members, and also respect for the community as such.

In human history, we have learned (I hope) that the conqueror role is eventually self-defeating. Why? Because it is implicit in such a role that the conqueror knows, *ex cathedra*, just what makes the community clock tick, and just what and who is valuable, and what and who is worthless, in community life. It always turns out that he knows neither, and this is why his conquests eventually defeat themselves.

In the biotic community, a parallel situation exists. Abraham knew exactly what the land was for: it was to drip milk and honey into Abraham's mouth. At the present moment, the assurance with which we regard this assumption is inverse to the degree of our education.

The ordinary citizen today assumes that science knows what makes the community clock tick; the scientist is equally sure that he does not. He knows that the biotic mechanism is so complex that its workings may never be fully understood.

That man is, in fact, only a member of a biotic team is shown by an ecological interpretation of history. Many historical events, hitherto explained solely in terms of human enterprise, were actually biotic interactions between people and land. The characteristics of the land determined the facts quite as potently as the characteristics of the men who lived on it.

Consider, for example, the settlement of the Mississippi valley. In the years following the Revolution, three groups were contending for its control: the native Indian, the French and English traders, and the American settlers. Historians wonder what would have happened if the English at Detroit had thrown a little more weight into the Indian side of those tipsy scales which decided the outcome of the colonial migration into the cane-lands of Kentucky. It is time now to ponder the fact that the cane-lands, when subjected to the particular mixture of forces represented by the cow, plow, fire, and axe of the pioneer, became bluegrass. What if the plant succession inherent in this dark and bloody ground had, under the impact of these forces, given us some worthless sedge, shrub, or weed? Would Boone and Kenton have held out? Would there have been any overflow into Ohio, Indiana, Illinois, and Missouri? Any Louisiana Purchase? Any transcontinental union of new states? Any Civil War?

Kentucky was one sentence in the drama of history. We are commonly told what the human actors in this drama tried to do, but we are seldom told that their success, or the lack of it, hung in large degree on the reaction of particular soils to the impact of the particular forces exerted by their occupancy. In the case of Kentucky, we do not even know where the bluegrass came from – whether it is a native species, or a stowaway from Europe.

Contrast the cane-lands with what hindsight tells us about the Southwest, where the pioneers were equally brave, resourceful, and persevering. The impact of occupancy here brought no bluegrass, or other plant fitted to withstand the bumps and buffetings of hard use. This region, when grazed by livestock, reverted through a series of more and more worthless grasses, shrubs, and weeds to a condition of unstable equilibrium. Each recession of plant types bred erosion; each increment to erosion bred a further recession of plants. The result today is a progressive and mutual deterioration, not only of plants and soils, but of the animal community subsisting thereon. The early settlers did not expect this: on the ciénegas of New Mexico some even cut ditches to hasten it. So subtle has been its progress that few residents of the region are aware of it. It is quite invisible to the tourist who finds this wrecked landscape colorful and charming (as indeed it is, but it bears scant resemblance to what it was in 1848).

This same landscape was "developed" once before, but with quite different results. The Pueblo Indians settled the Southwest in pre-Columbian times, but they happened *not* to be equipped with range live-stock. Their civilization expired, but not because their land expired.

In India, regions devoid of any sod-forming grass have been settled, apparently without wrecking the land, by the simple expedient of carrying the grass to the cow, rather than vice versa. (Was this the result of some deep wisdom, or was it just good luck? I do not know.)

In short, the plant succession steered the course of history; the pioneer simply demonstrated, for good or ill, what successions inhered in the land. Is history taught in this spirit? It will be, once the concept of land as a community really penetrates our intellectual life.

THE ECOLOGICAL CONSCIENCE

Conservation is a state of harmony between men and land. Despite nearly a century of propaganda, conservation still proceeds at a snail's pace; progress still consists largely of letterhead pieties and convention oratory. On the back forty we still slip two steps backward for each forward stride.

The usual answer to this dilemma is "more conservation education." No one will debate this, but is it certain that only the *volume* of education needs stepping up? Is something lacking in the *content* as well?

It is difficult to give a fair summary of its content in brief form, but, as I understand it, the content is substantially this: obey the law, vote right, join some organizations, and practice what conservation is profitable on your own land; the government will do the rest.

Is not this formula too easy to accomplish anything worthwhile? It defines no right or wrong, assigns no obligation, calls for no sacrifice, implies no change in the current philosophy of values. In respect of land use, it urges only enlightened self-interest. Just how far will such education take us? An example will perhaps yield a partial answer.

By 1930 it had become clear to all except the ecologically blind that southwestern Wisconsin's topsoil was slipping seaward. In 1933 the farmers were told that if they would adopt certain remedial practices for five years, the public would donate CCC labor to install them, plus the necessary machinery and materials. The offer was widely accepted, but the practices were widely forgotten when the five-year contract period was up. The farmers continued only those practices that yielded an immediate and visible economic gain for themselves.

This led to the idea that maybe farmers would learn more quickly if they themselves wrote the rules. Accordingly the Wisconsin Legislature in 1937 passed the Soil Conservation District Law. This said to farmers, in effect: *We, the public, will furnish you free technical service and loan you specialized machinery, if you will write your own rules for land-use. Each county may write its own rules, and these will have the force of law.* Nearly all the counties promptly organized to accept the proffered help, but after a decade of operation, *no county has yet written a single rule.* There has been visible progress in such practices as strip-cropping, pasture renovation, and soil liming, but none in fencing woodlots against grazing, and none in excluding plow and cow from steep slopes. The farmers, in short, have selected those remedial practices which were profitable anyhow, and ignored those which were profitable to the community, but not clearly profitable to themselves.

When one asks why no rules have been written, one is told that the community is not yet ready to support them; education must precede rules. But the education actually in progress makes no mention of obligations to land over and above those dictated by self-interest. The net result is that we have more education but less soil, fewer healthy woods, and as many floods as in 1937.

The puzzling aspect of such situations is that the existence of obligations over and above self-interest is taken for granted in such rural community enterprises as the betterment of roads, schools, churches, and baseball teams. Their existence is not taken for granted, nor as yet seriously discussed, in bettering the behavior of the water that falls on the land, or in the preserving of the beauty or diversity of the farm landscape. Land-use ethics are still governed wholly by economic self-interest, just as social ethics were a century ago.

To sum up: we asked the farmer to do what he conveniently could to save his soil, and he has done just that, and only that. The farmer who clears the woods off a 75 per cent slope, turns his cows into the clearing, and dumps its rainfall, rocks, and soil into the community creek, is still (if otherwise decent) a respected member of society. If he puts lime on his fields and plants his crops on contour, he is still entitled to all the privileges and emoluments of his Soil Conservation District. The District is a beautiful piece of social machinery, but it is coughing along on two cylinders because we have been too timid, and too anxious for quick success, to tell the farmer the true magnitude of his obligations. Obligations have no meaning without conscience, and the problem we face is the extension of the social conscience from people to land.

No important change in ethics was ever accomplished without an internal change in our intellectual emphasis, loyalties, affections, and convictions. The proof that conservation has not yet touched these foundations of conduct lies in the fact that philosophy and religion have not yet heard of it. In our attempt to make conservation easy, we have made it trivial.

SUBSTITUTES FOR A LAND ETHIC

When the logic of history hungers for bread and we hand out a stone, we are at pains to explain how much the stone resembles bread. I now describe some of the stones which serve in lieu of a land ethic.

One basic weakness in a conservation system based wholly on economic motives is that most members of the land community have no economic value. Wildflowers and songbirds are examples. Of the 22,000 higher plants and animals native to Wisconsin, it is doubtful whether more than 5 per cent can be sold, fed, eaten, or otherwise put to economic use. Yet these creatures are members of the biotic community, and if (as I believe) its stability depends on its integrity, they are entitled to continuance.

When one of these non-economic categories is threatened, and if we happen to love it, we invent subterfuges to give it economic importance. At the beginning of the century songbirds were supposed to be disappearing. Ornithologists jumped to the rescue with some distinctly shaky evidence to the effect that insects would eat us up if birds failed to control them. The evidence had to be economic in order to be valid.

It is painful to read these circumlocutions today. We have no land ethic yet, but we have at least drawn nearer the point of admitting that birds should continue as a matter of biotic right, regardless of the presence or absence of economic advantage to us.

A parallel situation exists in respect of predatory mammals, raptorial birds, and fish-eating birds. Time was when biologists somewhat overworked the evidence that these creatures preserve the health of game by killing weaklings, or that they control rodents for the farmer, or that they prey only on "worthless" species. Here again, the evidence had to be economic in order to be valid. It is only in recent years that we hear the more honest argument that predators are members of the community, and that no special interest has the right to exterminate them for the sake of a benefit, real or fancied, to itself. Unfortunately this enlightened view is still in the talk stage. In the field the extermination of predators goes merrily on: witness the impending erasure of the timber wolf by fiat of Congress, the Conservation Bureaus, and many state legislatures.

Some species of trees have been "read out of the party" by economics-minded foresters because they grow too slowly, or have too low a sale value to pay as timber crops: white cedar, tamarack, cypress, beech, and hemlock are examples. In Europe, where forestry is ecologically more advanced, the non-commercial tree species are recognized as members of the native forest community, to be preserved as such, within reason. Moreover some (like beech) have been found to have a valuable function in building up soil fertility. The interdependence of the forest and its constituent tree species, ground flora, and fauna is taken for granted.

Lack of economic value is sometimes a character not only of species or groups, but of entire biotic communities: marshes, bogs, dunes, and "deserts" are examples. Our formula in such cases is to relegate their conservation to government as refuges, monuments, or parks. The difficulty is that these communities are usually interspersed with more valuable private lands; the government cannot possibly own or control such scattered parcels. The net effect is that we have relegated some of them to ultimate extinction over large areas. If the private owner were ecologically minded, he would be proud to be the custodian of a reasonable proportion of such areas, which add diversity and beauty to his farm and to his community.

In some instances, the assumed lack of profit in these "waste" areas has proved to be wrong, but only after most of them had been done away with. The present scramble to reflood muskrat marshes is a case in point.

There is a clear tendency in American conservation to relegate to government all necessary jobs that private landowners fail to perform. Government ownership, operation, subsidy, regulation is now widely prevalent in forestry, range management, soil and watershed management, park and wilderness conservation, fisheries management, and migratory bird management, with more to come. Most of this growth in governmental conservation is proper and logical, some of it is inevitable. That I imply no disapproval of it is implicit in the fact that I have spent most of my life working for it. Nevertheless the question arises: What is the ultimate magnitude of the enterprise? Will the tax base carry its eventual ramifications? At what point will governmental conservation, like the mastodon, become handicapped by its own dimensions? The answer, if there is any, seems to be in a land ethic, or some other force which assigns more obligation to the private landowner.

Industrial landowners and users, especially lumbermen and stockmen, are inclined to wail long and loudly about the extension of government ownership and regulation to land, but (with notable exceptions) they show little disposition to develop the only visible alternative: the voluntary practice of conservation on their own lands.

When the private landowner is asked to perform some unprofitable act for the good of the community, he today assents only with outstretched palm. If the act costs him cash this is fair and proper, but when it costs only fore-thought, open-mindedness, or time, the issue is at least debatable. The overwhelming growth of land-use subsidies in recent years must be ascribed, in large part, to the government's own agencies for conservation education: the land bureaus, the agricultural colleges, and the extension services. As far as I can detect, no ethical obligation toward land is taught in these institutions.

To sum up: a system of conservation based solely on economic self-interest is hopelessly lopsided. It tends to ignore, and thus eventually to eliminate, many elements in the land community that lack commercial value, but that are (as far as we know) essential to its healthy functioning. It assumes, falsely, I think, that the economic parts of the biotic clock will function without the uneconomic parts. It tends to relegate to government many functions eventually too large, too complex, or too widely dispersed to be performed by government.

An ethical obligation on the part of the private owner is the only visible remedy for these situations.

THE LAND PYRAMID

An ethic to supplement and guide the economic relation to land presupposes the existence of some mental image of land as a biotic mechanism. We can be ethical only in relation to something we can see, feel, understand, love, or otherwise have faith in.

The image commonly employed in conservation education is "the balance of nature." For reasons too lengthy to detail here, this figure of speech fails to describe accurately what little we know about the land mechanism. A much truer image is the one employed in ecology: the biotic pyramid. I shall first sketch the pyramid as a symbol of land, and later develop some of its implications in terms of land-use.

Plants absorb energy from the sun. This energy flows through a circuit called the biota, which may be represented by a pyramid consisting of layers. The bottom layer is the soil. A plant layer rests on the soil, an insect layer on the plants, a bird and rodent layer on the insects, and so on up through various animal groups to the apex layer, which consists of the larger carnivores.

The species of a layer are alike not in where they came from, or in what they look like, but rather in what they eat. Each successive layer depends on those below it for food and often for other services, and each in turn furnishes food and services to those above. Proceeding upward, each successive layer decreases in numerical abundance. Thus, for every carnivore there are hundreds of his prey, thousands of their prey, millions of insects, uncountable plants. The pyramidal form of the system reflects this numerical progression from apex to base. Man shares an intermediate layer with the bears, raccoons, and squirrels which eat both meat and vegetables.

The lines of dependency for food and other services are called food chains. Thus soil–oak–deer–Indian is a chain that has now been largely converted to soil–corn–cow–farmer. Each species, including ourselves, is a link in many chains. The deer eats a hundred plants other than oak, and the cow a hundred plants other than corn. Both, then, are links in a hundred chains. The pyramid is a tangle of chains so complex as to seem disorderly, yet the stability of the system proves it to be a highly organized structure. Its functioning depends on the co-operation and competition of its diverse parts.

In the beginning, the pyramid of life was low and squat; the food chains short and simple. Evolution has added layer after layer, link after link. Man is one of thousands of accretions to the height and complexity of the pyramid. Science has given us many doubts, but it has given us at least one certainty: the trend of evolution is to elaborate and diversify the biota.

Land, then, is not merely soil; it is a fountain of energy flowing through a circuit of soils, plants, and animals. Food chains are the living channels which conduct energy upward; death and decay return it to the soil. The circuit is not closed; some energy is dissipated in decay, some is added by absorption from the air, some is stored in soils, peats, and long-lived forests; but it is a sustained circuit, like a slowly augmented revolving fund of life. There is always a net loss by downhill wash, but this is normally small and offset by the decay of rocks. It is deposited in the ocean and, in the course of geological time, raised to form new lands and new pyramids.

The velocity and character of the upward flow of energy depend on the complex structure of the plant and animal community, much as the upward flow of sap in a tree depends on its complex cellular organization. Without this complexity, normal circulation would presumably not occur. Structure means the characteristic numbers, as well as the characteristic kinds and functions, of the component species. This interdependence between the complex structure of the land and its smooth functioning as an energy unit is one of its basic attributes.

When a change occurs in one part of the circuit, many other parts must adjust themselves to it. Change does not necessarily obstruct or divert the flow of energy; evolution is a long series of self-induced changes, the net result of which has been to elaborate the flow mechanism and to lengthen the circuit. Evolutionary changes, however, are usually slow and local. Man's invention of tools has enabled him to make changes of unprecedented violence, rapidity, and scope.

One change is in the composition of floras and faunas. The larger predators are lopped off the apex of the pyramid; food chains, for the first time in history, become shorter rather than longer. Domesticated species from other lands are substituted for wild ones, and wild ones are moved to new habitats. In this world-wide pooling of faunas and floras, some species get out of bounds as pests and diseases, others are extinguished. Such effects are seldom intended or foreseen; they represent unpredicted and often untraceable readjustments in the structure. Agricultural science is largely a race between the emergence of new pests and the emergence of new techniques for their control.

Another change touches the flow of energy through plants and animals and its return to the soil. Fertility is the ability of soil to receive, store, and release energy. Agriculture, by overdrafts on the soil, or by too radical a substitution of domestic for native species in the superstructure, may derange the channels of flow or deplete storage. Soils depleted of their storage, or of the organic matter which anchors it, wash away faster than they form. This is erosion.

Waters, like soil, are part of the energy circuit. Industry, by polluting waters or obstructing them with dams, may exclude the plants and animals necessary to keep energy in circulation.

Transportation brings about another basic change: the plants or animals grown in one region are now consumed and returned to the soil in another. Transportation taps the energy stored in rocks, and in the air, and uses it elsewhere; thus we fertilize the garden with nitrogen gleaned by the guano birds from the fishes of seas on the other side of the Equator. Thus the formerly localized and self-contained circuits are pooled on a world-wide scale.

The process of altering the pyramid for human occupation releases stored energy, and this often gives rise, during the pioneering period, to a deceptive exuberance of plant and animal life, both wild and tame. These releases of biotic capital tend to becloud or postpone the penalties of violence.

[. . .]

This thumbnail sketch of land as an energy circuit conveys three basic ideas:

1 That land is not merely soil.
2 That the native plants and animals kept the energy circuit open; others may or may not.
3 That man-made changes are of a different order than evolutionary changes, and have effects more comprehensive than is intended or foreseen.

These ideas, collectively, raise two basic issues: Can the land adjust itself to the new order? Can the desired alterations be accomplished with less violence?

Biotas seem to differ in their capacity to sustain violent conversion. Western Europe, for example, carries a far different pyramid than Caesar found there. Some large animals are lost; swampy forests have become meadows or plowland; many new plants and animals are introduced, some of which escape as pests; the remaining natives are greatly changed in distribution and abundance. Yet the soil is still there and, with the help of imported nutrients, still fertile; the waters flow normally; the new structure seems to function and to persist. There is no visible stoppage or derangement of the circuit.

Western Europe, then, has a resistant biota. Its inner processes are tough, elastic, resistant to strain. No matter how violent the alterations, the pyramid, so far, has developed some new *modus vivendi* which preserves its habitability for man, and for most of the other natives.

Japan seems to present another instance of radical conversion without disorganization.

Most other civilized regions, and some as yet barely touched by civilization, display various stages of disorganization, varying from initial symptoms to advanced wastage. In Asia Minor and North Africa diagnosis is confused by climatic changes, which may have been either the cause or the effect of advanced wastage. In the United States the degree of disorganization varies locally; it is worst in the Southwest, the Ozarks, and parts of the South, and least in New England and the Northwest. Better land-uses may still arrest it in the less advanced regions. In parts of Mexico, South America, South Africa, and Australia a violent and accelerating wastage is in progress, but I cannot assess the prospects.

This almost world-wide display of disorganization in the land was to be similar to disease in an animal, except that it never culminates in complete disorganization or death. The land recovers, but at some reduced level of complexity, and with a reduced carrying capacity for people, plants, and animals. Many biotas currently regarded as "lands of opportunity" are in fact already subsisting on exploitative agriculture, i.e. they have already exceeded their sustained carrying capacity. Most of South America is overpopulated in this sense.

In arid regions we attempt to offset the process of wastage by reclamation, but it is only too evident that the prospective longevity of reclamation projects is often short. In our own West, the best of them may not last a century.

The combined evidence of history and ecology seems to support one general deduction: the less violent the man-made changes, the greater the probability of successful readjustment in the pyramid. Violence, in turn, varies with human population density; a dense population requires a more violent conversion. In this respect, North America has a better chance for permanence than Europe, if she can contrive to limit her density.

This deduction runs counter to our current philosophy, which assumes that because a small increase in density enriched human life, that an indefinite increase will enrich it indefinitely. Ecology knows of no density relationship that holds for indefinitely wide limits. All gains from density are subject to a law of diminishing returns.

Whatever may be the equation for men and land, it is improbable that we as yet know all its terms. Recent discoveries in mineral and vitamin nutrition reveal unsuspected dependencies in the up-circuit: incredibly minute quantities of certain substances determine the value of soils to plants, of plants to animals. What of the down-circuit? What of the vanishing species, the preservation of which we now regard as an esthetic luxury? They helped build the soil; in what unsuspected ways may they be essential to its maintenance? Professor Weaver proposes that we use prairie flowers to reflocculate the wasting soils of the dust bowl; who knows for what purpose cranes and condors, otters and grizzlies may some day be used? . . .

THE OUTLOOK

It is inconceivable to me that an ethical relation to land can exist without love, respect, and admiration for land, and a high regard for its value. By value, I of course mean something far broader than mere economic value; I mean value in the philosophical sense.

Perhaps the most serious obstacle impeding the evolution of a land ethic is the fact that our educational and economic system is headed away from, rather than toward, an intense consciousness of land. Your true modern is separated from the land by many middlemen, and by innumerable physical gadgets. He has no vital relation to it; to him it is the space between cities on which crops grow. Turn him loose for a day on the land, and if the spot does not happen to be a golf links or a "scenic" area, he is bored stiff. If crops could be raised by hydroponics instead of farming, it would suit him very well. Synthetic substitutes for wood, leather, wool, and other natural land products suit him better than the originals. In short, land is something he has "outgrown."

Almost equally serious as an obstacle to a land ethic is the attitude of the farmer for whom the land is still an adversary, or a taskmaster that keeps him in slavery. Theoretically, the mechanization of farming ought to cut the farmer's chains, but whether it really does is debatable.

One of the requisites for an ecological comprehension of land is an understanding of ecology, and this is by no means co-extensive with "education"; in fact, much higher education seems deliberately to avoid ecological concepts. An understanding of ecology does not necessarily originate in courses bearing ecological labels; it is quite as likely to be labeled geography, botany, agronomy, history, or economics. This is as it should be, but whatever the label, ecological training is scarce.

The case for a land ethic would appear hopeless but for the minority which is in obvious revolt against these "modern" trends.

The "key-log" which must be moved to release the evolutionary process for an ethic is simply this: quit thinking about decent land-use as solely an economic problem. Examine each question in terms of what is ethically and esthetically right, as well as what is economically expedient. A thing is right when it tends to preserve the integrity, stability, and beauty of the biotic community. It is wrong when it tends otherwise.

It of course goes without saying that economic feasibility limits the tether of what can or cannot be done for land. It always has and it always will. The fallacy the economic determinists have tied around our collective neck, and which we now need to cast off, is the belief that economics determines *all* land-use. This is simply not true. An innumerable host of actions and attitudes, comprising perhaps the bulk of all land relations, is determined by the land-users' tastes and predilections, rather than by his purse. The bulk of all land relations hinges on investments of time, forethought, skill, and faith rather than on investments of cash. As a land-user thinketh, so is he.

I have purposely presented the land ethic as a product of social evolution because nothing so important as an ethic is ever "written." Only the most superficial student of history supposes that Moses "wrote" the Decalogue; it evolved in the minds of a thinking community, and Moses wrote a tentative summary of it for a "seminar." I say tentative because evolution never stops.

The evolution of a land ethic is an intellectual as well as emotional process. Conservation is paved with good intentions which prove to be futile, or even dangerous, because they are devoid of critical understanding either of the land, or of economic land-use. I think it is a truism that as the ethical frontier advances from the individual to the community, its intellectual content increases.

The mechanism of operation is the same for any ethic: social approbation for right actions: social disapproval for wrong actions.

By and large, our present problem is one of attitudes and implements. We are remodeling the Alhambra with a steam-shovel, and we are proud of our yardage. We shall hardly relinquish the shovel has many good points, but we are in need of gentler and more objective criteria for its successful use.

WILDERNESS

Wilderness is the raw material out of which man has hammered the artifact called civilization.

Wilderness was never a homogeneous raw material. It was very diverse, and the resulting artifacts are very diverse. These differences in the end-product are known as cultures. The rich diversity of the world's cultures reflects a corresponding diversity in the wilds that gave them birth.

For the first time in the history of the human species, two changes are now impending. One is the exhaustion of wilderness in the more habitable portions of the globe. The other is the world-wide hybridization of cultures through modern transport and industrialization. Neither can be prevented, and perhaps should not be, but the question arises whether, by some slight amelioration of the impending changes, certain values can be preserved that would otherwise be lost.

To the laborer in the sweat of his labor, the raw stuff on his anvil is an adversary to be conquered. So was wilderness an adversary to the pioneer.

But to the laborer in repose, able for the moment to cast a philosophical eye on his world, that same raw stuff is something to be loved and cherished, because it gives definition and meaning to his life. This is a plea for the preservation of some tag-ends of wilderness, as museum pieces, for the edification of those who may one day wish to see, feel, or study the origins of their cultural inheritance.

ONE

"Orthodox Planning and The North End"

from the "Introduction" to the *The Death and Life of Great American Cities* (1961)

Jane Jacobs

Editors' Introduction

Despite many good intentions, urban planning in the twentieth century often proceeded in directions that were profoundly unsustainable. Planners promoted freeways and other automobile infrastructure without considering their sprawl-inducing impacts, authorized the bulldozing of vibrant older urban neighborhoods for redevelopment into bland, modernist apartment blocks, allowed the destruction of natural landscape features such as streams and wetlands, and aided in the segregation of racial or socioeconomic groups through zoning and red-lining. Such actions were often camouflaged behind an image of the planner as detached, objective expert, with the authority of scientific method justifying decisions that were essentially subjective or political in nature.

Writers such as Mumford and William H. Whyte complained vigorously against modernist urban planning during the 1940s and 1950s. Paul and Percival Goodman's 1947 book *Communitas: Ways of Livelihood and Means of Life* (New York: Columbia University Press, 1947) was an especially profound critique of modernist city-building, as well as a visionary exploration of alternatives, although this work did not receive wide circulation until it was reissued in the 1960s with a new introduction by Mumford. However, Jane Jacobs' 1961 book *The Death and Life of Great American Cities* was the bombshell that shocked many people around the world into questioning prevailing modes of urban planning. An editor at an architectural magazine, Jacobs had been living in New York City's Greenwich Village when plans were announced for an expressway through the community. At the time, the city's planning czar Robert Moses had bulldozed a network of expressways and bridges through dozens of other neighborhoods. However, Jacobs and her neighbors organized opposition and eventually defeated the city's plans. In the process she grew to appreciate even more the rich community life of this older urban neighborhood, and observed as well the effect of modernist redevelopment on other older neighborhoods such as Boston's West End, a tight-knit working-class community destroyed in order to create a cluster of bland apartment buildings for more affluent residents.

In *The Death and Life* Jacobs described in detail what makes dense urban neighborhoods work, and how modern city-building practices undermine many of the qualities that encourage pedestrian use of the street, neighborhood contacts, and a thriving local economy of small businesses. Her writing helped lay the groundwork for the field of environmental design research, in which later investigators carefully studied how people actually used buildings, streets, and neighborhoods, rather than simply following some abstract set of architectural criteria. Jacobs' specific emphasis on pedestrian-oriented urban form also served as an inspiration to many later urban activists, including the New Urbanists.

In her later writings Jacobs focused on questions related to local and regional economies. In her book *The Economy of Cities* (New York: Random House, 1969) she portrayed cities as the economic engines that have made possible broader development of societies. In *Cities and the Wealth of Nations: Principles of Economic Life* (New York: Random House, 1984), she expanded this argument, and argued particularly that to grow and be successful cities and urban regions need to gradually replace the goods that they had previously imported by producing them internally. In her emphasis on the importance of networks of small firms she foreshadowed much later work by academic economists, and comes close to the import-substitution method recommended by Shuman (see Part 2).

In this selection Jacobs describes her reaction to the proposed redevelopment of Boston's North End, a dense, tight-knit Italian-American community that she held up repeatedly as an example of a vibrant urban neighborhood. She contrasts in particular the attitude of planners with the reality of life in the neighborhood as she observed it. Related critiques have since been developed by many other writers. In her book *Dreaming the Rational City* (Cambridge, MA: MIT Press, 1983), Christine Boyer provides a detailed historical analysis of how urban planning has been used to enforce the dominant discourses and power of capitalist elites. Robert Goodman writes bitingly about urban renewal and planning as a vehicle for exploitation in his book *After the Planners* (New York: Simon and Schuster, 1971). And neo-Marxist authors such as David Harvey, Manuel Castells, and Robert Beauregard have argued that urban planners have often helped carry out the agendas of powerful class or economic interests, rather than working as a more independent force for social change.

The effect of all these analyses was to throw prevailing modes of urban planning into question during the second half of the twentieth century, to challenge the view of the planner as a detached, scientific expert, and to fuel calls for greater public participation and contextual, culturally informed understandings of urban problems. But Jacobs was the clearest, most down-to-earth and most biting of these critics, and her critique laid the groundwork for much re-evaluation of the field.

There is nothing economically or socially inevitable about either the decay of old cities or the fresh-minted decadence of the new unurban urbanization. On the contrary, no other aspect of our economy and society has been more purposefully manipulated for a full quarter of a century to achieve precisely what we are getting. Extraordinary governmental financial incentives have been required to achieve this degree of monotony, sterility and vulgarity. Decades of preaching, writing and exhorting by experts have gone into convincing us and our legislators that mush like this must be good for us, as long as it comes bedded with grass.

Automobiles are often conveniently tagged as the villains responsible for the ills of cities and the disappointments and futilities of city planning. But the destructive effects of automobiles are much less a cause than a symptom of our incompetence at city building. Of course planners, including the highwaymen with fabulous sums of money and enormous powers at their disposal, are at a loss to make automobiles and cities compatible with one another. They do not know what to do with automobiles in cities because they do not know how to plan for workable and vital cities anyhow – with or without automobiles.

The simple needs of automobiles are more easily understood and satisfied than the complex needs of cities, and a growing number of planners and designers have come to believe that if they can only solve the problems of traffic, they will thereby have solved the major problem of cities. Cities have much more intricate economic and social concerns than automobile traffic. How can you know what to try with traffic until you know how the city itself works, and what else it needs to do with its streets? You can't.

It may be that we have become so feckless as a people that we no longer care how things do work, but only what kind of quick, easy outer impression they give. If so, there is little hope for our cities or probably for much else in our society. But I do not think this is so.

Specifically, in the case of planning for cities, it is clear that a large number of good and earnest people do care deeply about building and renewing. Despite some corruption, and considerable greed for the other man's vineyard, the intentions going into the messes we make are, on the whole, exemplary. Planners, architects of city design, and those they have led along with them in their beliefs are not consciously disdainful of the importance of knowing how things work. On the contrary, they have gone to great pains to learn what the saints and sages of modern orthodox planning have said about how cities ought to work and what ought to be good for people and businesses in them. They take this with such devotion that when contradictory reality intrudes, threatening to shatter their dearly won learning, they must shrug reality aside.

Consider, for example, the orthodox planning reaction to a district called the North End in Boston. This is an old, low-rent area merging into the heavy industry of the waterfront, and it is officially considered Boston's worst slum and civic shame. It embodies attributes which all enlightened people know are evil because so many wise men have said they are evil. Not only is the North End bumped right up against industry, but worse still it has all kinds of working places and commerce mingled in the greatest complexity with its residences. It has the highest concentration of dwelling units, on the land that is used for dwelling units, of any part of Boston, and indeed one of the highest concentrations to be found in any American city. It has little parkland. Children play in the streets. Instead of super-blocks, or even decently large blocks, it has very small blocks; in planning parlance it is "badly cut up with wasteful streets." Its buildings are old. Everything conceivable is presumably wrong with the North End. In orthodox planning terms it is a three-dimensional textbook of "megalopolis" in the last stages of depravity. The North End is thus a recurring assignment for MIT and Harvard planning and architectural students, who now and again pursue, under the guidance of their teachers, the paper exercise of converting it into super-blocks and park promenades, wiping away its nonconforming uses, transforming it to an ideal of order and gentility so simple it could be engraved on the head of a pin.

Twenty years ago, when I first happened to see the North End, its buildings – town houses of different kinds and sizes converted to flats, and four- or five-story tenements built to house the flood of immigrants first from Ireland, then from Eastern Europe and finally from Sicily – were badly overcrowded, and the general effect was of a district taking a terrible physical beating and certainly desperately poor.

When I saw the North End again in 1959, I was amazed at the change. Dozens and dozens of buildings had been rehabilitated. Instead of mattresses against the windows there were Venetian blinds and glimpses of fresh paint. Many of the small, converted houses now had only one or two families in them instead of the old crowded three or four. Some of the families in the tenements (as I learned later, visiting inside) had uncrowded themelves by throwing two older apartments together, and had equipped these with bathrooms, new kitchens and the like. I looked down a narrow alley, thinking to find at least here the old, squalid North End, but no: more neatly repainted brickwork, new blinds, and a burst of music as a door opened. Indeed, this was the only city district I had ever seen – or have seen to this day – in which the sides of buildings around parking lots had not been left raw and amputated, but repaired and painted as neatly as if they were intended to be seen. Mingled all among the buildings for living were an incredible number of splendid food stores, as well as such enterprises as upholstery making, metal working, carpentry, food processing. The streets were alive with children playing, people shopping, people strolling, people talking. Had it not been a cold January day, there would surely have been people sitting.

The general street atmosphere of buoyancy, friendliness and good health was so infectious that I began asking directions of people just for the fun of getting in on some talk. I had seen a lot of Boston in the past couple of days, most of it sorely distressing, and this struck me, with relief, as the healthiest place in the city. But I could not imagine where the money had come from for the rehabilitation, because it is almost impossible today to get any appreciable mortgage money in districts of American cities that are not either high-rent, or else imitations of suburbs. To find out, I went into a bar and restaurant (where an animated conversation about fishing was in progress) and called a Boston planner I know.

"Why in the world are you down in the North End?" he said. "Money? Why, no money or work has gone into the North End. Nothing's going on down there. Eventually, yes, but not yet. That's a slum!"

"It doesn't seem like a slum to me," I said.

"Why, that's the worst slum in the city. It has two hundred and seventy-five dwelling units to the net acre! I hate to admit we have anything like that in Boston, but it's a fact."

"Do you have any other figures on it?" I asked.

"Yes, funny thing. It has among the lowest delinquency, disease and infant mortality rates in the city. It also has the lowest ratio of rent to income in the city. Boy, are those people getting bargains. Let's see . . . the child population is just about average for the city, on the nose. The death rate is low, 8.8 per thousand, against the average city rate of 11.2. The TB death rate is very low, less than 1 per ten thousand, can't understand it, it's lower even than Brookline's. In the old days the North End used to be the city's worst spot for tuberculosis, but all that has changed. Well, they must be strong people. Of course it's a terrible slum."

"You should have more slums like this," I said. "Don't tell me there are plans to wipe this out. You ought to be down here learning as much as you can from it."

"I know how you feel," he said. "I often go down there myself just to walk around the streets and feel that wonderful, cheerful street life. Say, what you ought to do, you ought to come back and go down in the summer if you think it's fun now. You'd be crazy about it in summer. But of course we have to rebuild it eventually. We've got to get those people off the streets."

Here was a curious thing. My friend's instincts told him the North End was a good place, and his social statistics confirmed it. But everything he had learned as a physical planner about what is good for people and good for city neighborhoods, everything that made him an expert told him the North End had to be a bad place.

The leading Boston savings banker, "a man 'way up there in the power structure,'" to whom my friend referred me for my inquiry about the money, confirmed what I learned, in the meantime, from people in the North End. The money had not come through the grace of the great American banking system, which now knows enough about planning to know a slum as well as the planners

do. "No sense in lending money into the North End," the banker said. "It's a slum! It's still getting some immigrants! Furthermore, back in the Depression it had a very large number of foreclosures; bad record." (I had heard about this too, in the mean time, and how families had worked and pooled their resources to buy back some of those foreclosed buildings.)

The largest mortgage loans that had been fed into this district of some 15,000 people in the quarter-century since the Great Depression were for $3,000, the banker told me, "and very, very few of those." There had been some others for $1,000 and for $2,000. The rehabilitation work had been almost entirely financed by business and housing earnings within the district, plowed back in, and by skilled work bartered among residents and relatives of residents.

By this time I knew that this inability to borrow for improvement was a galling worry to North Enders, and that furthermore some North Enders were worried because it seemed impossible to get new building in the area except at the price of seeing themselves and their community wiped out in the fashion of the students' dreams of a city Eden, a fate which they knew was not academic because it had already smashed completely a socially similar – although physically more spacious – nearby district called the West End. They were worried because they were aware also that patch and fix with nothing else could not do forever. "Any chance of loans for new construction in the North End?" I asked the banker.

"Absolutely not!" he said, sounding impatient at my denseness. "That's a slum!"

Bankers, like planners, have theories about cities on which they act. They have gotten their theories from the same intellectual sources as the planners. Bankers and government administrative officials who guarantee mortgages do not invent planning theories nor, surprisingly, even economic doctrine about cities. They are enlightened nowadays, and they pick up their ideas from idealists, a generation late. Since theoretical city planning has embraced no major new ideas for considerably more than a generation, theoretical planners, financers and bureaucrats are all just about even today.

And to put it bluntly, they are all in the same stage of elaborately learned superstition as medical science was early in the last century, when physicians put their faith in bloodletting, to draw out the evil humors which were believed to cause disease. With bloodletting, it took years of learning to know precisely which veins, by what rituals, were to be opened for what symptoms. A superstructure of technical complication was erected in such deadpan detail that the literature still sounds almost plausible. However, because people, even when they are thoroughly enmeshed in descriptions of reality which are at variance with reality, are still seldom devoid of the powers of observation and independent thought, the science of bloodletting, over most of its long sway, appears usually to have been tempered with a certain amount of common sense. Or it was tempered until it reached its highest peaks of technique in, of all places, the young United States. Bloodletting went wild here. It had an enormously influential proponent in Dr Benjamin Rush, still revered as the greatest statesman–physician of our revolutionary and federal periods, and a genius of medical administration. Dr Rush Got Things Done. Among the things he got done, some of them good and useful, were to develop, practice, teach and spread the custom of bloodletting in cases where prudence or mercy had heretofore restrained its use. He and his students drained the blood of very young children, of consumptives, of the greatly aged, of almost anyone unfortunate to be sick in his realms of influence. His extreme practices aroused the alarm and horror of European bloodletting physicians. And yet as late as 1851, a committee appointed by the State Legislature of New York solemnly defended the thoroughgoing use of bloodletting. It scathingly ridiculed and censured a physician, William Turner, who had the temerity to write a pamphlet criticizing Dr Rush's doctrines and calling "the practice of taking blood in diseases contrary to common sense, to general experience, to enlightened reason and to the manifest laws of the divine Providence." Sick people needed fortifying, not draining, said Dr Turner, and he was squelched.

Medical analogies, applied to social organisms, are apt to be farfetched, and there is no point in mistaking mammalian chemistry for what occurs in a city. But analogies as to what goes on in the brains of earnest and learned men, dealing with complex phenomena they do not understand at all and trying to make do with a pseudoscience, do have a point. As in the pseudoscience of bloodletting, just so in the pseudoscience of city rebuilding and planning, years of learning and a plethora of subtle and complicated dogma have arisen on a foundation of nonsense. The tools of technique have steadily been perfected. Naturally, in time, forceful and able men, admired administrators, having swallowed the initial fallacies and having been provisioned with tools and with public confidence, go on logically to the greatest destructive excesses, which prudence or mercy might previously have forbade. Bloodletting could heal only by accident or insofar as it broke the rules, until the time when it was abandoned in favor of the hard, complex business of assembling, using and testing, bit by bit, true descriptions of reality drawn not from how it ought to be, but from how it is. The pseudoscience of city planning and its companion, the art of city design, have not yet broken with the specious comfort of wishes, familiar superstitions, oversimplifications, and symbols and have not yet embarked upon the adventure of probing the real world. . . .

One principle emerges so ubiquitously, and in so many and such complex different forms, that [it] . . . becomes the heart of my argument. This ubiquitous principle is the need of cities for a most intricate and close-grained diversity of uses that give each other constant mutual support, both economically and socially. The components of this diversity can differ enormously, but they must supplement each other in certain concrete ways.

I think that unsuccessful city areas are areas which lack this kind of intricate mutual support, and that the science of city planning and the art of city design, in real life for real cities, must become the science and art of catalyzing and nourishing these close-grained working relationships.

"Plight and Prospect"

from *Design With Nature* (1969)

Ian L. McHarg

Editors' Introduction

Born in 1921 near Glasgow in Scotland, Ian McHarg grew up observing the spread of that industrial city with all its poverty and grime, and found solace in long rambles through the countryside. The experience convinced him that nature should be an essential element of human communities. After service in World War II and training in landscape architecture at Harvard, he returned home to find many of his pastoral childhood haunts obliterated by urban development. Saddened, he embarked on a teaching and consulting career in which he continually sought ways to reintegrate nature with the city. He founded and for thirty years directed the University of Pennsylvania's Department of Landscape Architecture and Regional Planning, and also founded a well-known consulting firm, Wallace, McHarg, Roberts & Todd, that was involved in creating regional plans and large-scale design of new communities.

McHarg's 1969 book *Design With Nature* (New York: Natural History Press, 1969) played a crucial role in bringing together environmental and urban planning concerns in the mid-twentieth century. Part personal statement, part clarion call against environmental destruction, and part description of a method of environmental analysis using overlay maps, the book sold more than 250,000 copies. Along with other works such as Rachel Carson's *Silent Spring* (New York: Fawcett Crest, 1962), calling attention to the dangers of pesticides and other toxic chemicals, and Barry Commoner's *The Closing Circle* (New York: Knopf, 1971), likewise warning against the pollution and resource consumption impacts of technological society, it helped catalyze the modern environmental movement. But it did so in a way that was related more directly to city planning, landscape architecture, and urban design, exhorting these professions to integrate elements of the natural world into their work. As such, McHarg may be seen as the inheritor of the ecological concerns of Mumford and the RPAA, trying at a more specific level to figure out how urban growth might co-exist with the natural landscape. To do so he used the available technology of the time in the form of hand-drawn mapping overlays of information derived from natural sciences to determine where development should go. This method was a precursor to today's Geographic Information Systems (GIS) computer-based methods.

In retrospect McHarg may be seen as taking an overly optimistic view of the power of ecological science to order urban development, a view better fitted with a technocratic approach toward sustainable development than Leopold's call for inner change on a spiritual level. McHarg may also be criticized for ignoring the role of powerful social, cultural, and economic forces in promoting unsustainable development, and for a detached, Olympian rhetoric far different than Jacobs' emphasis on firsthand experience of life in a city's streets. But whatever his shortcomings, in *Design With Nature* – a book filled with lovely black-and-white photographs of life in all its forms – McHarg succeeded in issuing an eloquent alarm call about the unsustainability of twentieth-century development patterns, especially the rising tide of suburban growth that was the dominant and most unsustainable aspect of the century's urbanization. That alarm call and McHarg's invitation to co-ordinate design and nature resonated deeply with a generation.

Clearly the problem of man and nature is not one of providing a decorative background for the human play, or even ameliorating the grim city: it is the necessity of sustaining nature as source of life, milieu, teacher, sanctum, challenge and, most of all, of rediscovering nature's corollary of the unknown in the self, the source of meaning.

There are still great realms of empty ocean, deserts reaching to the curvature of the earth, silent, ancient forests and rocky coasts, glaciers and volcanoes, but what will we do with them? There are rich contented farms, and idyllic villages, strong barns and white-steepled churches, tree-lined streets and covered bridges, but these are residues of another time. There are, too, the silhouettes of all the Manhattans, great and small, the gleaming golden windows of corporate images – expressionless prisms suddenly menaced by another of our creations, the supersonic transport whose sonic boom may reduce this image to a sea of shattered glass.[1]

But what do we say now, with our acts in city and countryside? When I first addressed this question to Scotland in my youth, today the world directs the same question to the United States. What is our performance and example? What are the visible testaments to the American mercantile creed – the hamburger stand, gas station, diner, the ubiquitous billboards, sagging wires, the parking lot, car cemetery and that most complete conjunction of land rapacity and human disillusion, the subdivision. It is all but impossible to avoid the highway out of town, for here, arrayed in all its glory, is the quintessence of vulgarity, bedecked to give the maximum visibility to the least of our accomplishments.

And what of the cities? Think of the imprisoning gray areas that encircle the center. From here the sad suburb is an unrealizable dream. Call them no-place although they have many names. Race and hate, disease, poverty, rancor and despair, urine and spit live here in the shadows. United in poverty and ugliness, their symbol is the abandoned carcasses of automobiles, broken glass, alleys of rubbish and garbage. Crime consorts with disease, group fights group, the only emancipation is the parked car.

What of the heart of the city, where the gleaming towers rise from the dirty skirts of poverty? Is it like midtown Manhattan where twenty per cent of the population was found to be indistinguishable

from the patients in mental hospitals? Both stimulus and stress live here with the bitch goddess success. As you look at the faceless prisms do you recognize the home of *anomie*?

Can you find the river that first made the city? Look behind the unkempt industry, cross the grassy railroad tracks and you will find the rotting piers and there is the great river, scummy and brown, wastes and sewage bobbing easily up and down with the tide, endlessly renewed....

[...]

The American dream envisioned only the single-family house, the smiling wife and healthy children, the two-car garage, eye-level oven, foundation planting and lawn, the school nearby and the church of your choice. It did not see that a subdivision is not a community, that the sum of subdivisions that make a suburb is not a community, that the sum of suburbs that compose the metropolitan fringe of the city does not constitute community nor does a metropolitan region. It did not see that the nature that awaited the subdivider was vastly different from the pockmarked landscape of ranch and split-level houses.

And so the transformation from city to metropolitan area contains all the thwarted hopes of those who fled the old city in search of clean government, better schools, a more beneficent, healthy and safe environment, those who sought to escape slums, congestion, crime, violence and disease.

There are many problems caused by the form of metropolitan growth – the lack of institution which diminishes the power to effect even local decisions, the trauma that is the journey to work, the increasingly difficult problem of providing community facilities. Perhaps the most serious is the degree to which the subdivision, the suburb and the metropolitan area deny the dream and have failed to provide the smiling image of the advertisements. The hucksters made the dream into a cheap thing, subdivided we fell, and the instinct to find more natural environments became the impulse that destroyed nature, an important ingredient in the social objective of this greatest of all population migrations.

Let us address ourselves to this problem. In earlier studies we saw that certain types of land are of such intrinsic value, or perform work for man best in a natural condition or, finally, contain such hazards to development that they should not be urbanized. Similarly, there are other areas that, for perfectly specific reasons, are intrinsically suitable for urban uses. . . .

[I]t transpires, as we have seen before, that if one selects eight natural features, and ranks them in order of value to the operation of natural process, then that group reversed will constitute a gross order of suitability for urbanization. These are: surface water, floodplains, marshes, aquifer recharge areas, aquifers, steep slopes, forests and woodlands, un-forested land. . . .

[. . .]

The application of this model requires elaborate ecological inventories. Happily, recent technological advances facilitate these. Earth satellites with remote scanning devices with high-level air photography and ground-level identification can provide rich data and time series information on the dynamism of many natural processes. When such inventories are completed they can be constituted into a value system. They can also be identified not only in degrees of value but of tolerance and intolerance. These data, together with the conception of fitness, constitute the greatest immediate utility of the ecological model. Ecosystems can be viewed as fit for certain prospective land uses in a hierarchy. It is then possible to identify environments as fit for ecosystems, organisms and land uses. The more intrinsically an environment is fit for any of these, the less work of adaptation is necessary. Such fitting is creative. It is then a maximum-benefit/minimum-cost solution.

These inventories would then constitute a description of the world, continent or ecosystem under study as phenomena, as interacting process, as a value system, as a range of environments exhibiting degrees of fitness for organisms, men and land use. It would exhibit intrinsic form. It could be seen to exhibit degrees of health and pathology. The inventories would include human artifacts as well as natural processes.

Certainly the most valuable application of such inventories is to determine locations for land uses and most particularly for urbanization. Urban growth in the United States today consists of emptying the continent toward its seaboard conurbations, which expand by accretion and coalesce. This offers the majority of future necropolitans the choice between the environments of Bedford Stuyvesant and Levittown. There must be other alternatives. Let us ask the land where are the best sites. Let us establish criteria for many different types of excellence responding to a wide range of choice. We seek not only the maximum range of differing excellences *between* city locations, but the maximum range of choice *within* each one. . . .

In the quest for survival, success and fulfillment, the ecological view offers an invaluable insight. It shows the way for the man who would be the enzyme of the biosphere – its steward, enhancing the creative fit of man–environment, realizing man's design with nature.

NOTE

1 *Ed. note*: When McHarg wrote in the late 1960s there were plans to develop and mass-produce supersonic passenger airplanes, which were bitterly opposed by environmentalists in the 1970s and ultimately succumbed to public opposition. Only one – the Concorde – was ever built, and that can only be used supersonically over water.

"The Development of Underdevelopment"

from *Capitalism and Underdevelopment in Latin America* (1967)

Andre Gunder Frank

Editors' Introduction

In the 1960s a number of writers worldwide began to developed sophisticated analyses of why poverty persisted and even worsened despite the huge post-World War II international effort to promote global development. Development assistance by wealthy nations, they argued, was in many ways an extension of earlier colonization, and whatever its stated rationale functioned to make poverty and inequality in the Third World worse rather than better. Many writers followed Marx in developing extensive "structuralist" critiques of the institutions and mechanisms of capitalist economics.

One of the most important strains of this growing chorus of criticism was "dependency theory," largely developed in the late 1960s by a group of economists living in Latin America – Andre Gunder Frank, Raul Prebisch, Fernando Cadoszo, and James Caporaso – known as the Economic Commission on Latin America. Their work emerged in part in reaction to the trend within mainstream economics to focus on indigenous factors as the reason for the failure of the least developed countries ("LDCs") to develop, thus "blaming the poor."

The German-born Frank had studied in the United States with noted Marxist economist Paul Baran, whose book *The Political Economy of Growth* (New York: Monthly Review Press, 1957) had begun to analyze global underdevelopment. But after moving to Latin America in 1962 Frank came to see conventional development economics in the United States as "widely used to defend social irresponsibility, pseudo-scientific scientism, and political reaction." "I had to learn from those who have been persecuted," he wrote.[1] He and others argued that the main effect of capitalist development efforts was to make Third World economies and societies increasingly dependent on wealthier nations in Europe and North America. Frank in particular argued that development created a hierarchy of "satellites," each dependent on others above it in the global system, which took physical form within the towns and metropolitan areas of developing nations.

At mid-century many other writers began to question prevailing theories of economic development around the world, laying the groundwork for later calls for more sustainable forms of development. The work of famed Swedish economist Gunnar Myrdahl during the 1950s and 1960s focused heavily on inequity within capitalist systems. American economist Kenneth Boulding sought as early as the late 1940s to develop an ethical economics that could promote peace, co-operation, and a one-world perspective. Greek historian L.S. Stavrianos described the gradual growth of global inequality in works such as *Global Rift: The Third World Comes of Age* (New York: Morrow, 1981). Egyptian-born writer Samir Amin promoted theories similar to Frank's in books such as *Accumulation on a World Scale* (New York: Monthly Review Press, 1974) and *Maldevelopment: Anatomy of a Global Failure* (London: Zed Books, 1990). Famed linguist Noam Chomsky, in books such as *The Culture of Terrorism* (Boston: South End Press, 1988) and *Deterring Democracy* (London:

Verso, 1991), has been a consistent critic of the role of powerful political and economic forces in creating violence and exploitation worldwide. Finally, Indian-born, Nobel Prize-winning economist Amartya Sen has studied the recurrence of famines and the role of women within development, and in his book *Development as Freedom* (New York: Knopf, 1999) views development as a societal transformation guaranteeing more liberty for a greater number of people, rather than as an increase in economic production.

The style of critics such as Frank is very different from the prose of either Jacobs or McHarg. Schooled in academic Marxism, they adopted a more theoretical, formal language that found an audience among European and Third World intellectuals. Not surprisingly, given their Marxist orientation and theoretical style, these writers were not read widely in North America, although they drew extensively upon North American Marxist analysts such as Paul Baran and Paul Sweezy. They have also received criticism, such as for arguing that there is a single world capitalist system and that particular stages in development are inevitably to be found within it. However, their arguments are profound, and set the stage for several generations of international critique of globalization.

Underdevelopment in Brazil, as elsewhere, is the result of capitalist development. The military coup of April 1964 and the political and economic events which followed are the logical consequences of this. My purpose here is to trace and to explain the capitalist development of underdevelopment in Brazil since its settlement by Portugal in the sixteenth century and to show how and why, within the metropolis-satellite colonialist and imperialist structure of capitalism, even the economic and industrial development that Brazil is capable of is necessarily limited to an underdeveloped development. My intent is not an exhaustive study of Brazil *per se*; it is rather an attempt to use the case of Brazil to study the nature of underdevelopment and the limitations of capitalist development.

To account for the underdevelopment and limited development of Brazil, and similar areas, it is common to resort to a dualist model of society. Thus the French geographer Jaques Lambert says in his book *Os Dois Brasis* (The Two Brazils):

The Brazilians are divided into two systems of economic and social organization. . . . These two societies did not evolve at the same rate. . . . The two Brazils are equally Brazilian, but they are separated by several centuries. . . . In the course of the long period of colonial isolation, an archaic Brazilian culture was formed, a culture which keeps in isolation the same stability which still exists in the indigenous cultures of Asia and the Near East. . . . The dual economy and the dual social structure

which accompanies it are neither new nor characteristically Brazilian – they exist in all unequally developed countries. (Lambert 1959, pp. 105–112)

The same view is shared by Arnold Toynbee (1962) and many others. Celso Furtado (1962), Brazil's Minister of Planning until the April 1964 coup, refers to the one Brazil, the modern capitalist and industrially more advanced Brazil, as an open society and to the archaic rural Brazil as a closed society.

The essential argument of all these students is that the modern Brazil is more developed because it is an open capitalist society; and the other archaic Brazil remains underdeveloped because it is not open, particularly to the industrial part and to the world as a while, and not sufficiently capitalist, but rather pre-capitalist, feudal or semi-feudal. Development is then often viewed as diffusion: "In Brazil, the motor of evolution is everywhere in the cities, from which it radiates change to the countryside" (Lambert, p. 108). The underdeveloped Brazil would develop if only it would open up, and the more developed Brazil would develop still more if the other Brazil would stop being a drag on it and would open its market to industrial goods. My analysis of Brazil's historical and contemporary experience contends that this dualist model is factually erroneous and theoretically inadequate and misleading.

An alternative model may be advanced instead. As a photograph of the world taken at a point in time, this model consists of a world metropolis (today the United States) and its governing class, and its national and international satellites and their leaders – national satellites like the Southern states of the United States, and international satellites like São Paulo. Since São Paulo is a national metropolis in its own right, the model consists further of its satellites: the provincial metropolises, like Recife or Belo Horizonte, and their regional and local satellites in turn. That is, taking a photograph of a slice of the world we get a whole chain of metropolises and satellites, which runs from the world metropolis down to the hacienda or rural merchant who are satellites of the local commercial metropolitan center but who in their turn have peasants as their satellites. If we take a photograph of the world as a whole, we get a whole series of such constellations of metropolises and satellites.

There are several important characteristics of this model: (1) Close economic, political, social and cultural ties between each metropolis and its satellites, which result in the total integration of the farthest outpost and peasant into the system as a whole. This contrasts with the supposed isolation and non-incorporation of large parts of society according to the dualist model. (2) Monopolistic structure of the whole system, in which each metropolis holds monopoly power over its satellites; the source of form of this monopoly varies from one case to another, but the existence of this monopoly is universal throughout the system. (3) As occurs in any monopolistic system, misuse and misdirection of available resources throughout the whole system and metropolis–satellite chain. (4) As part of this misuse, the expropriation and appropriation of a large part or even all of and more than the economic surplus or surplus value of the satellite by its local, regional, national or international metropolis.

Instead of a photograph at a point in time, the model may be viewed as a moving picture of the course of history. It then shows the following characteristics: (1) Expansion of the system from Europe until it incorporates the entire planet in one world system and structure. (If the socialist countries have managed to escape from this system, then there are now two worlds – but in no case are there three.) (2) Development of capitalism, at first commercial and later also industrial, on a world scale as a single system. (3) Polarizing tendencies generic to the structure of the system at world, national, provincial, local and sectoral levels, which generate the development of the metropolis and the underdevelopment of the satellite. (4) Fluctuations within the system, like booms and depressions, which are transmitted from metropolis to satellite, and like the substitution of one metropolis by another, such as the passing of the metropolis from Venice to the Iberian peninsula to Holland to Britain to the United States. (5) Transformations within the system, such as the so-called Industrial Revolution. Among these transformations we give special emphasis below to important historical changes in the source or mechanism of monopoly which the capitalist world metropolis exercises over its satellites.

From this model in which metropolitan status generates development and satellite status generates underdevelopment, we may derive hypotheses about metropolis–satellite relations and their consequences which differ in important respects from some theses generally accepted, in particular those associated with the dualist model:

1   A metropolis (for example, a national metropolis) which is at the same time a satellite (of the world metropolis) will find that its development is not autonomous; it does not itself generate or maintain its development; it is a limited or misdirected development; it experiences, in a word, underdeveloped development.

2   The relaxation, weakening or absence of ties between metropolis and satellite will lead to a turning in upon itself on the part of the satellite, an involution, which may take one of two forms:

(a)   Passive capitalist involution toward or into a subsistence economy of apparent isolation and of extreme underdevelopment, such as that of the North and Northeast of Brazil. Here there may arise the apparently feudal or archaic features of the "other" sector of the dualist model. But these features are not original to the region, and they are not due to the region's or country's lack of incorporation into the system, as in the dualist model. On the contrary, they are due to and reflect precisely the region's ultra-incorporation, its strong (usually export) ties, which are followed by the region's temporary or permanent abandonment by the metropolis and by the relaxation of these ties.

(b) Weakening of ties together with active capitalist involution which may lead to more or less autonomous development or industrialization of the satellite, which is based on the metropolis–satellite relations of internal colonialism or imperialism. Examples of such active capitalist involution are the industrialization drives of Brazil, Mexico, Argentina, India and others during the Great Depression and the Second World War, while the metropolis was otherwise occupied. Development of the satellites thus appears not as the result of stronger ties with the metropolis, as the dualist model suggests, but occurs on the contrary because of the weakening of these ties. In the history of Brazil we find many cases of the first type of involution – in Amazonia, the Northeast, Minas Gerais, and Brazil as a whole – and one major instance of the second kind of involution in the case of São Paulo.

3 The renewal of stronger metropolis–satellite ties may correspondingly produce the following consequences in the satellite:

(a) The renewal of underdeveloped development consequent upon the reopening of the market for the retrenched region's export products, such as has occurred periodically in Brazil's Northeast. This apparent development is just as disadvantageous in the long run as was the satellite's initial metropolis-sponsored export economy: underdevelopment continues to develop.

(b) The strangulation and misdirection of the autonomous development undertaken by the satellite during the period of lesser ties through the renewal of stronger metropolis–satellite ties as the result of the recuperation of the metropolis after a depression, war, or other kinds of ups and downs. The inevitable result is the renewal of the generation of underdevelopment in the satellite, such as that which took place in the above-named countries after the Korean War.

4 There is a close interconnection of the economy and the socio-political structure of the satellite with those of the metropolis. The closer the satellite's links with and dependence on the metropolis, the closer is the satellite bourgeoisie, including the so-called "national bourgeoisie," linked and dependent on the metropolis. . . .

5 These ties, this growing interconnection, is accompanied by – no, produces – increasing polarization between the two ends of the metropolis–satellite chain in the world capitalist system. A symptom of this polarization is the growing international inequality of incomes and the absolute decline of the real income of the lower income recipients. Yet there is even more acute polarization at the lower end of the chain, between the national and/or local metropolises and their poorest rural and urban satellites whose absolute real income is steadily declining. This increasing polarization sharpens political tension, not so much between the international metropolis with its imperialist bourgeoisie and the national metropolises with national bourgeoisies, as between both of these and their rural and city slum satellites. This tension between the poles becomes sharper until the initiative and generation of the transformation of the system passes from the metropolitan pole, where it has been for centuries, to the satellite pole.

NOTE

1 *Capitalism and Underdevelopment in Latin America*, p. xviii.

REFERENCES

Furtardo, C. (1962) *A Pre-revoluca Brasileiro*. Rio de Janeiro: Editora Fundo de Cultura.

Lambert, J. (1959) *Os/dois Brasis*. Rio de Janeiro: Ministerio da Educacaoe Cultura.

Taynbee, A.J. (1962) America and the World Revolution. NY: Oxford University Press.

"Perspectives, Problems, and Models"

from the "Introduction" to
The Limits to Growth (1972)

Donella H. Meadows, Dennis L. Meadows,
Jörgen Randers, and William W. Behrens III

Editors' Introduction

One of the most influential books of the 1970s – and as far as we have been able to determine the first work ever to use the term "sustainable development" in its current sense – was *The Limits to Growth* (New York: Universe Books, 1972). This paperback bestseller by a team of MIT scientists catalyzed discussions world-wide about the future of human society. The book resulted from a study which the Club of Rome – an *ad hoc* group of global industrialists and humanitarians led by Italian economic consultant Aurelio Peccei – commissioned from Professor Jay Forrester and his graduate students at MIT. Given the emergence of computers as a powerful analytic tool at this time, Peccei asked the researchers to use computer models for the first time to attempt to analyze the future of the world. The group developed a model known as World3, using a method of analysis they called "systems dynamics."

The Forrester team analyzed the basic factors most likely to limit growth: population, agricultural production, natural resources, industrial production, and pollution. They concluded that following then-current trends the limits to the growth of human society on the planet would be reached within a hundred years, followed by a steep decline in global population and industrial capacity. Essentially, growing problems of resource depletion, pollution (including carbon dioxide concentration), loss of arable land, and declining food production would converge to halt progress. However, they also stated that it would be possible to alter these trends "to establish a condition of ecological and economic stability that is sustainable far into the future" (p. 24). The first of these conclusions shocked millions around the world, but fit with what many in the growing environmental movement had already concluded – that human development trends were headed in unsustainable directions. Other writers such as Rene Dubos, a French molecular biologist and originator of the phrase "think globally, act locally," Paul Ehrlich, a Stanford environmental scientist and author of the bestseller *The Population Bomb* (New York: Ballantine Books, 1968), Rachel Carson, Barry Commoner, and Ian McHarg had been saying much the same thing. The 1972 Stockholm Conference on Environment and Development and other United Nations events in the years that followed helped spread such ideas. But the *Limits to Growth* work was unique in that for the first time it used computer technology and scientific method to analyze the human future as well as ask questions such as whether growing human population and resource consumption were sustainable.

Not surprisingly, other writers vigorously opposed the *Limits to Growth* position, arguing that this approach was an alarmist recapitulation of arguments advanced by Thomas Malthus around 1800, comparing the linear increase in agricultural production with the geometric increase of population. These opponents argued that technology, economics, and human ingenuity would be able to help humanity surmount growth-related problems. In his 1981 book *The Ultimate Resource* (Princeton, NJ: Princeton University Press, 1981), business and marketing professor Julian Simon advanced the view that the world would never run out of resources. Scarcity-fueled increases in resource prices would encourage conservation or resource substitution, in his view, thus avoiding any long-term problems. Simon even bet Ehrlich $5,000 that the prices of a certain set of metals would in fact fall over the next five years, and won the bet (Ehrlich later claimed that the time frame was too short).

The *Limits to Growth* team revisited their work two decades later in a second book *Beyond the Limits* (Post Mills, VT: Chelsea Green, 1992). After running an updated version of their model and considering additional evidence over the twenty-year period, Meadows and her co-authors concluded that the fundamental themes of *Limits to Growth* had held up relatively well, and that the world had entered a period of "overshoot" in which it was well beyond sustainable levels of resource consumption, pollution, and population. Other conclusions remained the same – that it was still possible for humanity to change course, and that the sooner it did so the better.

The language of this selection reflects the sudden sense of global environmental crisis that many experienced in the late 1960s and 1970s, illustrated by the quote with which the authors start out from United Nations Secretary-General U Thant. Previously, the end of World War II had ushered in a period of optimism based on faith that economic progress and the spread of new technologies would bring about continual human betterment. However, by the end of the 1960s the world was in the midst of a cold war and a nuclear arms race, problems of global pollution and ecosystem degradation were being discovered, overpopulation was being recognized as a serious problem, and resources such as petroleum were suddenly seen to be limited. Events such as the 1973 energy crisis seemed to confirm this sense of crisis. Certainly the dangers of global catastrophe were overplayed by some. But certainly also the world's attention needed to be called to the challenge of living sustainably on a small planet in the long term. More than any other single work, *The Limits to Growth* helped to do this.

The style of this selection – especially that of principal author Donella Meadows – also reflects a rising awareness of the extent to which different worldviews or paradigms affect how problems are seen. The supposedly objective viewpoint of modern science, so strong at that time, was under increasing attack from many directions, especially from Thomas Kuhn, whose book *The Structure of Scientific Revolutions* (Chicago: University of Chicago Press, 1962) had argued that science, far from being completely objective and rational, in fact proceeds by moving through different paradigms dominant at different times. What is needed, Meadows and others argued, is an increased awareness of how our cognitive "lenses" affect our beliefs about global development. In particular, they believed that a new focus on the role of values in determining our beliefs and worldviews – and a rethinking of values themselves – was necessary in order to bring about more sustainable development practices.

I do not wish to seem overdramatic, but I can only conclude from the information that is available to me as Secretary-General, that the Members of the United Nations have perhaps ten years left in which to subordinate their ancient quarrels and launch a global partnership to curb the arms race, to improve the human environment, to defuse the population explosion, and to supply the required momentum to development efforts. If such a global partnership is not forged within the next decade, then I very much fear that the problems I have mentioned will have reached such staggering proportions that they will be beyond our capacity to control. (U Thant, 1969)

The problems U Thant mentions – the arms race, environmental deterioration, the population explosion, and economic stagnation – are often cited as the central, long-term problems of modern man. Many people believe that the future course of human society, perhaps even the survival of human society, depends on the speed and effectiveness with which the world responds to these issues. And yet only a small fraction of the world's population is actively concerned with understanding these problems or seeking their solutions.

HUMAN PERSPECTIVES

Every person in the world faces a series of pressures and problems that require his attention and action. These problems affect him at many different levels. He may spend much of his time trying to find tomorrow's food for himself and his family. He may be concerned about personal power or the power of the nation in which he lives. He may worry about a world war during his lifetime, or a war next week with a rival clan in his neighborhood. These very different levels of human concern can be represented on a graph like that in Figure 1. The graph has two dimensions, space and time. Every

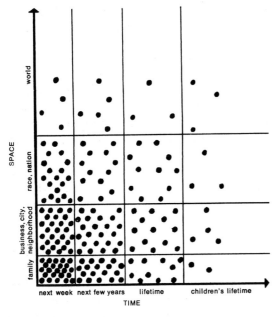

Figure 1. Human perspectives. Although the perspectives of the world's people vary in space and in time, every human concern falls somewhere on the space–time graph. The majority of the world's people are concerned with matters that affect only family or friends over a short period of time. Others look farther ahead in time or over a larger area – a city or a nation. Only a very few people have a global perspective that extends far into the future.

human concern can be located at some point on the graph, depending on how much geographical space it includes and how far it extends in time. Most people's worries are concentrated in the lower left-hand corner of the graph. Life for these people is difficult, and they must devote nearly all of their efforts to providing for themselves and their families, day by day. Other people think about and act on problems farther out on the space or time axes. The pressures they perceive involve not only themselves, but the community with which they identify. The actions they take extend not only days, but weeks or years into the future.

A person's time and space perspectives depend on his culture, his past experience, and the immediacy of the problems confronting him on each level. Most people must have successfully solved the problems in a smaller area before they move their concerns to a larger one. In general the larger the space and the longer the time associated with a problem, the smaller the number of people who are actually concerned with its solution.

There can be disappointments and dangers in limiting one's view to an area that is too small. There are many examples of a person striving with all his might to solve some immediate, local problem, only to find his efforts defeated by events occurring in a larger context. A farmer's carefully maintained fields can be destroyed by an international war. Local officials' plans can be overturned by a national policy. A country's economic development can be thwarted by a lack of world demand for its products. Indeed there is increasing concern today that most personal and national objectives may ultimately be frustrated by long-term, global trends such as those mentioned by U Thant.

Are the implications of these global trends actually so threatening that their resolution should take precedence over local, short-term concerns?

Is it true, as U Thant suggested, that there remains less than a decade to bring these trends under control?

If they are not brought under control, what will the consequences be?

What methods does mankind have for solving global problems, and what will be the results and the costs of employing each of them?

These are the questions that we have been investigating in the first phase of The Club of Rome's Project on the Predicament of Mankind. Our concerns thus fall in the upper right-hand corner of the space–time graph.

PROBLEMS AND MODELS

Every person approaches his problems, wherever they occur on the space–time graph, with the help of models. A model is simply an ordered set of assumptions about a complex system. It is an attempt to understand some aspect of the infinitely varied world by selecting from perceptions and past experience a set of general observations applicable to the problem at hand. A farmer uses a mental model of his land, his assets, market prospects, and past weather conditions to decide which crops to plant each year. A surveyor constructs a physical model – a map – to help in planning a road. An economist uses mathematical models to understand and predict the flow of international trade.

Decision-makers at every level unconsciously use mental models to choose among policies that will shape our future world. These mental models are, of necessity, very simple when compared with the reality from which they are abstracted. The human brain, remarkable as it is, can only keep track of a limited number of the complicated, simultaneous interactions that determine the nature of the real world.

We, too, have used a model. Ours is a formal, written model of the world.[1] It constitutes a preliminary attempt to improve our mental models of long-term, global problems by combining the large amount of information that is already in human minds and in written records with the new information-processing tools that mankind's increasing knowledge has produced – the scientific method, systems analysis, and the modern computer.

Our world model was built specifically to investigate five major trends of global concern – accelerating industrialization, rapid population growth, widespread malnutrition, depletion of nonrenewable resources, and a deteriorating environment. These trends are all interconnected in many ways, and their development is measured in decades or centuries, rather than in months or years. With the model we are seeking to understand the causes of these trends, their interrelationships, and their implications as much as one hundred years in the future.

The model we have constructed is, like every other model, imperfect, oversimplified, and unfinished. We are well aware of its shortcomings, but we believe that it is the most useful model now available for dealing with problems far out on the space–time graph. To our knowledge it is the only formal model in existence that is truly global in scope, that has a time horizon longer than thirty years, and that includes important variables such as population, food production, and pollution, not as independent entities, but as dynamically interacting elements, as they are in the real world.

Since ours is a formal, or mathematical, model it also has two important advantages over mental models. First, every assumption we make is written in a precise form so that it is open to inspection and criticism by all. Second, after the assumptions have been scrutinized, discussed, and revised to agree with our best current knowledge, their implications for the future behavior of the world system can be traced without error by a computer, no matter how complicated they become.

We feel that the advantages listed above make this model unique among all mathematical and mental world models available to us today. But there is no reason to be satisfied with it in its present form. We intend to alter, expand, and improve it as our own knowledge and the world data base gradually improve.

In spite of the preliminary state of our work, we believe it is important to publish the model and our findings now. Decisions are being made every day in every part of the world, that will affect the physical, economic, and social conditions of the world system for decades to come. These decisions cannot wait for perfect models and total understanding. They will be made on the basis of some model, mental or written, in any case. We feel that the model described here is already sufficiently developed to be of some use to decision-makers. Furthermore, the basic behavior modes we have already observed in this model appear to be so fundamental and general that we do not expect our broad conclusions to be substantially altered by further revisions.

It is not the purpose of this book to give a complete, scientific description of all the data and mathematical equations included in the world model. Such a description can be found in the final technical report of our project. Rather, in *The Limits to Growth* we summarize the main features of the model and our findings in a brief, nontechnical way. The emphasis is meant to be not on the equations or the intricacies of the model, but on what it tells us about the world. We have used a computer as a tool to aid our own understanding of the causes and consequences of the accelerating trends that characterize the modern world, but familiarity with computers is by no means necessary to comprehend or to discuss our conclusions. The implications of those accelerating trends raise issues that go far beyond the proper domain of a purely scientific document. They must be debated by a wider community than that of scientists alone. Our purpose here is to open that debate.

The following conclusions have emerged from our work so far. We are by no means the first group to have stated them. For the past several decades, people who have looked at the world with a global, long-term perspective have reached similar conclusions. Nevertheless, the vast majority of policy-makers seems to be actively pursuing goals that are inconsistent with these results.

Our conclusions are:

1   If the present growth trends in world population, industrialization, pollution, food production, and resource depletion continue unchanged, the limits to growth on this planet will be reached sometime within the next one hundred years. The most probable result will be a rather sudden and uncontrollable decline in both population and industrial capacity.

2   It is possible to alter these growth trends and to establish a condition of ecological and economic stability that is sustainable far into the future. The state of global equilibrium could be designed so that the basic material needs of each person on earth are satisfied and each person has an equal opportunity to realize his individual human potential.

3   If the world's people decide to strive for this second outcome rather than the first, the sooner they begin working to attain it, the greater will be their chances of success.

These conclusions are so far-reaching and raise so many questions for further study that we are quite frankly overwhelmed by the enormity of the job that must be done. We hope that this book will serve to interest other people, in many fields of study and in many countries of the world, to raise the space and time horizons of their concerns and to join us in understanding and preparing for a period of great transition – the transition from growth to global equilibrium.

NOTE

1   The prototype model on which we have based our work was designed by Professor Jay W. Forrester of the Massachusetts Institute of Technology. A description of that model has been published in his book *World Dynamics* (Cambridge, MA: Wright-Allen Press, 1971).

"The Steady-State Economy"
from *Toward a Steady-State Economy* (1973)

Herman E. Daly

Editors' Introduction

Although the field of economics provided much of the foundation for twentieth-century global development, a growing number of economists toward the middle and end of the century came to question its prevailing assumptions. Eloquent observers such as Boulding, Schumacher, and Hazel Henderson published works arguing that prevailing economic perspectives were deeply flawed, and that humanity needed to take environmental and social impacts of economic development into account as well as adopt a longer term perspective than that supplied by economics. Particularly influential were Shumacher's book *Small is Beautiful* (New York: Harper & Row, 1973), advocating "appropriate technology," and Boulding's call for an "economics for a spaceship earth" in books such as *Beyond Economics: Essays on Society, Religion, and Ethics* (Ann Arbor, MI: University of Michigan Press, 1968). In the 1980s and 1990s environmental economists and ecological economists worked in more specific ways to change the discipline. The former group has tried to integrate environmental and social concerns into existing economic tools, while the second group has gone further to try to put economics into the context of a broader global picture.

However, it is Herman Daly, a professor first at the University of Louisiana and later the University of Maryland, who has developed perhaps the most fundamental critique of traditional economics from a sustainability perspective. Picking up on an idea first mentioned by John Stuart Mill in the nineteenth century, Daly argued in the early 1970s that an economy based on endless growth in physical production was impossible, and called instead for a "steady-state economy" based on qualitative but not quantitative growth. Daly's preferred method for achieving this steady state was through depletion quotas on resources, through which the government would essentially auction off the right to consume basic resources. The amount allowed to be consumed would decrease over time, setting in place market mechanisms (through higher prices, conservation, better technologies, substitution and so on) to reduce consumption of that resource. Somewhat tongue-in-cheek, Daly even suggested that this mechanism be applied to the right to have children, so as to use economics to address the population problem.

Daly's ideas constituted a challenge to conventional economics that could not be ignored, and indeed questioning "growth" of all sorts – including population growth, metropolitan spatial growth, and growth in energy or resource consumption – has become a central feature of sustainability debates. For a time in the 1990s Daly was even hired by the World Bank as an in-house economic adviser, though his ideas appear to have had little impact on the Bank's overall policies. Although mainstream economists, politicians, and the global business community have yet to take the concept of a steady-state economy seriously, and assume continual growth in production and consumption to be essential for human welfare, Daly's ideas remain important as a leading philosophical alternative to endless growth, one that may eventually underpin a more sustainable society.

Like Leopold and Meadows, Daly embraced notions of ethical and spiritual change that went far beyond his professional field. He argued for changes in values and moral growth to serve as the underpinnings of a new economy, and developed these ideas further in collaboration with theologian John B. Cobb, Jr. in their book *For the Common Good* (Boston: Beacon Press, 1989). In that volume Daly and Cobb proposed an Index of Sustainable Economic Welfare that might serve as an alternative to the Gross Domestic Product (GDP) for measuring human well-being. Instead of a single measure of economic production, this indicator combines a wide range of measures of human welfare such as public expenditures on health and education, net capital growth, the value of household services, costs of pollution, depletion of non-renewable resources, and distributional inequality. The nongovernmental organization Redefining Progress, based in San Francisco, has produced a similar statistic called the Genuine Progress Indicator (for more information, see www.rprogress.org). Both measures show social welfare in the USA peaking in the early 1970s and declining ever since, due in part to rapidly growing inequities, environmental damage, and mounting personal costs for health and education.

GROWTHMANIA

The fragmentation of knowledge and people by excessive specialization, the disequilibrium between the human economy and the natural ecosystem, the congestion and pollution of our spatial dimension of existence, the congestion and pollution of our temporal dimension of existence with the resulting state of harried drivenness and stress – all these evils and more are symptomatic of the basic malady of growthmania.

"Growthmania" is an insufficiently pejorative term for the paradigm or mind-set that always puts growth in first place – the attitude that there is no such thing as enough, that cannot conceive of too much of a good thing. It is the set of unarticulated preconceptions which allows the President's Council of Economic Advisers to say, "If it is agreed that economic output is a good thing it follows by definition that there is not enough of it." As a sop to environmentalists the Council does admit that "growth of GNP has its costs, and beyond some point they are not worth paying." But instead of raising the obvious question – "What determines this point of optimal GNP, and how do we know when we have reached it?" – the Council merely pontificates that "the existing propensities of the population and the policies of the government constitute claims upon GNP itself that can only be satisfied by rapid economic growth." That of course is merely to restate the problem, not to give a solution. Apparently these "existing propensities and policies" are beyond discussion. That is growthmania. Brezhnev, Castro, and Franco receive much the same advice from their respective

Councils of Economic Advisers. Growthmania is ecumenical.

The answer to the avoided question "When do the costs of growth in GNP outweigh the benefits?" is contained in the question itself. This occurs when the decreasing marginal benefit of extra GNP becomes less than the increasing marginal cost. The marginal benefit is measured by the market value of extra goods and services – i.e., the increment in GNP itself in value units. But what statistical series measures the cost? Answer: *none!* That is growthmania; literally not counting the costs of growth.

But the worst is yet to come. We take the real costs of increasing GNP as measured by the defensive expenditures incurred to protect ourselves from the *unwanted* side effects of production, and *add* these expenditures to GNP rather than subtract them. We count the real costs as benefits – this is hyper-growthmania. Since the net benefit of growth can never be negative with this Alice-in-Wonderland accounting system, the rule becomes "grow forever" or at least until it kills you – and then count your funeral expenses as further growth. This is terminal hyper-growthmania. Is the water table falling? Dig deeper wells, build bigger pumps, and up goes GNP! Mines depleted? Build more expensive refineries to process lower grade ores, and up goes GNP! Soil depleted? Produce more fertilizer, etc. As we press against the carrying capacity of our physical environment, these "extra-effort" and "defensive expenditures" (which are really costs masquerading as benefits) will loom larger and larger. As more and more of the finite physical world is converted into wealth, less and less is left over as nonwealth – i.e. the nonwealth physical world becomes scarce, and in becoming scarce it gets a price and thereby becomes wealth. This creates the illusion of becoming better off, when in actuality we are becoming worse off. We may already have passed the point where the marginal cost of growth exceeds the marginal benefit. This suspicion is increased by looking at who absorb the costs and who receive the benefits. We all get some of each, but not equal shares. Who buys a second car or a third TV? Who lives in the most congested, polluted areas? The benefits of growth go mainly to the rich, the costs go mainly to the poor. That statement is based on casual empiricism – we do not have social accounts which allow us to say precisely who receive the benefits and who absorb the costs of growth, a fact which is itself very revealing. Ignorance, if not blissful, is often politically expedient.

Growthmania is the paradigm upon which stand the models and policies of our current political economy. The answer to every problem is growth.

For example:

1 Poverty? Grow more to provide more employment for the poor and more tax revenues for welfare programs.
2 Unemployment? Invest and grow to bolster aggregate demand and employment.
3 Inflation? Grow by raising productivity so that more goods will be chased by the same number of dollars and prices will fall.
4 Balance of payments? Grow more and increase productivity in order to increase exports. Cutting imports is seen only as a short-run stopgap, not a solution.
5 Pollution and depletion? Grow so we will be rich enough to afford the cost of cleaning up and of discovering new resources and technologies.
6 War? We must grow to be strong and have *both* guns and butter.

The list could be extended, but it can also be summarized in one sentence: The way to have your cake and eat it too is to make it grow.

Growthmania is the attitude in economic theory that begins with the theological assumption of infinite wants, and then with infinite hubris goes on to presume that the original sin of infinite wants has its redemption vouchsafed by the omnipotent savior of technology, and that the first commandment is to produce more and more goods for more and more people, world without end. And that this is not only possible, but desirable.

Environmental degradation is an iatrogenic disease induced by economic physicians who treat the basic malady of unlimited wants by prescribing unlimited economic growth. We experience environmental degradation in the form of increased scarcity of clean air, pure water, relaxed moments, etc. But the only way the growthmania paradigm knows to deal with scarcity is to recommend growth. Yet one certainly does not cure a treatment-induced disease by increasing the treatment dosage! Nevertheless the usual recommendation for combating pollution is to grow more because "a rising GNP will enable the nation more easily to bear the costs of eliminating pollution." Such a view is patently absurd.

The growth paradigm has outlived its usefulness. It is a senile ideology that should be unceremoniously retired into the history of economic doctrines. In the terminology of Thomas Kuhn's book, *The Structure of Scientific Revolutions*, the growth paradigm has been more than exhausted by the normal science puzzle-solving research done within its confines. Political economy must enter a period of revolutionary science to establish a new paradigm to guide a new period of normal science. Just as mercantilism gave way to physiocracy, physiocracy to classical *laissez-faire*, *laissez-faire* to Keynesianism, Keynesianism to the neoclassical growth synthesis – so the current neoclassical growthmania must give way to a new paradigm. What will the new paradigm be? I submit that it must be very similar to an idea from classical economics that never attained the status of a paradigm, except for a brief chapter in John Stuart Mill's *Principles of Political Economy*. This idea is that of the steady-state economy.

THE STEADY STATE

What is meant by a "steady-state economy"?

Why is it necessary?

How can it be attained?

The first two questions are relatively easy and have been dealt with elsewhere. Hence they will be treated rapidly. The third question is extremely difficult, and will be the main focus of attention.

The steady state is defined as an economy in which the total population and the total stock of physical wealth are maintained constant at some desired levels by a "minimal" rate of maintenance throughput (i.e., by birth and death rates that are equal at the lowest feasible level, and by physical production and consumption rates that are equal at the lowest feasible level). The first part of the definition (constant stocks) goes back to John Stuart Mill, and the second part ("minimal" flow of throughput) goes back to 1949 vintage Kenneth Boulding. Minimizing throughput implies maximizing the average life expectancy of a member of the stock.

Why is the steady state necessary? Not for the reasons given by the classical economists who saw increasing rent and interest eliminating profit and thus the incentive for "progress." Rather, the necessity follows immediately from physical first principles. The world is finite, the ecosystem is a steady state. The human economy is a subset of the steady-state ecosystem. Therefore at some level and over some time period the subsystem must also become a steady state, at least in its physical dimensions of people and physical wealth. The steady-state economy is therefore a physical necessity....

When we raise the third question, how to attain the steady state, things become more difficult. First, we must give operational definitions to the specific goals contained in the definition of steady state. Second, we must specify the technologies, social institutions, and moral values which are in harmony with and supportive of the steady state.

To define more clearly the goal of the steady state we must face four questions.

1 At what levels should the stocks of wealth and people be maintained constant? Specifying the stock of wealth and of people also specifies the wealth per person or standard of living. In other words the question becomes the old one of what is the optimum population? So far no one has given a definite answer, and I certainly cannot.... [T]he optimum population is more likely to be discovered by experience than by a priori thought. We should attain a stationary population at some feasible nearby level. After experiencing it we could then decide whether the optimum level is above or below the current level....

2 What is the optimal level of maintenance throughput for a given level of stocks? For the time being the answer is probably "as low as possible" or at least "less than at present"....

3 What is the optimal time horizon or accounting period over which population and wealth are required to be constant? Obviously we cannot mean day-to-day constancy and probably not even year-to-year constancy. Related to this is the question of the optimum amplitude of fluctuation around the steady-state mean during the accounting period.

Again, I cannot pretend to be able to answer this question.... But somehow [in standard economic theory] we manage to choose an accounting period and muddle through, and so we could also in a steady state....

4 What is the optimal rate of transition from the growing economy to the steady state? We can never attain a steady state in the long run if our efforts to do so kill us in the short run. In the case of population there are interesting trade-offs between speed of attainment of a stationary population versus size of the stationary population and the amplitude of fluctuations in the birth rate induced by the current nonequilibrium age structure.

Once again I do not know the optimum rate of transition. But I think we are very unlikely to exceed it. In any case the sooner we begin deceleration to zero growth the longer we can afford to take and the less disruptive that adjustment will be. The important thing from all points of view is to begin the deceleration now. Later we can argue about the optimum rate....

[...]

CONSTANT PHYSICAL WEALTH

.... Let quotas be set on new depletion of each of the basic resources, both renewable and non-renewable, during a given time. The legal right to deplete to the amount of the quota for each resource would be auctioned off by the government at the beginning of each time period, in conveniently divisible units, to private firms, individuals, and public enterprises. After puchase from the government the quota rights would be freely transferable by sale or gift. As population growth and economic growth press against resources, the prices of the depletion quotas would be driven higher and higher. In the interests of conserving nonrenewable resources and optimal exploitation of renewable resources, quotas could then be reduced to lower levels, thereby driving the price of the quotas still higher. In this way, the increasing windfall rents resulting from increasing pressure of demand on a fixed supply would be collected by the government through the auctioning of the depletion rights. The government spends the revenues, let us say, by paying a social dividend. Even though the monetary flow is therefore undiminished, the real flow has been physically limited by the resource quotas. All prices of resources and of goods then increase, the prices of resource-intensive goods increase relatively more, and total resource consumption (depletion) is reduced. Moreover, in accordance with the law of conservation of matter-energy, reduction of initial inputs will result in reduction of ultimate outputs (pollution), reducing the aggregate throughput and with it the stress it puts on the ecosystem.

With depletion now made more expensive and with higher prices on final goods, recycling becomes more profitable. As recycling increases pollution is reduced even more. Higher prices make consumers more interested in durability and careful maintenance of wealth. Most importantly, prices now provide a strong incentive to develop new technologies and patterns of consumption that are resource saving. If there is any static inefficiency incurred in setting the rate of depletion outside the market (a doubtful point), it is likely to be more than offset by the dynamic benefits of greater inducements to develop resource-saving technology.

Adjustment of the throughput of depletion and pollution flows to long-run ecologically sustainable levels can be effected gradually. At first depletion quotas could be set at the preceding year's levels, and if necessary gradually reduced by, say, two per cent per year until we reach the "optimal" throughput. Stocks will then adjust to equilibrium with the new throughput. Thereafter the constant stock would be maintained by the constant throughput. As we gradually exhaust nonrenewable resources, quotas for their depletion will approach zero and recycling will become the only source of inputs, at which time, presumably, the ever rising

price of the resource will have led to the development of a recycling technology. . . .

[. . .]

Here only still rests some hopes on new technology ads.

CONTROL OF DISTRIBUTION

Distribution is the rock upon which most ships of state, including the steady state, are very likely to run aground. Currently we seek to improve distribution by establishing a minimum standard of living guaranteed by a negative income tax. In the growthmania paradigm there is no upper limit to the standard of living. In the steady-state paradigm there is an upper limit. Furthermore the higher the lower limit below which no one is allowed to fall, the lower must be the upper limit above which no one is allowed to rise. The lower limit has considerable political acceptance, the upper limit does not. But in the steady state the upper limit is a logical necessity. It implies confiscation and redistribution of wealth above a certain limit per person or per family. What does one say to the cries of "destruction of incentive"? Remember – we are no longer anxious to grow in the first place! Also one recalls Jonathan Swift's observation:

> In all well-instituted commonwealths, care has been taken to limit men's possessions; which is done for many reasons, and, among the rest, for one which, perhaps, is not often considered; that when bounds are set to men's desires, after they have acquired as much as the laws will permit them, their private interest is at an end, and they have nothing to do but to take care of the public.[1]

The basic institution for controlling distribution is very simple: set maximum and minimum limits on wealth and income, the maximum limit on wealth being the most important. Such a proposal is in no way an attack on private property. Indeed, as John Stuart Mill argues, it is really a defense of private property:

> Private property, in every defense made of it, is supposed to mean the guarantee to individuals of the fruits of their own labor and abstinence. The guarantee to them of the fruits of the labor

and abstinence of others, transmitted to them without any merit or exertion of their own, is not of the essence of the institution, but a mere incidental consequence which, when it reaches a certain height, does not promote, but conflicts with, the ends which render private property legitimate.[2]

According to Mill, private property is legitimated as a bastion against exploitation. But this is true only if everyone owns some minimum amount. Otherwise, when some own a great deal of it and others have very little, private property becomes an *instrument* of exploitation rather than a guarantee against it. It is implicit in this view that private property is legitimate only if there is some distributist institution (like, for example, the Jubilee year of the Old Testament) that keeps inequality of wealth within some tolerable limits. Such an institution is now lacking. The proposed institution of maximum and minimum wealth and income limits would remedy this severe defect and make private property legitimate again. Also it would go a long way toward legitimating the free market, since most of our blundering interference with the price system (e.g., the farm program, the minimum wage) has as its goal an equalizing alteration in the distribution of income and wealth. . . .

[. . .]

ON MORAL GROWTH

Is the above sketch of a steady state unrealistic and idealistic? On the contrary, it is in broad characteristics the only realistic possibility. The present economy is literally unrealistic because in its disregard for natural laws it is attempting the impossible. The steady-state paradigm, unlike growthmania, is realistic because it takes the physical laws of nature as its first premise.

Let us assume for a moment that the necessity of the steady state and the above outline of its appropriate technologies and social institutions are accepted. Logic and necessity are not sufficient to bring about social reform. The philosopher Leibnitz observed that,

> If geometry conflicted with our passions and interests as much as do ethics, we would contest it and violate it as much as we do ethics now, in spite of all the demonstrations of Euclid and Archimedes, which would be labeled paralogisms and dreams.[3]

Leibnitz is surely correct. However logical and necessary the above outline of the steady state, it is, on the assumption of static morality, nothing but a dream. The physically steady economy absolutely requires moral growth beyond the present level.

NOTES

1 Jonathan Swift, "Thoughts on Various Subjects," reprinted in G.B. Woods *et al.* (eds), *The Literature of England*, New York: Scott, Foresman and Company, 1958, p. 1003.
2 John Stuart Mill, "Of Property," in *Principles of Political Property*, Volume II, London: John Parker and Son, 1857, Chapter 1.
3 Leibnitz quoted in A. Sauvy, *The General Theory of Population*, New York: Basic Books, 1970, p. 270.

"Towards Sustainable Development"

excerpted from *Our Common Future* (1987)

World Commission on Environment and Development (The Brundtland Commission)

Editors' Introduction

No event did more to push sustainable development into the mainstream of worldwide policy debates than the 1987 release of the report of the World Commission on Environment and Development, commonly known as the Brundtland Commission. Widely distributed as a trade paperback entitled *Our Common Future* (New York: Norton, 1987), this volume formulated what has become the standard definition of sustainable development ("development that meets the needs of the present without jeopardizing the ability of future generations to meet their own needs"). A directive of the United Nations General Assembly established the commission in 1983, in the tradition of several previous influential UN-affiliated international commissions, the Palme Commission on Security and Disarmament and the Brandt Commission on North–South Issues. Chaired by former Norwegian Prime Minister Gro Harlem Brundtland, the Commission consisted of leading citizens from twenty-one nations. The group held public hearings on five continents, reviewed 10,000 pages of testimony, sought advice from numerous experts and advisory panels, and commissioned more than seventy-five studies and reports.

A rich compendium of analysis and strategies, the Brundtland Commission report succeeded remarkably well at calling global attention to the need for sustainable development and developing a common formulation of this concept. The Commission helped establish a strong foundation for the United Nations Conference on Environment and Development held in Rio de Janeiro in 1992 and many other subsequent events and programs. However, it has been criticized on many grounds as well, particularly for accepting conventional notions of continued economic growth as the path to improved human welfare, for insufficiently incorporating an analysis of global power relations, and for developing a definition of sustainable development that is highly anthropocentric and dependent on the difficult-to-define concept of "needs."

Other leading reports that called attention to the need for sustainable development in general, and sustainable urban development in particular, included the Worldwatch Reports from the Worldwatch Institute (www.worldwatch.org), an influential series of pamphlet-style analyses that began in 1975, the annual *State of the World* books published by the same organization beginning in 1984, the Global 2000 Report to US President Jimmy Carter in 1980, and annual *World Conservation Strategy* reports from the World Conservation Union beginning in 1980. The establishment of national groups such as Canada's National Roundtable on the Environment and Economy has emulated the work of the Brundtland Commission on a smaller scale, helping to place sustainability issues on public agendas.

A CALL FOR ACTION

Over the course of this century, the relationship between the human world and the planet that sustains it has undergone a profound change.

When the century began, neither human numbers nor technology had the power radically to alter planetary systems. As the century closes, not only do vastly increased human numbers and their activities have that power, but major, unintended changes are occurring in the atmosphere, in soils, in waters, among plants and animals, and in the relationships among all of these. The rate of change is outstripping the ability of scientific disciplines and our current capabilities to assess and advise. It is frustrating the attempts of political and economic institutions, which evolved in a different, more fragmented world, to adapt and cope. It deeply worries many people who are seeking ways to place those concerns on the political agendas.

The onus lies with no one group of nations. Developing countries face the obvious life-threatening challenges of desertification, deforestation, and pollution, and endure most of the poverty associated with environmental degradation. The entire human family of nations would suffer from the disappearance of rain forests in the tropics, the loss of plant and animal species, and changes in rainfall patterns. Industrial nations face the life-threatening challenges of toxic chemicals, toxic wastes, and acidification. All nations may suffer from the releases by industrialized countries of carbon dioxide and of gases that react with the ozone layer, and from any future war fought with the nuclear arsenals controlled by those nations. All nations will have a role to play in changing trends, and in righting an international economic system that increases rather than decreases inequality, that increases rather than decreases numbers of poor and hungry.

The next few decades are crucial. The time has come to break out of past patterns. Attempts to maintain social and ecological stability through old approaches to development and environmental protection will increase instability. Security must be sought through change. The Commission has noted a number of actions that must be taken to reduce risks to survival and to put future development on paths that are sustainable. Yet we are aware that such a reorientation on a continuing basis is simply beyond the reach of present decision-making structures and institutional arrangements, both national and international.

This Commission has been careful to base our recommendations on the realities of present institutions, on what can and must be accomplished today. But to keep options open for future generations, the present generation must begin now, and begin together.

To achieve the needed changes, we believe that an active follow-up of this report is imperative. It is with this in mind that we call for the UN General Assembly, upon due consideration, to transform this report into a UN Programme on Sustainable Development. Special follow-up conferences could be initiated at the regional level. Within an appropriate period after the presentation of this report to the General Assembly, an international conference could be convened to review progress made, and to promote follow-up arrangements that will be needed to set benchmarks and to maintain human progress.

First and foremost, this Commission has been concerned with people – of all countries and all walks of life. And it is to people that we address our report. The changes in human attitudes that we call for depend on a vast campaign of education, debate, and public participation. This campaign must start now if sustainable human progress is to be achieved.

The members of the World Commission on Environment and Development came from 21 very different nations. In our discussions, we disagreed often on details and priorities. But despite our widely differing backgrounds and varying national and international responsibilities, we were able to agree to the lines along which change must be drawn.

We are unanimous in our conviction that the security, well-being, and very survival of the planet depend on such changes, now.

A THREATENED FUTURE

The Earth is one but the world is not. We all depend on one biosphere for sustaining our lives. Yet each community, each country, strives for survival and prosperity with little regard for its impact on others. Some consume the Earth's resources at a rate that would leave little for future generations. Others, many more in number, consume far too little and live with the prospect of hunger, squalor, disease, and early death.

Yet progress has been made. Throughout much of the world, children born today can expect to live longer and be better educated than their parents. In many parts, the new-born can also expect to attain a higher standard of living in a wider sense. Such progress provides hope as we contemplate the improvements still needed, and also as we face our failures to make this Earth a safer and sounder home for us and for those who are to come.

The failures that we need to correct arise both from poverty and from the short-sighted way in which we have often pursued prosperity. Many parts of the world are caught in a vicious downwards spiral: Poor people are forced to overuse environmental resources to survive from day to day, and their impoverishment of their environment further impoverishes them, making their survival ever more difficult and uncertain. The prosperity attained in some parts of the world is often precarious, as it has been secured through farming, forestry, and industrial practices that bring profit and progress only over the short term.

Societies have faced such pressures in the past and, as many desolate ruins remind us, sometimes succumbed to them. But generally these pressures were local. Today the scale of our interventions in nature is increasing and the physical effects of our decisions spill across national frontiers. The growth in economic interaction between nations amplifies the wider consequences of national decisions. Economics and ecology bind us in ever-tightening networks. Today, many regions face risks of irreversible damage to the human environment that threaten the basis for human progress.

These deepening interconnections are the central justification for the establishment of this Commission. We traveled the world for nearly three years, listening. At special public hearings organized by the Commission, we heard from government leaders, scientists, and experts, from citizens' groups concerned about a wide range of environment and development issues, and from thousands of individuals – farmers, shanty-town residents, young people, industrialists, and indigenous and tribal peoples.

We found everywhere deep public concern for the environment, concern that has led not just to protests but often to changed behaviour. The challenge is to ensure that these new values are more adequately reflected in the principles and operations of political and economic structures.

We also found grounds for hope: that people can cooperate to build a future that is more prosperous, more just, and more secure; that a new era of economic growth can be attained, one based on policies that sustain and expand the Earth's resource base; and that the progress that some have known over the last century can be experienced by all in the years ahead. But for this to happen, we must understand better the symptoms of stress that confront us, we must identify the causes, and we must design new approaches to managing environmental resources and to sustaining human development.

SYMPTOMS AND CAUSES

Environmental stress has often been seen as the result of the growing demand on scarce resources and the pollution generated by the rising living standards of the relatively affluent. But poverty itself pollutes the environment, creating environmental stress in a different way. Those who are poor and hungry will often destroy their immediate environment in order to survive: They will cut down forests, their livestock will overgraze grasslands; they will overuse marginal land; and in growing numbers they will crowd into congested cities. The cumulative effect of these changes is so far-reaching as to make poverty itself a major global scourge.

On the other hand, where economic growth has led to improvements in living standards, it has sometimes been achieved in ways that are globally damaging in the longer term. Much of the improvement in the past has been based on the use of increasing amounts of raw materials, energy, chemicals, and synthetics and on the creation of pollution that is not adequately accounted for in figuring the costs of production processes. These trends have had unforeseen effects on the environment. Thus today's environmental challenges arise both from the lack of development and from the unintended consequences of some forms of economic growth. . . .

[. . .]

Sustainable development is development that meets the needs of the present without compromising the ability of future generations to meet their own needs. It contains within it two key concepts:

▪ the concept of 'needs', in particular the essential needs of the world's poor, to which overriding priority should be given; and
▪ the idea of limitations imposed by the state of technology and social organization on the environment's ability to meet present and future needs.

Thus the goals of economic and social development must be defined in terms of sustainability in all countries – developed or developing, market-oriented or centrally planned. Interpretations will vary, but must share certain general features and must flow from a consensus on the basic concept of sustainable development and on a broad strategic framework for achieving it.

Development involves a progressive transformation of economy and society. A development path that is sustainable in a physical sense could theoretically be pursued even in a rigid social and political setting. But physical sustainability cannot be secured unless development policies pay attention to such considerations as changes in access to resources and in the distribution of costs and benefits. Even the narrow notion of physical sustainability implies a concern for social equity between generations, a concern that must logically be extended to equity within each generation.

THE CONCEPT OF SUSTAINABLE DEVELOPMENT

The satisfaction of human needs and aspirations is the major objective of development. The essential needs of vast numbers of people in developing countries – for food, clothing, shelter, jobs – are not being met, and beyond their basic needs these people have legitimate aspirations for an improved quality of life. A world in which poverty and inequity are endemic will always be prone to ecological and other

crises. Sustainable development requires meeting the basic needs of all and extending to all the opportunity to satisfy their aspirations for a better life.

Living standards that go beyond the basic minimum are sustainable only if consumption standards everywhere have regard for long-term sustainability. Yet many of us live beyond the world's ecological means, for instance in our patterns of energy use. Perceived needs are socially and culturally determined, and sustainable development requires the promotion of values that encourage consumption standards that are within the bounds of the ecologically possible and to which all can reasonably aspire.

Meeting essential needs depends in part on achieving full growth potential, and sustainable development clearly requires economic growth in places where such needs are not being met. Elsewhere, it can be consistent with economic growth, provided the content of growth reflects the broad principles of sustainability and nonexploitation of others. But growth by itself is not enough. High levels of productive activity and widespread poverty can coexist, and can endanger the environment. Hence sustainable development requires that societies meet human needs both by increasing productive potential and by ensuring equitable opportunities for all.

An expansion in numbers can increase the pressure on resources and slow the rise in living standards in areas where deprivation is widespread. Though the issue is not merely one of population size but of the distribution of resources, sustainable development can only be pursued if demographic developments are in harmony with the changing productive potential of the ecosystem.

A society may in many ways compromise its ability to meet the essential needs of its people in the future – by overexploiting resources, for example. The direction of technological developments may solve some immediate problems but lead to even greater ones. Large sections of the population may be marginalized by ill-considered development.

Settled agriculture, the diversion of watercourses, the extraction of minerals, the emission of heat and noxious gases into the atmosphere, commercial forests, and genetic manipulation are all examples of human intervention in natural systems during the course of development. Until recently, such interventions were small in scale and their impact limited. Today's interventions are more drastic in scale and impact, and more threatening to life-support systems both locally and globally. This need not happen. At a minimum, sustainable development must not endanger the natural systems that support life on Earth: the atmosphere, the waters, the soils, and the living beings.

Growth has no set limits in terms of population or resource use beyond which lies ecological disaster. Different limits hold for the use of energy, materials, water, and land. Many of these will manifest themselves in the form of rising costs and diminishing returns, rather than in the form of any sudden loss of a resource base. The accumulation of knowledge and the development of technology can enhance the carrying capacity of the resource base. But ultimate limits there are, and sustainability requires that long before these are reached, the world must ensure equitable access to the constrained resource and reorient technological efforts to relieve the pressure.

Economic growth and development obviously involve changes in the physical ecosystem. Every ecosystem everywhere cannot be preserved intact. A forest may be depleted in one part of a watershed and extended elsewhere, which is not a bad thing if the exploitation has been planned and the effects on soil erosion rates, water regimes, and genetic losses have been taken into account. In general, renewable resources like forests and fish stocks need not be depleted provided the rate of use is within the limits of regeneration and natural growth. But most renewable resources are part of a complex and interlinked ecosystem, and maximum sustainable yield must be defined after taking into account system-wide effects of exploitation.

As for nonrenewable resources, like fossil fuels and minerals, their use reduces the stock available for future generations. But this does not mean that such resources should not be used. In general the rate of depletion should take into account the criticality of that resource, the availability of technologies for minimizing depletion, and the likelihood of substitutes being available. Thus land should not be degraded beyond reasonable recovery. With minerals and fossil fuels, the rate of depletion and the emphasis on recycling and economy of use should be calibrated to ensure that the resource does not run out before acceptable substitutes are available. Sustainable development requires that the rate of depletion of nonrenewable resources should foreclose as few future options as possible.

Development tends to simplify ecosystems and to reduce their diversity of species. And species, once extinct, are not renewable. The loss of plant and animal species can greatly limit the options of future generations; so sustainable development requires the conservation of plant and animal species.

So-called free goods like air and water are also resources. The raw materials and energy of production processes are only partly converted to useful products. The rest comes out as wastes. Sustainable development requires that the adverse impacts on the quality of air, water, and other natural elements are minimized so as to sustain the ecosystem's overall integrity.

In essence, sustainable development is a process of change in which the exploitation of resources, the direction of investments, the orientation of technological development, and institutional change are all in harmony and enhance both current and future potential to meet human needs and aspirations.

"The Rio Declaration on Environment and Development", "Introduction to Chapter 7" from *Agenda 21* (1992),[1] and the "Istanbul Declaration on Human Settlements"

United Nations

Editors' Introduction

The 1992 United Nations Conference on Environment and Development held in Rio de Janeiro – the "Earth Summit" – was like the Brundtland Commission a watershed event that helped alert the global media and the public to the need for sustainable development. This conference, attended by leaders of 178 nations, produced a lengthy declaration known as *Agenda 21* that laid out sustainable development principles in many different areas. The Rio Declaration on Environment and Development below summarizes some of its major themes. The Earth Summit also led to the creation of a United Nations Commission on Sustainable Development which meets annually to review international implementation efforts, and a UN Division for Sustainable Development to co-ordinate the agency's work in this field. More information on the latter is available at www.un.org/esa/sustdev.

Chapter 7 of *Agenda 21* laid out directions for sustainable urban development, and, at the urging of non-governmental organizations (NGOs) such as the International Council on Local Environmental Initiatives (ICLEI), Chapter 28 on implementation stipulates that "by 1996, most local authorities in each country should have undertaken a consultative process with their population and achieved a consensus on a local Agenda 21 for the(ir) communities." This mandate for "Local Agenda 21" planning has stimulated a large number of local planning initiatives, especially in the United Kingdom, northern Europe, and a number of developing nations.

The Earth Summit was followed in the early and mid-1990s by other United Nations conferences on global population, social development, women, and urban development. The 1996 Habitat II "City Summit," held in Istanbul, produced a lengthy consensus document on urban development principles, and perhaps just as importantly featured a huge exhibition of global best practices in sustainable city-building. The 2002 World Summit on Sustainable Development, held in Johannesburg, South Africa, sought to review global progress in the ten years following the Rio Conference. Largely ignored by the US administration of George W. Bush, which successfully persuaded delegates to avoid specific timetables for change, this event nonetheless produced some additional financial commitments to sustainable development programs from a number of European nations and the European Union.

Despite optimistic rhetoric from some agencies, international declarations such as *Agenda 21* have often been downplayed or ignored by national governments. It is debatable to what extent the UN treaties, conferences, and declarations represent an important global process of education, consensus-building, and action, and to what extent they are ineffectual gestures or window-dressing for the status quo. Some might argue that the global trade institutions set up during the 1990s, such as the World Trade Organization, represent a second, more powerful global governance mechanism entirely bypassing the United Nations – one promoting the economic goals of powerful nations at the expense of sustainable development for all countries. However, it is clear that the Earth Summit and other UN-related events have indeed played at least some role in promoting global debates about sustainability, and, even if ignored by nations such as the USA in the short term, have helped lay the groundwork for international action in the long run.

THE RIO DECLARATION ON ENVIRONMENT AND DEVELOPMENT

The United Nations Conference on Environment and Development, having met at Rio de Janeiro from 3 to 14 June 1992, reaffirming the Declaration of the United Nations Conference on the Human Environment, adopted at Stockholm on 16 June 1972,[2] and seeking to build upon it, with the goal of establishing a new and equitable global partnership through the creation of new levels of cooperation among States, key sectors of societies and people, working towards international agreements which respect the interests of all and protect the integrity of the global environmental and developmental system, recognizing the integral and interdependent nature of the Earth, our home, proclaims that:

Principle 1

Human beings are at the center of concerns for sustainable development. They are entitled to a healthy and productive life in harmony with nature.

Principle 2

States have, in accordance with the Charter of the United Nations and the principles of international law, the sovereign right to exploit their own resources pursuant to their own environmental and developmental policies, and the responsibility to ensure that activities within their jurisdiction or control do not cause damage to the environment of other States or of areas beyond the limits of national jurisdiction.

Principle 3

The right to development must be fulfilled so as to equitably meet developmental and environmental needs of present and future generations.

Principle 4

In order to achieve sustainable development, environmental protection shall constitute an integral part of the development process and cannot be considered in isolation from it.

Principle 5

All States and all people shall cooperate in the essential task of eradicating poverty as an indispensable requirement for sustainable development, in order to decrease the disparities in standards of living and better meet the needs of the majority of the people of the world.

Principle 6

The special situation and needs of developing countries, particularly the least developed and those most environmentally vulnerable, shall be given special priority. International actions in the field of environment and development should also address the interests and needs of all countries.

Principle 7

States shall cooperate in a spirit of global partnership to conserve, protect and restore the health and integrity of the Earth's ecosystem. In view of the different contributions to global environmental degradation, States have common but differentiated responsibilities. The developed countries acknowledge the responsibility that they bear in the international pursuit of sustainable development in view of the pressures their societies place on the global environment and of the technologies and financial resources they command.

Principle 8

To achieve sustainable development and a higher quality of life for all people, States should reduce and eliminate unsustainable patterns of production and consumption and promote appropriate demographic policies.

Principle 9

States should cooperate to strengthen endogenous capacity-building for sustainable development by improving scientific understanding through exchanges of scientific and technological knowledge, and by enhancing the development, adaptation, diffusion and transfer of technologies, including new and innovative technologies.

Principle 10

Environmental issues are best handled with the participation of all concerned citizens, at the relevant level. At the national level, each individual shall have appropriate access to information concerning the environment that is held by public authorities, including information on hazardous materials and activities in their communities, and the opportunity to participate in decision-making processes. States shall facilitate and encourage public awareness and participation by making information widely available. Effective access to judicial and administrative proceedings, including redress and remedy, shall be provided.

Principle 11

States shall enact effective environmental legislation. Environmental standards, management objectives and priorities should reflect the environmental and developmental context to which they apply. Standards applied by some countries may be inappropriate and of unwarranted economic and social cost to other countries, in particular developing countries.

Principle 12

States should cooperate to promote a supportive and open international economic system that would lead to economic growth and sustainable development in all countries, to better address the problems of environmental degradation. Trade policy measures for environmental purposes should not constitute a means of arbitrary or unjustifiable discrimination or a disguised restriction on international trade. Unilateral actions to deal with environmental challenges outside the jurisdiction of the importing country should be avoided. Environmental measures addressing transboundary or global environmental problems should, as far as possible, be based on an international consensus.

Principle 13

States shall develop national law regarding liability and compensation for the victims of pollution and other environmental damage. States shall also cooperate in an expeditious and more determined manner to develop further international law regarding liability and compensation for adverse effects of environmental damage caused by activities within their jurisdiction or control to areas beyond their jurisdiction.

Principle 14

States should effectively cooperate to discourage or prevent the relocation and transfer to other States of any activities and substances that cause severe environmental degradation or are found to be harmful to human health.

Principle 15

In order to protect the environment, the precautionary approach shall be widely applied by States according to their capabilities. Where there are threats of serious or irreversible damage, lack of full scientific certainty shall not be used as a reason for postponing cost-effective measures to prevent environmental degradation.

Principle 16

National authorities should endeavor to promote the internalization of environmental costs and the use of economic instruments, taking into account the approach that the polluter should, in principle, bear the cost of pollution, with due regard to the public interest and without distorting international trade and investment.

Principle 17

Environmental impact assessment, as a national instrument, shall be undertaken for proposed activities that are likely to have a significant adverse impact on the environment and are subject to a decision of a competent national authority.

Principle 18

States shall immediately notify other States of any natural disasters or other emergencies that are likely to produce sudden harmful effects on the environment of those States. Every effort shall be made by the international community to help States so afflicted.

Principle 19

States shall provide prior and timely notification and relevant information to potentially affected States on activities that may have a significant adverse transboundary environmental effect and shall consult with those States at an early stage and in good faith.

Principle 20

Women have a vital role in environmental management and development. Their full participation is therefore essential to achieve sustainable development.

Principle 21

The creativity, ideals and courage of the youth of the world should be mobilized to forge a global partnership in order to achieve sustainable development and ensure a better future for all.

Principle 22

Indigenous people and their communities and other local communities have a vital role in environmental management and development because of their knowledge and traditional practices. States should recognize and duly support their identity, culture and interests and enable their effective participation in the achievement of sustainable development.

Principle 23

The environment and natural resources of people under oppression, domination and occupation shall be protected.

Principle 24

Warfare is inherently destructive of sustainable development. States shall therefore respect international law providing protection for the environment in times of armed conflict and cooperate in its further development, as necessary.

Principle 25

Peace, development and environmental protection are interdependent and indivisible.

Principle 26

States shall resolve all their environmental disputes peacefully and by appropriate means in accordance with the Charter of the United Nations.

Principle 27

States and people shall cooperate in good faith and in a spirit of partnership in the fulfillment of the principles embodied in this Declaration and in the further development of international law in the field of sustainable development.

CHAPTER 7 OF *AGENDA 21* (EXCERPT FROM INTRODUCTION)

Promoting sustainable human settlement development

7.1. In industrialized countries, the consumption patterns of cities are severely stressing the global ecosystem, while settlements in the developing world need more raw material, energy, and economic development simply to overcome basic economic and social problems. Human settlement conditions in many parts of the world, particularly the developing countries, are deteriorating mainly as a result of the low levels of investment in the sector attributable to the overall resource constraints in these countries. In the low-income countries for which recent data are available, an average of only 5.6 per cent of central government expenditure went to housing, amenities, social security and welfare. Expenditure by international support and finance organizations is equally low. For example, only 1 per cent of the United Nations

system's total grant-financed expenditures in 1988 went to human settlements, while in 1991, loans from the World Bank and the International Development Association (IDA) for urban development and water supply and sewerage amounted to 5.5 and 5.4 per cent, respectively, of their total lending.

7.2. On the other hand, available information indicates that technical cooperation activities in the human settlement sector generate considerable public and private sector investment. For example, every dollar of UNDP technical cooperation expenditure on human settlements in 1988 generated a follow-up investment of $122, the highest of all UNDP sectors of assistance.

7.3. This is the foundation of the "enabling approach" advocated for the human settlement sector. External assistance will help to generate the internal resources needed to improve the living and working environments of all people by the year 2000 and beyond, including the growing number of unemployed – the no-income group. At the same time the environmental implications of urban development should be recognized and addressed in an integrated fashion by all countries, with high priority being given to the needs of the urban and rural poor, the unemployed and the growing number of people without any source of income.

Human settlement objective

7.4. The overall human settlement objective is to improve the social, economic and environmental quality of human settlements and the living and working environments of all people, in particular the urban and rural poor. Such improvement should be based on technical cooperation activities, partnerships among the public, private and community sectors and participation in the decision-making process by community groups and special interest groups such as women, indigenous people, the elderly and the disabled. These approaches should form the core principles of national settlement strategies. In developing these strategies, countries will need to set priorities among the eight programme areas in this chapter in accordance with their national plans and objectives, taking fully into account their social and cultural capabilities. Furthermore, countries should make appropriate provision to monitor the impact of their strategies on marginalized and disenfranchised groups, with particular reference to the needs of women.

7.5. The programme areas included in this chapter are:

(a) Providing adequate shelter for all.
(b) Improving human settlement management.
(c) Promoting sustainable land-use planning and management.
(d) Promoting the integrated provision of environmental infrastructure: water, sanitation, drainage and solid-waste management.
(e) Promoting sustainable energy and transport systems in human settlements.
(f) Promoting human settlement planning and management in disaster-prone areas.
(g) Promoting sustainable construction industry activities.
(h) Promoting human resource development and capacity-building for human settlement development.

THE "ISTANBUL DECLARATION ON HUMAN SETTLEMENTS"

1. We, the Heads of State or Government and the official delegations of countries assembled at the United Nations Conference on Human Settlements (Habitat II) in Istanbul, Turkey from 3 to 14 June 1996, take this opportunity to endorse the universal goals of ensuring adequate shelter for all and making human settlements safer, healthier and more liveable, equitable, sustainable and productive. Our deliberations on the two major themes of the Conference – adequate shelter for all and sustainable human settlements development in an urbanizing world – have been inspired by the Charter of the United Nations and are aimed at reaffirming existing and forging new partnerships for action at the international, national and local levels to improve our living environment. We commit ourselves to the objectives, principles and recommendations contained in the Habitat Agenda and pledge our mutual support for its implementation.

2. We have considered, with a sense of urgency, the continuing deterioration of conditions of shelter and human settlements. At the same time, we recognize cities and towns as centres of civilization, generating economic development and social, cultural, spiritual and scientific advancement. We must take advantage of the opportunities presented by our settlements and preserve their diversity to promote solidarity among all our peoples.

3. We reaffirm our commitment to better standards of living in larger freedom for all humankind. We recall the first United Nations Conference on Human Settlements, held at Vancouver, Canada, the celebration of the International Year of Shelter for the Homeless and the Global Strategy for Shelter to the Year 2000, all of which have contributed to increased global awareness of the problems of human settlements and called for action to achieve adequate shelter for all. Recent United Nations world conferences, including, in particular, the United Nations Conference on Environment and Development, have given us a comprehensive agenda for the equitable attainment of peace, justice and democracy built on economic development, social development and environmental protection as interdependent and mutually reinforcing components of sustainable development. We have sought to integrate the outcomes of these conferences into the Habitat Agenda.

4. To improve the quality of life within human settlements, we must combat the deterioration of conditions that in most cases, particularly in developing countries, have reached crisis proportions. To this end, we must address comprehensively, inter alia, unsustainable consumption and production patterns, particularly in industrialized countries; unsustainable population changes, including changes in structure and distribution, giving priority consideration to the tendency towards excessive population concentration; homelessness; increasing poverty; unemployment; social exclusion; family instability; inadequate resources; lack of basic infrastructure and services; lack of adequate planning; growing insecurity and violence; environmental degradation; and increased vulnerability to disasters.

5. The challenges of human settlements are global, but countries and regions also face specific problems which need specific solutions. We recognize the need to intensify our efforts and cooperation

to improve living conditions in the cities, towns and villages throughout the world, particularly in developing countries, where the situation is especially grave, and in countries with economies in transition. In this connection, we acknowledge that globalization of the world economy presents opportunities and challenges for the development process, as well as risks and uncertainties, and that achievement of the goals of the Habitat Agenda would be facilitated by, inter alia, positive actions on the issues of financing of development, external debt, international trade and transfer of technology. Our cities must be places where human beings lead fulfilling lives in dignity, good health, safety, happiness and hope.

6. Rural and urban development are interdependent. In addition to improving the urban habitat, we must also work to extend adequate infrastructure, public services and employment opportunities to rural areas in order to enhance their attractiveness, develop an integrated network of settlements and minimize rural-to-urban migration. Small- and medium-sized towns need special focus.

7. As human beings are at the centre of our concern for sustainable development, they are the basis for our actions as in implementing the Habitat Agenda. We recognize the particular needs of women, children and youth for safe, healthy and secure living conditions. We shall intensify our efforts to eradicate poverty and discrimination, to promote and protect all human rights and fundamental freedoms for all, and to provide for basic needs, such as education, nutrition and life-span health care services, and, especially, adequate shelter for all. To this end, we commit ourselves to improving the living conditions in human settlements in ways that are consonant with local needs and realities, and we acknowledge the need to address the global, economic, social and environmental trends to ensure the creation of better living environments for all people. We shall also ensure the full and equal participation of all women and men, and the effective participation of youth, in political, economic and social life. We shall promote full accessibility for people with disabilities, as well as gender equality in policies, programmes and projects for shelter and sustainable human settlements development. We make these commitments with particular reference to the more than one billion people living in absolute poverty and to the members of vulnerable and disadvantaged groups identified in the Habitat Agenda.

8. We reaffirm our commitment to the full and progressive realization of the right to adequate housing as provided for in international instruments. To that end, we shall seek the active participation of our public, private and non-governmental partners at all levels to ensure legal security of tenure, protection from discrimination and equal access to affordable, adequate housing for all persons and their families.

9. We shall work to expand the supply of affordable housing by enabling markets to perform efficiently and in a socially and environmentally responsible manner, enhancing access to land and credit and assisting those who are unable to participate in housing markets.

10. In order to sustain our global environment and improve the quality of living in our human settlements, we commit ourselves to sustainable patterns of production, consumption, transportation and settlements development; pollution prevention; respect for the carrying capacity of ecosystems; and the preservation of opportunities for future generations. In this connection, we shall cooperate in a spirit of global partnership to conserve, protect and restore the health and integrity of the Earth's ecosystem. In view of different contributions to global environmental degradation, we reaffirm the principle that countries have common but differentiated responsibilities. We also recognize that we must take these actions in a manner consistent with the precautionary principle approach, which shall be widely applied according to the capabilities of countries. We shall also promote healthy living environments, especially through the provision of adequate quantities of safe water and effective management of waste.

11. We shall promote the conservation, rehabilitation and maintenance of buildings, monuments, open spaces, landscapes and settlement patterns of historical, cultural, architectural, natural, religious and spiritual value.

12. We adopt the enabling strategy and the principles of partnership and participation as the most democratic and effective approach for the realization of our commitments. Recognizing local authorities as our closest partners, and as essential, in the implementation of the Habitat Agenda, we must, within the legal framework of each country, promote decentralization through democratic local authorities and work to strengthen their financial and institutional capacities in accordance with the conditions of countries, while ensuring their transparency, accountability and responsiveness to the needs of people, which are key requirements for Governments at all levels. We shall also increase our cooperation with parliamentarians, the private sector, labour unions and non-governmental and other civil society organizations with due respect for their autonomy. We shall also enhance the role of women and encourage socially and environmentally responsible corporate investment by the private sector. Local action should be guided and stimulated through local programmes based on *Agenda 21*, the Habitat Agenda, or any other equivalent programme, as well as drawing upon the experience of worldwide cooperation initiated in Istanbul by the World Assembly of Cities and Local Authorities, without prejudice to national policies, objectives, priorities and programmes. The enabling strategy includes a responsibility for Governments to implement special measures for members of disadvantaged and vulnerable groups when appropriate.

13. As the implementation of the Habitat Agenda will require adequate funding, we must mobilize financial resources at the national and international levels, including new and additional resources from all sources – multilateral and bilateral, public and private. In this connection, we must facilitate capacity-building and promote the transfer of appropriate technology and know-how. Furthermore, we reiterate the commitments set out in recent United Nations conferences, especially those in *Agenda 21* on funding and technology transfer.

14. We believe that the full and effective implementation of the Habitat Agenda will require the strengthening of the role and functions of the United Nations Centre for Human Settlements (Habitat), taking into account the need for the Centre to focus on well-defined and thoroughly developed objectives and strategic issues. To this end, we pledge our support for the successful implementation of the Habitat Agenda and its global plan of action. Regarding the implementation of the Habitat Agenda, we fully recognize the contribution of the regional and national action plans prepared for this Conference.

15. This Conference in Istanbul marks a new era of cooperation, an era of a culture of solidarity. As we move into the twenty-first century, we offer a positive vision of sustainable human settlements, a sense of hope for our common future and an exhortation to join a truly worthwhile and engaging challenge, that of building together a world where everyone can live in a safe home with the promise of a decent life of dignity, good health, safety, happiness and hope.

NOTES

1 New York: United Nations. 1992. (Report of the United Nations Conference on Environment and Development, Rio de Janeiro, 3–14 June 1992, Annex I; adopted by more than 178 governments.) Available at http://www.un.org/esa/sustdev/agenda21text.htm.
2 Report of the United Nations Conference on the Human Environment, Stockholm, 5–16 June 1972 (United Nations publication, Sales No. E.73.II.A.14 and corrigendum), ch. I.

PART TWO

Dimensions of urban sustainability

INTRODUCTION TO PART TWO

After this historical overview, we turn to some of the specific subject areas addressed within sustainable urban development. Land use, urban design, transportation, environmental planning, resource use, environmental justice, local economic development, architecture and building construction practices – these are among the dimensions of urban planning in which thoughtful action can make a difference. The following selections address these and other topics in turn. The editors' introductions to each selection suggest further directions for readers interested in exploring each topic at greater length.

Several points deserve emphasis as we enter Part 2.

First, even while exploring particular dimensions of urban planning, it is important to keep in mind an overall sense of how these planning topics fit together. How do transportation systems link to land use, housing, or environmental planning? How can particular economic development strategies reinforce local land-use planning visions or community livability? How can better urban design or architecture promote environmental justice and equity? Many such questions could be asked. We encourage readers to think about the links between different readings, and how action in each area can help create more balanced and sustainable urban environments overall. Indeed, a leading reason for unsustainable urban development in the past is that planners, elected leaders, and citizens often have not made such linkages. Engineers have planned freeway systems, yet have paid little attention to the ways they promote sprawling suburban land use. Land-use planners and developers have often mapped out new subdivisions without considering whether they have promoted an equitable distribution of affordable housing in the region. Architects have designed buildings as though no ecological context existed. Economic development planners have sought rapid expansion of jobs without asking whether housing was available for the new workers or whether this development would benefit existing business and residents. As Jacobs, McHarg, and Mumford have suggested, such compartmentalized thinking has helped create current urban problems.

Second, planning actions may take place at different scales of government. It is important to be aware of which level is appropriate for action at any given time, and to integrate initiatives across different scales. Urban land-use decisions, for example, are made primarily by local governments, but are also profoundly influenced by action at national, state or provincial, and regional levels. Transportation planning and regulation of air and water pollution are often handled regionally, but take place within the context of national, state, or provincial policy. Some tasks such as urban design and site planning tend to be carried out primarily in particular neighborhoods or at specific sites, but policies adopted at broader scales, such as municipal zoning codes and state or national building codes, determine what it is possible to do. Thus, as we consider specific topics of urban sustainability planning, it is important to keep in mind how actions at different scales can reinforce one another and help create an overall context in which sustainable urban development can occur.

Finally, urban planning and development actions are undertaken by a wide variety of different actors, not just by city governments. Nonprofit organizations, neighborhood associations, private developers, the news media, and concerned individuals all influence the character and growth of cities, in addition to local governments, regional agencies, state agencies, utility companies, and many other public sector actors. Even within local government many different players are involved, such as planning offices and staffs, public works departments, redevelopment agencies, appointed city commissions, and elected mayors and city council members. Each of these actors has many opportunities to make a difference. For each one of us there are multiple avenues for professional or activist involvement aimed at creating more sustainable communities. Learning the landscape of "who does what" is important in order to be able to best understand how change may be brought about.

It is important to bear in mind these three themes – how individual issues fit together, how actions at different scales relate, and who does what at each level – as we explore the various dimensions of sustainable urban development.

LAND USE AND URBAN DESIGN

"The Next American Metropolis"

from *The Next American Metropolis: Ecology, Community, and the American Dream* (1993)

Peter Calthorpe

Editors' Introduction

In the early twentieth century, urban thinkers often focused on improving the physical form of the city, with specific proposals for new towns, improved neighborhoods, and dispersion of population from overcrowded industrial cities into regional constellations of communities. This tradition of visionary physical planning never entirely disappeared during the middle of the century – figures such as Ian McHarg, American planning consultant Victor Gruen, and Greek visionary Constantine Doxiotis continued to explore new directions – but by and large urban planning became a more pragmatic field built on a foundation of scientific or economic analysis. Planning documents themselves no longer had as many maps, drawings, or graphic visions in them. Instead, many planners opted for the collection of quantitative data on economics, housing, or transportation, and relied on computer models and policy analysis. Some theorists such as University of California at Los Angeles urban geographer Edward Soja have argued that the dimension of "space" itself disappeared from planning discourses. Normative statements about what constitutes good city form also became scarce.

Toward the end of the century the pendulum began to swing back the other way, toward a renewed appreciation of the role of physical planning and urban design. Many observers came to see the need for new types of urban form that would make cities and towns more habitable and ecologically oriented. Strong public movements to manage outward urban expansion ("growth management") and to create more coherent systems of parks, greenways, and open space also emerged. Jane Jacobs helped lay the groundwork for a renewed emphasis on "place-making" with her critique of the sterile, automobile-oriented urban landscapes created by much mid-twentieth-century modernist architecture and urban renewal. What was important, in her view, was the day-to-day life and vitality of urban places. MIT planning professor Kevin Lynch also helped catalyze a new interest in normative urban design values with books such as *Good City Form* (Cambridge, MA: MIT Press, 1981), which analyzed the physical form of human settlements throughout history and arrived at a set of design principles that Lynch argued were important for livable cities. University of California at Berkeley architecture professor Christopher Alexander and his colleagues likewise sought to determine features of what they called "the timeless way of building," and in their book *A Pattern Language* (New York: Oxford University Press, 1977) set forth a list of fifty characteristics of good urban form throughout history that they argued could be combined to produce livable places.

These and other writers helped lay the groundwork for renewed attention to ways of creating livable, walkable places, but the leading movement in terms of actually changing community form came to be called the

New Urbanism. This philosophy emerged in the 1980s and 1990s as a number of architects and planners sought ways to create neighborhoods that emulated features of the traditional American small town. Early on, leaders of the movement used terms such as "traditional neighborhood design" to describe their work, and adopted many design concepts from towns laid out a hundred years before such as grid-like street networks, mid-block alleys, village centers with small shops and workplaces, front porches, and garages at the rear of houses rather than in the front. (If these designers had used European small towns as a model instead, they might well have gravitated toward more winding, organic street patterns and more urban housing forms.)

Miami-based architects Andres Duany and Elizabeth Plater-Zyberk (designers of new communities such as Seaside and Kentlands), Bay Area-based designer Peter Calthorpe (designer of Laguna West and regional planning consultant for Portland, Salt Lake City, Minneapolis-St. Paul, and Chicago), and Los Angeles-based designers Stefanos Polyzoides and Elizabeth Moule were among the founders of the new movement. By taking the name Congress for the New Urbanism (CNU), they consciously positioned themselves as an alternative to the 1930s modernist architectural movement known as the Congrès Internationaux d'Architecture Moderne (CIAM). The CNU held its first annual Congress in Alexandria, Virginia in 1993, and issued a Charter for the New Urbanism in 1996 (San Francisco: Congress for the New Urbanism, 2000). By the turn of the millennium several hundred New Urbanist-inspired neighborhoods were under construction in North America, both on infill locations (within existing urban areas) and greenfield sites (unbuilt open land at the urban fringe). Equally importantly, New Urbanist design principles were diffusing into planning and design professions throughout the world. In Britain, Prince Charles' Prince of Wales Institute served as a vehicle for promoting similar types of urban design, and on the continent architects such as Rob and Leon Krier designed relatively dense new urban additions to existing cities. Many New Urbanist projects may be seen as promoting sustainability, in that they help produce more compact, pedestrian-oriented, resource-efficient urban communities. However, they can also be criticized on various grounds, such as for not providing enough affordable housing, not using green architecture or landscaping principles, or at times for being built on inappropriate locations outside of existing urban areas.

The move to rethink land-use planning and urban design has been strengthened by a wide variety of urban growth management efforts in North America, Europe, and elsewhere. These land-use planning initiatives have sought to deal with a problem unforeseen by early twentieth-century urban thinkers – rapid suburban sprawl made possible in large part by the automobile. In the United States, states such as Oregon, Vermont, Florida, and New Jersey first passed growth management legislation in the 1960s and 1970s, is some cases requiring local governments to plan urban growth boundaries (UGBs) or to limit expansion of urban services such as water and sewer utilities. Additional states such as Washington, Maryland, Massachusetts, Maine, and Pennsylvania launched initiatives in the 1980s and 1990s, often under the banner of "smart growth." The smart growth movement borrowed many principles from the New Urbanism but focusing also on reducing infrastructure costs and creating a fairer distribution of affordable housing. Smart growth efforts have been resisted by many local governments, landowners, developers, and property rights advocates. Libertarians and free-market economists have argued that people choose to live in automobile-oriented, sprawling suburbs, that compact development is not a cure for traffic congestion, and that supposedly sprawling cities such as Los Angeles actually have higher residential densities than do growth management models such as Portland. Growth management proponents reply that citizens have little choice but to live in sprawl, that the housing market has been distorted for many years by public and private subsidies for sprawl, that traffic can be reduced only through a combination of policies including better pricing and transportation alternatives as well as better land use, and that "sprawl" consists of many factors beyond sheer population density. For a good example of this debate see Peter Gordon and Harry Richardson's article "Are compact cities a desirable planning goal?" (*Journal of the American Planning Association*, 63(1), 1997, pp. 95–107) and Reid Ewing's response "Is Los Angeles-style sprawl desirable?" (same issue, pp. 107–127). Whatever the exact outcome of these arguments, it is clear to many these days that new approaches to physical planning are necessary for sustainable urban development.

Calthorpe, one of the leading New Urbanists, may be seen as an heir to Howard and Mumford in that through his regional and neighborhood planning work he has sought to develop a new version of the city–country balance. The co-editor (with ecological architect Sim Van der Ryn) of an earlier book entitled *Sus-

tainable Communities (San Francisco: Sierra Club Books, 1986), Calthorpe later sought a more pragmatic synthesis of pedestrian-oriented planning principles that could be adopted by the mainstream development industry. In works such *as The Next American Metropolis* (Princeton, NJ: Princeton Architectural Press, 1993) and *The Regional City* (Washington, DC: Island Press, 2001; with William Fulton), he has sought to promote co-ordinated physical planning changes on neighborhood, city, and regional scales. Calthorpe has also been a leading proponent of "transit-oriented development," clustering communities around a regional network of rail transit stations.

One of the greatest contributions of Calthorpe and other New Urbanists has been to develop consensus on specific design guidelines and place-making strategies. Calthorpe's graphics in this book represent some of these principles. More are provided by other New Urbanist designers such as Duany, Plater-Zyberk, and Jeff Speck in their book *Suburban Nation* (New York: North Point Press, 2000), and by organizations such as the Congress for the New Urbanism (www.cnu.org), the Sacramento-based Local Government Commission (www.lgc.org), and the Smart Growth Network (www.smartgrowth.org).

Although he speaks primarily to an American audience and talks of redefining the "American Dream," it is important to realize that Calthorpe is talking about a mode of development which has become common the world over – a suburban world of cul-de-sacs, detached single-family houses, single-use zoning, and dependence on automobiles. This "dream" is now sought with increasing frequency in Indonesia, South Africa, The Netherlands, Mexico, eastern Europe, and countless other locations. Reasons for this include omnipresent American television, movies, and popular culture, the power of multinational corporations and their advertising to promote materialist lifestyles, and the employment of American planning consultants throughout the world.

The American Dream is an evolving image and the American Metropolis is its ever-changing reflection. The two feed one another in a complex, interactive cycle. At one point a dream moves us to a new vision of the city and community, at another the reflection of the city transforms that dream with harsh realities or alluring opportunities. We are at a point of transformation once again and the two, city and dream, are changing together. World War II created a distinct model for each: the nuclear family in the suburban landscape. That model and its physical expression is now stressed beyond retention. The family has grown more complex and diverse, while the suburban form has grown more demanding and less accessible. The need for change is blatant, with sprawl reaching its limits, communities fracturing into enclaves, and families seeking more inclusive identities. Clearly we need a new paradigm of development; a new vision of the American Metropolis and a new image for the American Dream.

The old suburban dream is increasingly out of sync with today's culture. Our household makeup has changed dramatically, the work place and work force have been transformed, average family wealth is shrinking, and serious environmental concerns have surfaced. But we continue to build post-World War II suburbs as if families were large and had only one breadwinner, as if the jobs were all downtown, as if land and energy were endless, and as if another lane on the freeway would end traffic congestion.

Over the last 20 years these patterns of growth have become more and more dysfunctional. Finally they have come to produce environments which often frustrate rather than enhance everyday life. Suburban sprawl increases pollution, saps inner-city development, and generates enormous costs – costs which ultimately must be paid by taxpayers, consumers, businesses, and the environment. These problems are not to be solved by limiting the scope, program, or location of development – they must be resolved by rethinking the nature and quality of growth itself, in every context.

This book attempts to map out a new direction for growth in the American Metropolis. It borrows from many traditions and theories: from the romantic environmentalism of Ruskin to the City Beautiful Movement, from the medieval urbanism of Sitte to the Garden Cities of Europe, from streetcar suburbs to the traditional towns of America, and from the theories of Jane Jacobs to those of Leon Krier. It is a work which has evolved from theory to practice in some of our fastest growing cities and regions. It is a search for a paradigm that combines the utopian ideal of an integrated and heterogeneous community with the realities of our time – the imperatives of ecology, affordability, equity, technology, and the relentless force of inertia. The work asserts that our communities must be designed to reestablish and reinforce the public domain, that our districts must be human-scaled, and that our neighborhoods must be diverse in use and population. And finally, that the form and identity of the metropolis must integrate historic context, unique ecologies, and a comprehensive regional structure.

The net result is that we need to start creating neighborhoods rather than subdivisions; urban quarters rather than isolated projects; and diverse communities rather than segregated master plans. Quite simply, we need towns rather than sprawl.

Settlement patterns are the physical foundation of our society and, like our society, they are becoming more and more fractured. Our developments and local zoning laws segregate age groups, income groups, and ethnic groups, as well as family types. Increasingly they isolate people and activities in an inefficient network of congestion and pollution – rather than joining them in diverse and human scaled communities. Our faith in government and the fundamental sense of commonality at the center of any vital democracy is seeping away in suburbs designed more for cars than people, more for market segments than communities. Special interest groups have now replaced citizens in the political landscape, just as gated subdivisions have replaced neighborhoods.

REDEFINING THE AMERICAN DREAM

It is time to redefine the American Dream. We must make it more accessible to our diverse population: singles, the working poor, the elderly, and the pressed middle-class families who can no longer afford the "Ozzie and Harriet" version of the good life. Certain traditional values – diversity, community, frugality, and human scale – should be the foundation of a new direction for both the American Dream and the American Metropolis. These values are not a retreat to nostalgia or imitation, but a recognition that certain qualities of culture and community are timeless. And that these timeless imperatives must be married to the modern condition in new ways.

The alternative to sprawl is simple and timely: neighborhoods of housing, parks, and schools placed within walking distance of shops, civic services, jobs, and transit – a modern version of the traditional town. The convenience of the car and the opportunity to walk or use transit can be blended in an environment with local access for all the daily needs of a diverse community. It is a strategy which could preserve open space, support transit, reduce auto traffic, and create affordable neighborhoods. Applied at a regional scale, a network of such mixed-use neighborhoods could create order in our balkanized metropolis. It could balance inner-city development with suburban investment by organizing growth around an expanding transit system and setting defensible urban limit lines and greenbelts. The increments of growth in each neighborhood would be small, but the aggregate could accommodate regional growth with minimal environmental impacts; less land consumed, less traffic generated, less pollution produced.

Such neighborhoods, called Pedestrian Pockets or Transit-Oriented Developments, ultimately could be more affordable for working families, environmentally responsible, and cost-effective for business and government. But such a growth strategy will mean fundamentally changing our preconceptions and local regulatory priorities, as well as redesigning the federal programs that shape our cities.

At the core of this alternative, philosophically and practically, is the pedestrian. Pedestrians are the catalyst which makes the essential qualities of communities meaningful. They create the place and the time for casual encounters and the practical integration of diverse places and people. Without the pedestrian, a community's common ground – its parks, sidewalks, squares, and plazas – become useless obstructions to the car. Pedestrians are the lost measure of a community, they set the scale for both center and edge of our neighborhoods. Without the pedestrian, an area's focus can be easily lost. Commerce and civic uses are easily decentralized into distant chain store destinations and government centers. Homes and jobs are isolated in subdivisions and office parks.

Although pedestrians will not displace the care anytime soon, their absence in our thinking and planning is a fundamental source of failure in our new developments. To plan as if there were pedestrians may be a self-fulfilling act; it will give kids some autonomy, the elderly basic access, and others the choice to walk again. To plan as if there were pedestrians will turn suburbs into towns, projects into neighborhoods, and networks into communities.

If we are now to reinvest in America, careful consideration should be given to what kind of America we want to create. Our investments in transit must be supported by land use patterns which put riders and jobs within an easy walk of stations. Our investments in affordable housing should place families in neighborhoods where they can save dollars by using their autos less. Our investments in open space should reinforce regional greenbelts and urban limit lines. Our investments in highways should not unwittingly support sprawl, inner-city disinvestments, or random job decentralization. Our investments in inner-cities and urban businesses ought to be linked by transit to the larger region, not isolated by gridlock. Our planning and zoning codes should help create communities, not sprawl.

Is such as transformation possible? Americans love their cars, they love privacy and independence, and they are evolving ever larger institutions. The goal of community planning for the pedestrian or transit is not to eliminate the car, but to balance it. In the 1970s the national love affair with the car was certainly hot, but we traveled on average 50 per cent fewer miles per year than we do now. It *is* possible to accommodate the car and still free pedestrians. Practically, it means narrowing local roads and placing parking to the rear of buildings, not eliminating access for the car. Similarly, the suburban goals of privacy and independence do not have to be abandoned in the interests of developing communities with vital urban centers and neighborly streets. In fact, a walkable neighborhood may produce increased independence for growing segments of the population, the elderly and kids. The scale of our institutions may no longer fit the human scale proportions of an old village, but with careful design they could be integrated into mixed-use communities. Large businesses are quickly becoming aware of the benefits of being part of a neighborhood rather than an office park, with shared amenities and local services topping the list.

This new balance calls for the integration of seemingly opposing forces. Community and privacy,

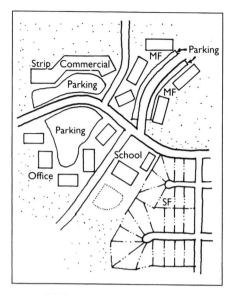

CONVENTIONAL SUBURBAN
DEVELOPMENT

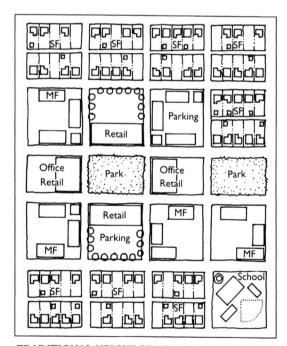

TRADITIONAL NEIGHBORHOOD DEVELOPMENT

Figure 1. Conventional suburban development vs. traditional neighborhood development.

auto and pedestrian, large institution and small business, suburban and urban; these are the poles that must be fused in a new pattern of growth. The design imperatives of creating the post-suburban metropolis are complex and challenging. They are to develop a regional growth strategy which integrates social diversity, environmental protection, and transit; create an architecture that reinforces the public domain without sacrificing the variety and character of individual buildings; advance a planning approach that reestablishes the pedestrian in mixed-use, livable communities; and evolve a design philosophy that is capable of accommodating modern institutions without sacrificing human scale and memorable places.

DEFINITIONS

Transit-Oriented Development (TOD)

A Transit-Oriented Development (TOD) is a mixed-use community within an average 2,000-foot

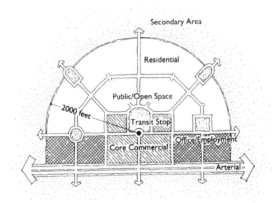

Figure 2. Transit-Oriented Development.

walking distance of a transit stop and core commercial area. TODs mix residential, retail, office, open space, and public uses in a walkable environment, making it convenient for residents and employees to travel by transit, bicycle, foot, or car.

Residential areas

TOD residential areas include housing that is within a convenient walking distance from core commercial areas and transit stops. Residential density requirements should be met with a mix of housing types, including small lot single-family, townhomes, condominiums, and apartments.

Secondary areas

Each TOD may have a Secondary Area adjacent to it, including areas across and arterial, which are no further than one mile from the core commercial area. The Secondary Area street network must provide multiple direct street and bicycle connections to the transit stop and core commercial area, with a minimum of arterial crossings. Secondary Areas may have lower density single-family housing, public schools, large community parks, low intensity employment-generating uses, and park-and-ride lots.

Relationship to transit and circulation

The site must be located on an existing or planned trunk transit line or on a feeder bus route within 10 minutes transit travel time from a stop on the trunk line. Where transit may not occur for a period

Small-lot single-family

Townhouses

Duplexes

Apartments & condominiums

Figure 3. Housing types.

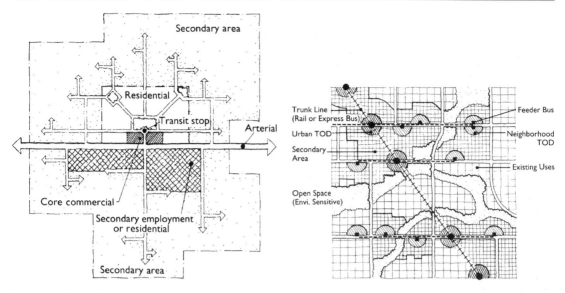

Figure 4. Secondary areas.   **Figure 5.** Relationship to transit.

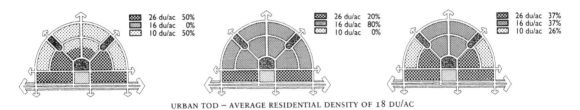

URBAN TOD – AVERAGE RESIDENTIAL DENSITY OF 18 DU/AC

Figure 6. Residential density mix.

of time, the land use and street patterns within a TOD must function effectively in the interim.

Residential mix

A mix of housing densities, ownership patterns, price, and building types is desirable in a TOD. Average minimum densities should vary between 10 and 25 dwelling units/net residential acre (25 to 62 units/hectare), depending on the relationship to surrounding existing neighborhoods and location within the urban area.

Street and circulation system

The local street system should be recognizable, formalized, and inter-connected, converging to transit stops, core commercial areas, schools, and

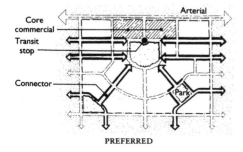

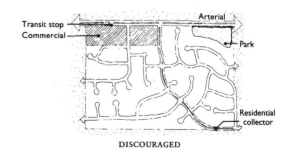

Figure 7. Street and circulation system.

parks. Multiple and parallel routes must be provided between the core commercial area, residential, and employment uses so that local trips are not forced onto arterial streets. Streets must be pedestrian friendly; sidewalks, street trees, building entries, and parallel parking must shelter and enhance the walking environment.

Regional form

Regional form should be the product of transit accessibility and environmental constraints. Major natural resources, such as rivers, bays, ridgelands, agriculture, and sensitive habitat should be preserved and enhanced. An Urban Growth Boundary should be established that provides adequate area for growth while honoring these criteria.

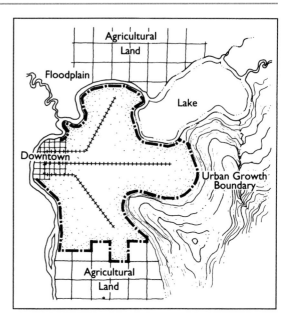

Figure 8. Regional form.

"Outdoor Space and Outdoor Activities"

from *Life Between Buildings* (1980)

Jan Gehl

Editors' Introduction

Beginning in the 1960s writers such as Jacobs, Lynch, William H. Whyte, Clare Cooper Marcus, and Danish designer Jan Gehl emphasized the need to base urban design on study of how people actually experience and use urban environments. A new discipline of environmental design emerged, devoted to researching how built environments work for people. Researchers developed methods using behavior observation, time-lapse photography, post-occupancy evaluation surveys, and cognitive mapping (in which people were asked to draw maps or images of how they perceived their urban environments) to provide factual information for improved urban design.

In his pioneering book *Life Between Buildings: Using Public Space* (New York: Van Nostrand Reinhold, 1980), Gehl took a remarkably perceptive look at different types of outdoor spaces and their social uses. What is most needed, he argued, is an increase in optional activities taking place in the public realm. The number and variety of human interactions, especially chance meetings in public spaces, was in his view the way to a healthier urban community. Analyzing public spaces within Copenhagen, he found places such as the Stroget (one of Europe's pioneering pedestrian streets) and the Tivoli Gardens particularly conducive to social life. Although many of Gehl's observations may seem common sense today, they represented a major departure from modernist urban design practices in which abstract architectural principles, rather than careful observation of how people actually use places, often dictated urban form. Other books in this vein include Whyte's *The Social Life of Small Urban Spaces* (Washington, DC: The Conservation Foundation, 1980), Marcus and Wendy Sarkissian's *Housing as if People Mattered*, Marcus and Carolyn Francis' *People Places* (New York: Van Nostrand Reinhold, 1990), and Lynch's *The Image of the City* (Cambridge, MA: MIT Press, 1960).

THREE TYPES OF OUTDOOR ACTIVITIES

An ordinary day on an ordinary street. Pedestrians pass on the sidewalks, children play near front doors, people sit on benches and steps, the postman makes his rounds with the mail, two passersby greet on the sidewalk, two mechanics repair a car, groups engage in conversation. This mix of outdoor activities is influenced by a number of conditions. Physical environment is one of the factors: a factor that influences the activities to a varying degree and in many different ways. Outdoor activities, and a number of the physical conditions that influence them, are the subject of this book.

Greatly simplified, outdoor activities in public spaces can be divided into three categories, each of which places very different demands on the physical environment: *necessary activities*, *optional activities*, and *social activities*.

Necessary activities include those that are more or less compulsory – going to school or to work, shopping, waiting for a bus or a person, running errands, distributing mail – in other words, all activities in which those involved are to a greater or lesser degree required to participate.

In general, everyday tasks and pastimes belong to this group. Among other activities, this group includes the great majority of those related to walking.

Because the activities in this group are necessary, their incidence is influenced only slightly by the physical framework. These activities will take place throughout the year, under nearly all conditions, and are more or less independent of the exterior environment. The participants have no choice.

Optional activities – that is, those pursuits that are participated in if there is a wish to do so and if time and place make it possible – are quite another matter.

This category includes such activities as taking a walk to get a breath of fresh air, standing around enjoying life, or sitting and sunbathing.

These activities take place only when exterior conditions are optimal, when weather and place invite them. This relationship is particularly important in connection with physical planning because most of the recreational activities that are especially pleasant to pursue outdoors are found precisely in this category of activities. These activities are especially dependent on exterior physical conditions.

When outdoor areas are of poor quality, only strictly necessary activities occur.

When outdoor areas are of high quality, necessary activities take place with approximately the same frequency – though they clearly tend to take a longer time, because the physical conditions are better. In addition, however, a wide range of optional activities will also occur because place and situation now invite people to stop, sit, eat, plan, and so on.

In streets and city spaces of poor quality, only the bare minimum of activity takes place. People hurry home.

In a good environment, a completely different, broad spectrum of human activities is possible.

	Quality of the physical environment	
	Poor	Good
Necessary activities	●	●
Optional activities	·	⬤
"Resultant" activities (Social activities)	·	●

Figure 1. Graphic representation of the relationship between the quality of outdoor spaces and the rate of occurrence of outdoor activities. When the quality of outdoor areas is good, optional activities occur with increasing frequency. Furthermore, as levels of optional activity rise, the number of social activities usually increases substantially.

Social activities are all activities that depend on the presence of others in public spaces. Social activities include children at play, greetings and conversations, communal activities of various kinds, and finally – as the most widespread social activity – passive contacts, that is simply seeing and hearing other people.

Different kinds of social activities occur in many places: in dwellings; in private outdoor spaces, gardens, and balconies; in public buildings; at places of work; and so on; but in this context only those activities that occur in publicly accessible spaces are examined.

These activities could also be termed "resultant" activities, because in nearly all instances they evolve from activities linked to the other two activity categories. They develop in connection with the other activities because people are in the same space, meet, pass by one another, or are merely within view. . . .

[. . .]

LIFE BETWEEN BUILDINGS

It is difficult to pinpoint precisely what life between buildings means in relation to the *need for contact*.

Opportunities for meetings and daily activities in the public spaces of a city or residential area enable one to be among, to see, and to hear others, to experience other people functioning in various situations.

These modest "see and hear contacts" must be considered in relation to other forms of contact and as part of the whole range of social activities, from very simple and noncommittal contacts to complex and emotionally involved connections.

The concept of varying degrees of contact intensity is the basis of the following simplified outline of various contact forms:

High intensity	↑	Close friendships
		Friends
		Acquaintances
		Chance contacts
Low intensity		Passive contacts ("see and hear" contacts)

In terms of this outline life between buildings represents primarily the low-intensity contacts located at the bottom of the scale. Compared with the other contact forms, these contacts appear insignificant, yet they are valuable both as independent contact forms and as prerequisites for other, more complex interactions.

Opportunities related to merely being able to meet, see, and hear others include:

- contact at a modest level
- a possible starting point for contact at other levels
- a possibility for maintaining already established contacts
- a source of information about the social world outside
- a source of inspiration, an offer of stimulating experience.

The possibilities related to the low-intensity contact forms offered in public spaces perhaps can best be described by the situation that exists if they are lacking.

If activity between buildings is missing, the lower end of the contact scale also disappears. The varied transitional forms between being alone and being together have disappeared. The boundaries between isolation and contact become sharper – people are either alone or else with others on a relatively demanding and exacting level.

Life between buildings offers an opportunity to be with others in a relaxed and undemanding way. One can take occasional walks, perhaps make a detour along a main street on the way more or pause at an inviting bench near a front door to be among people for a short while. One can take a long bus ride every day, as many retired people have been found to do in large cities. Or one can do daily shopping, even that it would be more practical to do it once a week. Even looking out of the window now and then, if one is fortunate to have something to look at, can be rewarding. Being among others, seeing and hearing others, receiving impulses from others, imply positive experiences, alternatives to being alone. One is not necessarily with a specific person, but one is, nevertheless, with others.

As opposed to being a passive observer of other people's experiences on television or video or film, in public spaces the individual himself is present, participating in a modest way, but most definitely participating.

Low-intensity contact is also a situation from which other forms of contact can grow. It is a medium for the unpredictable, the spontaneous, the unplanned. . . .

[. . .]

The trend from living in lifeless cities and residential areas that has accompanied industrialization, segregation of various city functions, and reliance on the automobile also has caused cities to become duller and more monotonous. This points up another important need, namely *the need for stimulation.*

Experiencing other people represents a particularly colorful and attractive opportunity for stimulation. Compared with experiencing buildings and other inanimate objects, experiencing people, who speak and move about, offers a wealth of sensual variation. No moment is like the previous or the following when people circulate among people. The number of new situations and new stimuli is limitless. Furthermore it concerns the most important subject in life: people.

Living cities, therefore, ones in which people can act with one another, are always stimulating because they are rich in experiences, in contrast to lifeless cities, which can scarcely avoid being poor in experiences and thus dull, no matter how many colors and variations of shape in buildings are introduced. . . .

[. . .]

OUTDOOR ACTIVITIES AND THE QUALITY OF OUTDOOR SPACE

Life between buildings is discussed here because the extent and character of outdoor activities are greatly influenced by physical planning. Just as it is possible through choice of materials and colors to create a certain palette in a city, it is equally possible through planning decisions to influence patterns of activities, to create better or worse conditions for outdoor events, and to create lively or lifeless cities.

The spectrum of possibilities can be described by two extremes. One extreme is the city with multistory buildings, underground parking facilities, extensive automobile traffic, and long distances between buildings and functions. This type of city can be found in a number of North American and "modernized" European cities and in many suburban areas.

In such cities one sees buildings and cars, but few people, if any, because pedestrian traffic is more or less impossible, and because conditions for outdoor stays in the public areas near buildings are very poor. Outdoor spaces are large and impersonal. With great distances in the urban plan, there is nothing much to experience outdoors, and the few activities that do take place are spread out in time and space. Under these conditions most residents prefer to remain indoors in front of the television or on their balcony or in other comparably private outdoor spaces.

Another extreme is the city with reasonably low, closely spaced buildings, accommodation for foot traffic, and good areas for outdoor stays along the streets and in direct relation to residences, public buildings, places of work, and so forth. Here it is possible to see buildings, people coming and going, and people stopping in outdoor areas near the buildings because the outdoor spaces are easy and inviting to use. This city is a living city, one in which spaces inside buildings are supplemented with usable outdoor areas, and where public spaces are allowed to function. . . .

In a survey recording all activities occurring in the center of Copenhagen during the spring and summer of 1986, it was found that the number of pedestrian streets and squares in the city center had tripled between 1968 and 1986. Parallel to this improvement of the physical conditions, a tripling in the number of people standing and sitting was recorded.

In cases where neighboring cities offer varying conditions for city activities, great differences can also be found.

In Italian cities with pedestrian streets and automobile-free squares, the outdoor city life is often much more pronounced than in the car-oriented neighboring cities, even though the climate is the same.

A 1978 survey of street activities in both trafficked and pedestrian streets in Sydney, Melbourne, and Adelaide, Australia, carried out by architectural students from the University of Melbourne and the Royal Melbourne Institute of Technology found a direct connection between street quality and street activity. In addition, an experimental improvement of increasing the number of seats by 100 per cent on the pedestrian street in Melbourne resulted in an 88 per cent increase in seated activities.

William H. Whyte, in his book *The Social Life of Small Urban Spaces*, describes the close connection between qualities of city space and city activities and documents how often quite simple physical alterations can improve the use of the city space noticeably.

Comparable results have been achieved in a number of improvement projects executed in New York and other US cities by the Project for Public Spaces.

In residential areas as well, both in Europe and the United States, traffic reduction schemes, courtyard clearing, laying out of parks, and comparable outdoor improvements have had a marked effect.

Plate 1. A sociable street in Copenhagen.

TRANSPORTATION

"Transit and the Metropolis: Finding Harmony"

from *The Transit Metropolis: A Global Inquiry* (1998)

Robert Cervero

Editors' Introduction

Rising traffic volume and congestion are leading citizen concerns in most cities and towns the world over, and of course produce other sustainability-related problems such as air pollution, greenhouse gas emissions, depletion of nonrenewable fossil fuels, destruction of open space by roads and suburban sprawl, and degradation of local neighborhood quality of life. Vehicle ownership continues to grow rapidly in most countries, and the number of miles driven per capita has doubled in nations such as the USA over the last generation. How can this situation ever be changed? While there is no easy answer to this question, a number of combined strategies involving land use, public transit, other alternative travel modes, and pricing are likely to make the difference. This chapter explores some of these areas crucial to improving urban sustainability.

University of California at Berkeley professor Robert Cervero has studied relationships between transportation and land use the world over and is a leading authority on strategies to reduce automobile use. In this selection from his book *The Transit Metropolis: A Global Inquiry* (Washington, DC: Island Press, 1998), he asks why automobile use continues to grow and public transit use decline, and what characteristics can lead urban regions to buck this trend. Solutions, he believes, can be of several sorts. Regions may adapt their land use to fit around major transit systems such as subways or light rail lines ("adaptive cities"). Or they might adapt their transit systems to fit their low-density land use by employing on-demand shuttles and vans and/or flexible bus systems ("adaptive transit"). Or various hybrid options are possible. Pricing of transportation and other "transportation demand management" policies will play a role as well. The long-term goal, in Cervero's view, is the "transit metropolis" where strong public transit alternatives exist to balance private vehicle use.

Other resources on the subject of reducing automobile use include Peter Newman and Jeffrey Kenworthy's *Sustainability and Cities: Overcoming Automobile Dependence* (Washington, DC: Island Press, 1999; excerpted later in Part 2), Anthony Downs' *Stuck in Traffic: Coping With Peak-Hour Traffic Congestion* (Washington, DC: The Brookings Institution, 1992), and David Engwicht's *Reclaiming Our Cities & Towns: Better Living with Less Traffic* (Philadelphia: New Society Publishers, 1993). Two excellent internet resources on transportation are the Surface Transportation Policy Project (www.transact.org) and the Victoria Transportation Policy Institute (www.vtpi.org), both of which offer an impressive array of materials on transportation policy and how it might be reformed.

Public transit systems are struggling to compete with the private automobile the world over. Throughout North America, in much of Europe, and even in most developing countries, the private automobile continues to gain market shares of motorized trips at the expense of public transit systems. In the United States, just 1.8 per cent of all person trips were by transit in 1995, down from 2.4 per cent in 1977 and 2.2 per cent in 1983.[1] Despite the tens of billions of dollars invested in new rail systems and the underwriting of more than 75 per cent of operating expenses, ridership figures for transit's bread-and-butter market – the work trip – remain flat. Nationwide, 4.5 per cent of commutes were by transit in 1983; by 1995, this share had fallen to 3.5 per cent.

The declining role of transit has been every bit as alarming in Europe, prompting some observers to warn that it is just a matter of time before cities like London and Madrid become as automobile-oriented as Los Angeles and Dallas. England and Wales saw the share of total journeys by transit fall from 33 per cent in 1971 to 14 per cent in 1991.[2] Since 1980, transit's market shares of trips have plummeted in Italy, Poland, Hungary, and former East Germany. Eroding market shares have likewise been reported in such megacities as Buenos Aires, Bangkok, and Manila.

Numerous factors have fueled these trends. Part of the explanation for the decline in Europe has been sharp increases in fares resulting from government deregulation of the transit sector. Public disinvestment has left the physical infrastructure of some transit systems in shambles in Italy and parts of Eastern Europe. However, transit's decline has been more an outcome of powerful spatial and economic trends that have been unfolding over the past several decades than of overt government actions (or inaction). Factors that have steadily chipped away at transit's market share worldwide include rising personal incomes and car ownership, declining real-dollar costs for motoring and parking, and the decentralization of cities and regions. Of course, these forces have partly fed off each other. Rising wealth and cheaper motoring, for instance, have prompted firms, retailers, and households to exit cities in favor of less dense environs. Spread-out development has proven to be especially troubling for mass transit. With trip origins and destinations today spread all over the map, mass transit is often no match for the private automobile and its flexible, door-to-door, no-transfer features.

Suburbanization has not crippled transit systems everywhere, however. Some cities and regions have managed to buck the trend, offering transit services that are holding their own against the automobile's ever-increasing presence, and in some cases even grabbing larger market shares of urban travel. These are places, I contend, that have been superbly adaptive, almost in a Darwinian sense. Notably, they have found a harmonious fit between mass transit services and their cityscapes.

Some, like Singapore and Copenhagen, have adapted their settlement patterns so that they are more conducive to transit riding, mainly by rail transit, whether for reasons of land scarcity, open space preservation, or encouraging what are viewed as more sustainable patterns of growth and travel. This has often involved concentrating offices, homes, and shops around rail nodes in attractive, well-designed, pedestrian-friendly communities. Other places have opted for an entirely different approach, accepting their low-density, often market-driven lay of the land, and in response adapting mass transit services and technologies to better serve these spread-out environs. These are places, such as Karlsruhe in Germany and Adelaide, Australia, that have introduced flexible forms of mass transit that begin to emulate the speedy, door-to-door service features of the car.

Still other places, like Ottawa, Canada, and Curitiba, Brazil, have struck a middle ground, adapting their urban landscapes so as to become more transit-supportive while at the same time adapting their transit services so as to deliver customers closer to their destinations, minimize waits, and expedite transfers. It is because these places have found a workable nexus between their mass transit services and urban settlement patterns that they either are or are on the road to becoming great transit metropolises.

What these areas have in common – adaptability – is first and fundamentally a calculated process of making change by investing, reinvesting, organizing, reorganizing, inventing, and reinventing. Adaptability is about self-survival in a world of limited resources, tightly stretched budgets, and ever-changing cultural norms, lifestyles, technologies, and personal values. In the private sector, any business that resists adapting to changing consumer wants and preferences is a short-lived business. More and more, the public sector is being held to similar standards. There is no longer the public largesse or patience to allow business as usual. Transit authorities must adapt to change, as must city and regional governments. Trends like suburbanization, advances in telecommunications, and chained trip-making require that transit agencies refashion how they configure and deliver services and that builders and planners adjust their designs of communities and places. In the best of worlds, these efforts are closely coordinated. This will most likely occur when and where there is the motivation and the means to break out of traditional, entrenched practices, which, of course, is no small feat in the public realm. Yet even transit's most ardent defenders now concede that steadily eroding shares of metropolitan travel are a telltale sign that fresh, new approaches are needed. Places that appropriately adapt to changing times, I contend, are places where transit stands the best chance of competing with the car well into the next millennium.

It bears noting that a functional and sustainable transit metropolis is not equated with a region whereby transit largely replaces the private automobile or even captures the majority of motorized trips. Rather, the transit metropolis represents a built form and a mobility environment where transit is a far more respectable alternative to traveling than currently is the case in much of the industrialized world. It is an environment where transit and the built environment harmoniously co-exist, reinforcing and enhancing each other in the process. Thus, while automobile travel might still predominate, a transit metropolis is one where enough travelers opt for transit riding, by virtue of the workable transit–land use nexus, to place a region on a sustainable course.

It is also important to emphasize . . . connections between transit and urbanization at the regional scale versus the local one. While considerable attention has been given to transit-oriented development (TOD) and the New Urbanism movement in recent years, both by scholars and the popular press, much of this focus has been at the neighborhood and community levels. Micro-scale designs that encourage walking and promote community cohesion have captivated the attention of many proponents of TODs and New Urbanism. While good quality designs are without question absolutely essential to creating places that are physically conducive to transit riding, they are clearly not sufficient in and of themselves. Islands of TOD in a sea of freeway-oriented suburbs will do little to change fundamental travel behavior or the sum quality of regional living. The key to making TOD work is to make sure that it is well coordinated across a metropolis. While land use planning and urban design are local prerogatives, their impacts on travel are felt regionally. . . .

[. . .]

TYPES OF TRANSIT METROPOLISES

[There are] four classes of transit metropolises:

- *Adaptive cities.* These are transit-oriented metropolises that have invested in rail systems to guide urban growth for purposes of achieving larger societal objectives, such as preserving open space and producing affordable housing in rail-served communities. All feature compact, mixed-use suburban communities and new towns concentrated around rail nodes . . . . examples are Stockholm, Copenhagen, Tokyo, and Singapore.
- *Adaptive transit.* These are places that have largely accepted spread-out, low-density patterns of growth and have sought to appropriately adapt transit services and new technologies to best serve these environs. [Models include] technology-based examples (e.g. dual-track systems in Karlsruhe, Germany), service innovations (e.g. track-guided buses in Adelaide, Australia), and small-vehicle, entrepreneurial services (e.g. colectivos in greater Mexico City).
- *Strong-core cities.* [Cities such as] Zurich and Melbourne have successfully integrated transit and urban development within a more confined, central city context. They have done so by providing integrated transit services centered around mixed-traffic tram and light rail systems. In these places, trams designed into streetscapes co-exist nicely with pedestrians and bicyclists. These cities' primacies (high

shares of regional jobs and retail sales in their cores) and healthy transit patronage are testaments to the success of melding together the renewal of both central city districts and traditional tramways.

- *Hybrids: adaptive cities and adaptive transit.* [Cities such as] Munich, Ottawa, and Curitiba are best viewed as hybrids, in the sense that they have struck a workable balance between concentrating development along mainline transit corridors and adapting transit to efficiently serve their spread-out suburbs and exurbs. Greater Munich's hybrid of heavy rail trunkline services and light rail and conventional bus feeders – all coordinated through a regional transit authority – has strengthened the central city while also serving suburban growth axes. Both Ottawa and Curitiba have introduced flexible transit centered around dedicated busways, and at the same time have targeted considerable shares of regional commercial growth around key busway stations. The combination of flexible bus-based services and mixed-use development along busway corridors has given rise to unusually high per capita transit ridership rates in both cities.

[. . .]

TRANSIT SERVICES AND TECHNOLOGIES

I have opted for the term *transit* to describe generically the collective forms of passenger-carrying transportation services – ranging from vans and minibuses serving multiple origins and destinations (many-to-many) over nonfixed routes to modern, heavy rail trains operating point to point (one-to-one) over fixed guideways. *Transit* is the catchall used in the United States and Canada; however, almost everywhere else, *public transport* is the vernacular. And while in much of North America, *public transport* or *public transit* is associated with mass transit services provided by the public sector, almost everywhere else it means services that are available to the public at large, whether publicly or privately deployed. It is this broader, more inclusive definition of public transport that is adopted [here].

Types or classes of transit services can be defined along a continuum according to types of vehicles, passenger-carrrying capacities, and operating environments. The following sections elaborate on the forms of common-carrier transit services – i.e., those available to the general public. . . .

Paratransit

The smallest carriers often go by the name of *paratransit*, representing the spectrum of vans, jitneys, shuttles, microbuses, and minibuses that fall between the private automobile and conventional bus in terms of capacities and service features. Often owned and operated by private companies and individuals, paratransit services tend to be flexible and highly market-responsive, connecting multiple passengers to multiple destinations within a region, sometimes door-to-door and, because of multiple occupants, at a price below a taxi (but enough to more than cover full operating costs). Driven by the profit motive, paratransit entrepreneurs aggressively seek out new and expanding markets, innovating when and where necessary. Much of their success lies in their flexibility and adaptability. Unencumbered by strict operating rules, jitney drivers will sometimes make a slight detour to deliver someone hauling groceries to his or her front door in return for an extra charge. Besides being more human-scale, jitneys and minibuses can offer service advantages over bigger buses – often, they take less time to load and unload, arrive more frequently, stop less often, and are more maneuverable in busy traffic, and, studies show, passengers tend to feel more secure since each one is closer to the driver.[3]

In many parts of the developing world, jitneys and minibuses are the mainstays of the transit network. The archetypal service consists of a constellation of loosely regulated owner-operated collective-ride vehicles that follow more or less fixed routes with some deviations as custom, traffic, and hour of day permit. Jitney drivers respond to curbside hails pretty much anywhere along a route. Every paratransit system, however – whether the 2,000 *matatus* of Nairobi, the 15,000 *carros por puesto* minibuses in Caracas, or the 40,000-plus jeepneys of Manila – differs in some way. Some load customers in the rear of vehicles and others on the side; some are governed by federations of jitney owners while others engage in daily head-to-head competition; some have comfortable padded seats and others have hard wooden benches. Manila's jeepneys (converted US army jeeps that serve up to twelve riders on semifixed routes) carry about 60 per cent of all peak-period trips in the region. They cost 16 per cent less per seat mile than standard buses and generally provide a higher quality service (e.g., greater reliability, shorter waits) at a lower fare. Jeepney operations have historically been the last to petition for fare increases.[4]

Although banned in most wealthy countries, a handful of US cities today allow private minibus and jitney operators to ply their trade as long as they meet minimum safety and insurance requirements. New York City has the largest number of privately operated van services of any American city – an estimated 3,000 to 5,000 vehicles (seating 14 to 20 passengers) operate, both legally and illegally, on semifixed routes and variable schedules to subway stops and as connectors to Manhattan. Surveys show that more than three-quarters of New York's commuter van customers are former transit riders who value having a guaranteed seat and speedy, dependable services. Miami also has a thriving paratransit sector that caters mainly to recent immigrants from Cuba and the West Indies who find jitney-vans a more familiar and congenial form of travel than buses. Today, virtually all US cities allow private shuttle vans to serve airports.

Studies consistently show that jitneys and minibuses, whether in United States or Southeast Asia, confer substantial economic and financial benefits, both to the public sector and to private operators – namely, they are more effective at coaxing motorists out of cars than conventional transit in many settings, and do so without costly public subsidies.[5] However, as passenger volumes rise above a certain threshold (usually 4,000 or more per direction per hour), the economic advantages of paratransit begin to plummet, reflecting the limitations of smaller vehicles in carrying large line-haul loads. In both the developing and developed worlds, paratransit best operates in a supporting and supplement rather than substituting, role.

Bus transit

Urban *bus transit* services come in all shapes and sizes, but in most places they are characterized by 45- to 55-passenger pneumatic-tire coaches that ply fixed routes on fixed schedules. Buses are usually diesel propelled, though in some larger metropolises (e.g., Mexico City, Toronto), electric trolley buses powered by overhead wires also operate. Because they share road space, buses tend to be cheaper and more adaptive than rail services. However, on a per passenger kilometer basis, bus transit is generally a less efficient user of energy and emits more pollution than urban rail services. It is partly because of environmental concerns, as well as image consciousness, that some cities have sought to trade in their bus routes for urban rail services.

Bus transit is particularly important in developing countries, such as India, where some 40 per cent of all urban trips are by bus. In the Third World, the private sector serves more than 75 per cent of bus trips. In Karachi, Pakistan, private enterprises operating medium-size buses handle 82 per cent of transit journeys.[6] Because they are highly vulnerable to traffic congestion, buses are notoriously slow in megacities such as Shanghai, China, where it is generally faster to pedal a bike for trips under 14 kilometers in length.[7] One remedy is to reward high-occupancy travel through preferential treatment, such as reserved bus lanes and traffic signal preemptions. Bangkok, Thailand, has opened some 200 kilometers of reserved, contra-flow bus lanes to expedite bus flows in a city where rush-hour speeds often fall below 10 kilometers per hour.

In most developed countries, bus transit falls largely under the domain of the public sector, though concerns over rising subsidies have prompted more and more public transit agencies to competitively tender services to private contractors. In much of the United Kingdom and Scandinavia, public bus services have been turned over to the private sector outright. For many small to medium-size metropolitan areas of the United States, Canada, and Europe, conventional coaches (operating over fixed routes on published schedules) are the predominant transit carriers; in larger areas, buses often function mainly as feeders into mainline rail corridors. Providing exclusive busways can allow buses to integrate feeder and line-haul functions in a single vehicle. In . . . Ottawa and Curitiba, dedicated passageways are provided for buses, enabling rubber-tire vehicles to emulate the speed advantages of conventional steel-wheel trains on line-haul segments, yet perform as regular buses on surface streets as well. Guided busways, or O-Bahns, introduced so far in Essen, Germany; Adelaide, Australia; and two British cities, Leeds and Ipswich, are particularly suited to corridors (such as freeway medians) with restricted right-of-ways. Because of faster operating speeds, the theoretical maximum passenger throughputs of busways are as high as 20,000 persons per direction per hour, more than twice that of conventional surface-street buses.[8]

Trams and light rail transit

Rail transit systems are mass transit's equivalents to motorized expressways, providing fast, trunkline connections between central business districts, secondary activity centers, and suburban corridors. The oldest and slowest rail services – *streetcars* in the United States and *tramways* in Europe – functioned as mainline carriers in an earlier era, but as metropolitan areas grew outward, those that remained intact were relegated to the role of central city circulators. In cities such as Zurich, Munich, and Melbourne, aging tramways have been refurbished in recent times to improve vehicle comfort, safety, and maneuverability. Trams are enjoying a renaissance in a number of European cities because their slower speeds, street-scale operations, and Old World character blend nicely with a pedestrian-oriented, car-free central city.

The modern-day version of the electric streetcar, *light rail transit* (LRT), has gained popularity as a more affordable alternative to expensive heavy rail systems, particularly in medium-size metropolitan areas of under 3 million population. Compared to tram services, LRT generally operates along exclusive or semi-exclusive right-of-ways using modern, automated train controls and technologies. The LRT vehicles tend to be roomier and more comfortable than tram cars, with more head clearance and lower floors. In the United States, where the most LRT trackage has been laid since the early 1980s, costs are often saved by building along disused railroad corridors. Medium-size US cities with fairly low densities, such as Sacramento, California, have managed to build LRT for as low as US\$ 10 million per route mile; in Sacramento's case, costs were slashed by sharing a freight railroad right-of-way, building no-frills side-platform stations, and relying predominantly on single-track services. Light rail transit is generally considered safer than heavy rail because electricity comes from an overhead wire instead of a middle third rail. There is thus no need to fence in the track, not only saving costs but also allowing LRT cars to mix with traffic on city streets.

Today there are more than 100 tramways and LRT systems worldwide (mostly in Europe and North America), with the number continually rising. Among the factors behind the growing popularity of LRT and refurbished tramways are their lower costs relative to heavy rail investments and their ability to adapt to the streetscapes of built-up areas without much disruption. Other advantages include: they operate relatively quietly, thus are fairly environmentally benign and unobtrusive; they are electrically propelled, thus are less dependent than buses on the availability of petrochemical fuels; and they can be developed incrementally, a few miles at a time, eliminating the need for the long lead times associated with heavy rail construction.

. . . With four-car trains running as closely as three minutes apart, LRT can carry some 11,000 passengers per direction per hour; cutting the headways to ninety seconds (as found in some German cities, including Karlsruhe), maximum capacity can be doubled to more than 20,000. Advanced light rail transit (ALRT) systems – such as the skytrains in Vancouver, Toronto, and London's Docklands propelled by linear induction motors – can accommodate more than 25,000 passengers per direction per hour because of their higher engineering and design standards (though automated train control in lieu of on-board drivers constrains carrying capacities). It is for this reason they are also called intermediate capacity transit systems (ICTS).

Heavy rail and metros

In the world's largest cities, the big-volume transit carriers are the *heavy rail* systems, also called *rapid rail transit*, and known as *metros* in Europe,

Asia, and Latin America. Metros . . . work best in large, dense cities. Indeed, the relationship is symbiotic. The densities found on Hong Kong's Victoria Island and New York's Manhattan Island could not be sustained without heavy rail services. And heavy rail service could not be sustained without very high densities. Presently, more than 90 per cent of all peak-period trips to and from central London are by transit, mainly via the underground "tube"; for the remainder of greater London, transit serves fewer than a quarter of all peak-hour trips.[9]

Today, worldwide, there are some 80 metro systems, including 27 in Europe, 17 in Asia, 17 in the former Soviet Union, 12 in North America, seven in Latin America, and one in Africa. Some metros have been enormously successful, including Moscow's and Tokyo's, each of which carries 2.6 billion to 2.8 billion customers a year, more than twice as many as London's or Paris's metro systems, both of which are double the size of Moscow's and Tokyo's. On a riders per track kilometer basis, the world's most intensively used metros are, in order, São Paulo, Moscow, Tokyo, St Petersburg, Osaka, Hong Kong, and Mexico City. Most Western European, Canadian, and US metros have one-third to one-quarter the passenger throughput per track kilometer of these cities, in large part because more of their residents own cars and the cost of driving is relatively low.

In contrast to light rail systems, few new metros are being built today, partly for fiscal reasons and partly because most areas that can economically justify the costly outlays already have them. Except for Southern California, no new heavy rail lines or extensions are being planned, designed, or constructed in North America. The World Bank lending for metro systems ceased completely in 1980 and has resumed again only recently. The Bank generally frowns on funding rail projects, even in megacities paralyzed by traffic congestion, viewing them as cost-ineffective means of achieving the Bank's principal missions of alleviating poverty and stimulating economic growth.[10]

The niche market of heavy rail services is high-volume, mainline corridors. Accommodating more than 50,000 passengers per hour in each direction, heavy rail services provide high-speed, high-performance connections within built-up cities as well as between outlying areas and central business districts. In city cores, heavy rail systems almost always operate below ground, thus the names undergrounds (in Great Britain and its former colonies) and subways. To justify the high costs for right-of-way acquisitions, relocations, and excavation, undergrounds require very high traffic volumes (toward the upper end of the capacity threshold). Outside the core, metro lines are normally either above ground (called elevated or aerial alignments) or at-grade within expressway medians. Most heavy rail stations are far more substantial and sited farther apart than LRT stops, usually two or more kilometers from each other, except in downtowns, where they might be three or four blocks away. Because heavy rail systems are often the most expansive metropolitan rail services and operate at the highest speeds, their impacts on accessibility, and accordingly on urban development, tend to be the greatest.[11]

Heavy rail systems are almost universally electrically propelled, usually from a third rail, and each car has its own motor. Since contact with the high-voltage third rail can be fatal, rapid rail stations usually have high platforms and at-grade tracks are fenced.

Commuter and suburban railways

In terms of operating speed and geographic reach, *commuter rail* or *suburban rail*, stands at the top of the rail transit hierarchy. In Germany and central Europe, where suburb-to-city rail links are widespread, these services go by the name *S-Bahn*. Today, commuter rail services can be found on five continents in over 100 cities in more than 100 countries. Japan dominates the world's commuter rail market. In 1994, Tokyo carried almost six times the number of suburban rail commuters as Bombay, the largest commuter rail market outside Japan. Metropolitan New York's suburban rail is today only 2 per cent of Tokyo's. Nevertheless, metropolitan New York, along with a dozen or so other North American metropolises, is in the midst of a commuter rail renaissance. More commuter rail tracks are currently being planned, designed, and constructed in the United States and Canada than any form of rail transit. In all, twenty-one US and Canadian cities either have commuter rail services or hope to have them within the next decade. This would raise the total US and Canadian commuter rail trackage to some 8,000 kilometers, more than five times as long as LRT and seven times as long as heavy rail.

Commuter rail services typically link outlying towns and suburban communities to the edge of a region's central business district. They are most common in big metropolitan areas or along highly urbanized corridors and conurbations, such as the Richmond–Boston axis in the northeastern United States. Commuter rail is characterized by heavy equipment (e.g., locomotives that pull passenger coaches), widely spaced stations (e.g., 5 to 10 kilometers apart), and high maximum speeds that compete with cars on suburban freeways (although trains are slow in acceleration and deceleration). Services tend to be of a high quality, with every passenger getting a comfortable seat and ample leg room. Routes are typically 40 to 80 kilometers long and lead to a stub-end downtown terminal. Outlying depots are normally surrounded by surface parking lots that enable suburbanites and exurbanites to access stations conveniently by car. With the exception of the greater New York area (along the MetroNorth corridor to Connecticut), relatively little land-use concentration or redevelopment can be found around US commuter rail stations – after all, the very premise of commuter rail is to serve the low-density lifestyle preferences of well-off suburban professionals who work downtown. Serving commuter trips almost exclusively also means that ridership is highly concentrated in peak hours, more so than any other form of mass transit service.

NOTES

1 Urban Mobility Corporation. 1997. The 1995 Nationwide Personal Transportation Survey, *Innovation Briefs*, 8(7), p. 1; Pisarski, A. 1992. *Travel Behavior Issues in the 90's*. Washington, DC: Federal Highway Administration, US Department of Transportation.

2 Pucher J. and Lefèvre, C. 1996. *The Urban Transport Crisis in Europe and North America*. Basingstoke, UK: Macmillan Press.

3 Cervero, R. 1997. *Paratransit in America: Redefining Mass Transportation*. Westport, CT: Praeger.

4 Roth, G. and Wynne, G. 1982. *Learning from Abroad: Free Enterprise Urban Transportation*. New Brunswick, NJ: Transaction Books.

5 Roth and Wynne, op. cit.; Walters, A. 1979. The Benefits of Minibuses, *Journal of Transport Economics and Policy*, 13, pp. 320–334; Takyi, I. 1990. An Evaluation of Jitney Systems in Developing Countries, *Transportation Planning and Technology*, 44(1), pp. 163–177.

6 Armstrong-Wright, A. 1993. *Public Transport in Third World Cities*. London: HMSO Publications.

7 Shen, Q. 1997. Urban Transportation in Shanghai, China: Problems and Planning Implications, *International Journal of Urban and Regional Research*, 21(4), pp. 589–606.

8 Under ideal conditions (e.g., very light traffic, flat terrain, straight lanes, no interruptions to flow such as traffic signals), buses operating on a conventional highway can move as many as 9,000 passengers per lane per direction. Sources: Trolley, R. and Turton, B. 1995. *Transport Systems and Policy Planning: A Geographical Approach*. Harlow, Essex: Longman Scientific & Technical; Vuchic, V. 1992. Urban Passenger Transportation Modes, in G. Gray and L. Hoel (eds), *Public Transportation in the United States*. Englewood Cliffs, NJ: Prentice-Hall, pp. 79–113.

9 Dasgupta, P. and Bly, P. 1995. *Managing Urban Travel Demand: Perspectives on Sustainability*. London: Department of Transportation.

10 The International Institute for Energy Conservation. 1996. *The World Bank & Transportation*. Washington, DC: The International Institute for Energy Conservation; Gutman, J. and Scurfield, R. 1990. Towards a More Realistic Assessment of Urban Mass Transit, in *Rail Mass Transit for Developing Countries*, Institute of Civil Engineers, London: Thomas Telford, pp. 327–338.

11 Knight, R. and Trygg, L. 1977. Evidence of Land Use Impacts of Rapid Transit Systems, *Transportation*, 6(3), pp. 231–247; Cervero, R. 1984. Light Rail Transit and Urban Development, *Journal of the American Planning Association*, 50(2), pp. 133–147; and Huang, H. 1996. The Land-Use Impacts of Urban Rail Transit Systems, *Journal of Planning Literature*, 11(1), pp. 17–30.

"Traffic Calming"

from *Sustainability and Cities: Overcoming Automobile Dependence* (1999)

Peter Newman and Jeffrey Kenworthy

Editors' Introduction

Australian researchers Peter Newman and Jeffrey Kenworthy touched off an international debate in 1989 with their analysis of the relation between urban density and petroleum consumption in their book *Cities and Automobile Dependence* (Brookfield, VT: Gower Technical, 1989). This work showed both the enormous range of urban densities worldwide and the very strong correlation between higher densities and decreased resource use. In their later book *Sustainability and Cities* (Washington, DC: Island Press, 1999), they place transportation squarely at the center of the urban sustainability challenge, and outline various strategies for moving away from automobile dependence. Newman and Kenworthy argue that five key policies are needed to overcome automobile dependence:

1 Traffic calming "to slow auto traffic and create more urban humane environments better suited to other transportation modes,"
2 Improved transit, bicycling, and walking "to provide genuine options to the car,"
3 Improved land use, especially "urban villages" that can "create multinodal centers with mixed, dense land use that reduce the need to travel,"
4 Growth management "to prevent sprawl and redirect development into urban villages," and
5 Economic incentives, such as "taxing transportation better."

In this selection Newman and Kenworthy discuss approaches to calming traffic, and provide historical background on the global traffic-calming movement that began in Europe in the 1970s. This effort to reclaim automobile-dominated streets for human use is now worldwide and goes far beyond simply improving public safety. It may be seen as part of an effort to humanize public space and reclaim cities for people instead of cars. Other authors have made this point as well, such as Engwicht in his books *Reclaiming our Cities and Towns: Better Living with Less Traffic* (Gabriola Island, BC: New Society Press, 1993) and *Street Reclaiming: Creating Livable Streets and Vibrant Communities* (Gabriola Island, BC: New Society Press, 1999), Donald Appleyard in *Livable Streets* (Berkeley: University of California Press, 1981), and Bernard Rudofsky in his classic *Streets for People: A Primer for Americans* (New York: Doubleday, 1969).

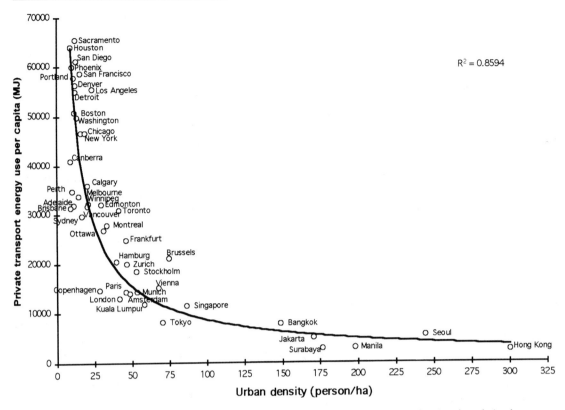

Figure 1. In a previous book, Newman and Kenworthy developed this classic diagram showing the relation between urban density and energy consumption for transportation.

Traffic calming (from the German *Verkehrsberuhigung*) is the process of slowing down traffic so that the street environment is safer and more conducive to pedestrians, cyclists, shoppers, and residential life. Traffic calming is best done by physically altering the street environment through different road textures; changing the geometry of the road through chicanes (also known as S-shaped diverters), neck-downs (also known as chokers), speed plateaus and bumps, and other traffic engineering devices; introducing new street furniture designed to create a more human, safe environment; and planting attractive landscaping.

Together, these changes make drivers slow down by causing them to see less open black-top and to perceive the road as a space that is to be shared with pedestrians, cyclists, and transit vehicles. Through the avenues of trees and street gardens that accompany good traffic-calming schemes, urban wildlife habitats and corridors through cities can be created and soft surfaces can be increased so there is less stormwater pollution.

Traffic calming has the potential not only to lessen the direct negative impacts of road traffic but to foster urban environments that are more human and interactive, more beautiful, and more economically successful due to the greater social vitality possible in a city's public spaces.

It is not known exactly where or when the concept of traffic calming originated, but the German term is believed to have first been used in German federal government reports in the early 1970s. The late John Roberts of Transport and Environment Studies[1] in London was the first to translate the word into English and to bring the concept to the attention of transportation planners in other parts of the world. The idea of traffic calming, however, has its roots in earlier movements to protect city environments from the worst excesses of the automobile. This reached a watershed in the early 1960s with the publication of the major report entitled "Traffic in Towns," by Colin Buchanan.[2] Although the British approach was to create more calmed city centers and protected residential precincts, the Buchanan report was used mostly to build large ring roads and bypasses that helped create automobile dependence. The report was used to justify major road proposals in Australian and North American cities as well. However, the European approach is based more on the organic integrity of the urban street and this approach is now gaining currency in the United Kingdom.[3]

Traffic calming emerged in Europe in the late 1960s from a number of sources: the Dutch *woonerf* or "living yard," created streets that had one shared surface with much planting to slow speeding traffic through inner-city streets and the original pedestrianization schemes in cities such as central Copenhagen.[4] Traffic calming gained rapid growth and acceptance in Europe in the 1980s through the successful action of many environmental groups trying to curtail the impacts of the automobile on European cities.[5]

Traffic calming's major objectives are to:

- Reduce the severity and number of accidents in urban areas;
- Reduce local air and noise pollution and vehicle fuel consumption;
- Improve the urban street environment for non-car-users;
- Reduce the car's dominance on roads by reclaiming road space for living space;
- Reduce the barrier effects of motor traffic on pedestrian and cycle movement; and
- Enhance local economic activity by creating a better environment for people.

With these broad objectives, traffic calming can also be of benefit to urban regeneration, housing renovation schemes, and city beautification programs (e.g., Freiburg, in southern Germany). These assist more deeply in reducing automobile dependence by bringing urban activity back to areas of the city that are inherently less dependent on the automobile (i.e., denser central and inner areas of cities built more around transit and nonmotorized modes). Traffic calming in Germany was in fact pioneered and promoted much more aggressively by the housing and urban development ministries than by the transportation ministry. This was primarily because of the positive impact traffic calming can have on the character and environmental quality

of neighborhoods, making them much more desirable urban redevelopment and residential areas, while a significant number of transportation planners viewed traffic calming changes with suspicion.[6]

TECHNIQUES OF TRAFFIC CALMING AND THEIR IMPLEMENTATION

Traffic calming was originally restricted mainly to improving residential streets, and this is still a major focus. Traffic calming seeks to alter road layout and design without actually totally rebuilding a street system. It does this through a whole suite of possible techniques such as narrowed entries to streets, plantings of trees with strong vertical elements, variable street surfaces, speed restricting devices, and visual barriers that encourages cautious driving. . . . However, it has been recognized that to be really effective and to not just shift traffic problems from one area to another, traffic calming must be applied more on an area-wide basis,[7] which means involving arterial or main roads.

There are now many examples of traffic calming on through roads and in other busy areas throughout Europe (e.g., Frankfurt, Hamburg, Nürnberg, Berlin, and Copenhagen). Denmark has a nationwide program of traffic calming on main roads called Environmentally Adapted Through Roads.[8]

The approach to traffic calming has to be somewhat different on main roads because of the volumes of traffic involved, although there is overlap in the basic techniques used. In busier areas where there is a need to better balance the needs of motor vehicles with the needs of pedestrians and cyclists, the main goal is to be able to reclaim road space for other uses by reducing the speed of traffic and its impact. In most cases roads are simply reduced from six to four traffic lanes, or from four to two lanes, through critical areas of a city. . . .

In some cases the reductions in road space are accompanied by significant improvements to transit such as new rail links (e.g., Nürnberg), and in others no major changes are made but incremental improvements are implemented. Road capacity is not necessarily reduced because the loss of lanes is offset by slower speeds that reduce vehicle headways and enable more vehicles to pass. Similarly, parking supply is not necessarily reduced and in some cases may be increased nominally. Often, parallel parking on two sides of a road is converted to angle parking on alternate sides separated by landscaped strips.

The implementation of traffic calming, however, is not just a technical process but a wide-ranging community process whereby local residents can have a strong input into identifying the problems and helping to find the solutions. It has been repeatedly shown that consultation with and involvement of the community are essential to the widespread acceptance of traffic-calming schemes. In fact, an important aspect of traffic calming is the way it has been able to provide a focal point for mobilizing and galvanizing many communities around the world into developing and fighting for a vision of a more sustainable and socially acceptable solution to the problem of traffic in urban environments.[9]

EFFECTS OF TRAFFIC-CALMING SCHEMES

Many of the major traffic-calming schemes in Europe have been formally sponsored by national and local governments as demonstration projects, and one of the aims has been to test the effects of the traffic-calming schemes on key environmental indicators and safety factors. Much of the available evidence about the effects of traffic-calming schemes comes from before-and-after studies of these projects.

The following is a brief summary of the general effects of traffic-calming schemes, along with some specific examples:

Reduced accidents. Accidents, particularly the severity of accidents, are generally significantly reduced with traffic calming because speed is the most critical factor in road accidents – particularly regarding the risk of serious injury and the danger to pedestrians and cyclists. In Berlin, for example, an area-wide scheme resulted in the reductions shown in Table 1.

Most other schemes report similar kinds of data, such as in Heidelberg, which experienced average accident reductions of 31 per cent and a

Type of traffic	Accident measure	Percent reduction
All traffic	Fatal accidents	−57
	Serious accidents	−45
	Slight accidents	−40
	Accident costs	−16
Nonmotorized	Pedestrians	−43
	Cyclists	−16
	Children	−66

Table 1 Accident reductions in Berlin Moabit (neighborhood) using comparable before and after periods

Source: Reported in Pharoah, T. and Russell, J. 1989. *Traffic Calming: Policy Evaluation in Three European Countries.* Occasional Paper 2/89, Department of Planning, Housing and Development. London: South Bank Polytechnic.

44 per cent reduction for casualties after thirty-kilometer-per-hour [eighteen-mile-per-hour] residential speed limits were introduced along with selected physical traffic-calming measures.[10] Area-wide schemes in The Netherlands have reduced accidents involving injury by 50 per cent in residential areas and 20 per cent overall (measured per million vehicle kilometers) and no increase in accidents has occurred in surrounding areas.[11]

The Center for Livable Communities, in their Livable Places Update for March 1998, summarized some of the best US examples of traffic calming, and in relation to accidents, found the following:

- The City of Seattle, where traffic-calming projects have been carried out for 20 years, surveyed the results of 119 completed projects and found an overwhelming 94 per cent reduction in accidents.
- In Portland, Oregon, 70 traffic circles and 300 speed bumps have been introduced and the number of reported accidents decreased by 50 per cent.
- A 1997 study of US street typology and accidents by Swift and Associates showed that as street width increases, accidents per mile per year increase exponentially. The safest residential street (curb to curb) turned out to be 24 feet (7.2 meters). Present US street regulations require

36 feet, primarily for access by fire vehicles, though the study found that fire vehicles can access 24-foot-wide roads when required. New Urbanism design guidelines are for 24-foot roads.

Noise reduced. Traffic calming generally results in a reduction in vehicle noise. Pharoah and Russell report that noise changes result from five factors: changes in traffic volume and composition, changes in carriageway layout, changes in carriageway surface, changes in vehicle speed, and changes in driving style.[12]

Air pollution benefits. Research in central Europe shows that in built-up areas, the higher the vehicle speed the more will be the proportion of acceleration, deceleration, and braking, and this increases air pollution. By contrast, traffic-calming schemes in some German residential areas have shown that idle times are reduced by 15 per cent, gear changing by 12 per cent, brake use by 14 per cent, and fuel use by 12 per cent.[13]

Evidence of the air pollution benefits of a slower, calmer style of driving comes from detailed work in Buxtehude, a German demonstration project (population 33,000). Table 2 shows the changes in the different types of emissions with a reduction of speed from 50 kilometers per hour (30 miles per hour) to 30 kilometers per hour (18 miles per hour) under two types of driving. In both aggressive and calm driving, emissions are reduced at the 30 kilometers per hour level, though the calm driving has a generally greater reduction and fuel use is lower.

	Driving style	
	Second gear, aggressive (%)	Third gear, calm (%)
Carbon monoxide	−17	−13
Hydrocarbons	−10	−22
Nitrogen oxides	−32	−48
Fuel consumption	+7	−7

Table 2 Changes in vehicle emissions and fuel use from 50 km/h to 30 km/h

Source: Reported in Pharoah, T. and Russell, J. 1989, op. cit.

It is also worth noting that even in instances when individual vehicles may experience an increase in fuel use and emissions (e.g., drivers do indulge in more acceleration, braking, and greater use of second gear), this may not result in an overall increase in local pollution and fuel use if the traffic-calming scheme has also resulted in lower traffic volumes.

Enhanced pedestrian and street activity. Traffic calming seeks to make the public environment safer and more attractive, so it is to be expected that traffic calming will result in a greater level of pedestrian and cycling activity in the area affected. In general, it can be expected that the results will be more noticeable in busier areas with a mix of land uses and the potential for people to make good use of reclaimed areas, such as for outdoor cafes and markets, children's facilities, etc.

Some formal measurements of the benefits are available from a summary of European experience by Pharoah and Russell (1989), such as in Berlin's federal demonstration project, where nonmotorized traffic on a wide range of streets in the scheme increased by between 27 per cent and 114 per cent; in Vinderup, a village in Denmark, where the main through route was traffic-calmed and outdoor activities increased by up to 47 per cent; and in Copenhagen, where traffic calming has led to immediate increases of pedestrian activity of between 20 per cent to 40 per cent, and in the long term, where central area activity is now 80 per cent pedestrian and 14 per cent by bike.[14] Where traffic calming reduces road capacity there is an overall decrease in traffic[15] and therefore better conditions are created for pedestrians.

Traffic calming also tends to increase the area used by pedestrians and cyclists and the extent to which streets are crossed by these users, since the severance effects of traffic are reduced. Pedestrians and cyclists tend not to confine themselves purely to walkways, but rather they extend their territory to the roadway in some instances.

Reduced crime rates. Appleyard (1981) showed that visiting among neighbors decreases when traffic increases,[16] and when neighboring ceases and people stop watching out for one another, then criminal activity can occur. The *Livable Places Update* (March 1998) overview on traffic calming quotes a Harvard University study that showed violent crimes in communities where residents willingly worked together were as much as 40 per cent lower than in neighborhoods where such relationships were not as strong. Race and income were not factors in people's willingness to take part in such community activity. An example of a place where crime rates diminished after traffic calming is Weinland Park in Columbus, Ohio.

Positive economic implications. As pointed out in the objectives of traffic calming, economic revitalization of an area is an explicit aim in some schemes. A study by TEST (1989) attempted to confirm the hypothesis that "A good physical environment is a good economic environment" and examined ten European cities in detail. Roberts sums up the work by saying, "the message is simple: there is a strong likelihood that traffic restraint in all its forms, and environmental improvement, and a healthy economy, are causally related."[17]

The basis of this finding would appear to involve at least the following factors:

- People like to come to humanly attractive, green cities.
- Businesses like to locate in areas with a high quality urban environment.
- Car access is not banned, but it is not facilitated to the point of dominating everything else.
- Other modes are generally facilitated.

Hass-Klau (1993) shows conclusively that pedestrianization and traffic calming both have positive effects on the economic performance of an area; the more aggressive is the traffic calming, the more pronounced is the positive economic effect.[18]

In the United States, a West Palm Beach, Florida, neighborhood was economically depressed and bisected by fast-moving traffic. A traffic-calming scheme slowed the traffic through road narrowing and construction of speed bumps, traffic circles, and pedestrian islands. Then the city raised intersections, made sidewalks level with the street, and added a fountain, benches, and an amphitheater for "block parties." The development spurred new private investment and the cost of commercial space rapidly moved from five dollars per square foot to twenty-five dollars per square foot.[19] Similar case studies are given in the UK Friends of the Earth publication.[20]

TRAFFIC CALMING: A BROADER APPROACH

Traffic calming can be viewed as a broader transportation planning philosophy and not merely as a series of physical changes to roads.[21] Traffic calming in this broader sense is aimed at reducing total dependence on the automobile and promoting a more self-sufficient community with a transportation system more oriented to pedestrian, cycle, and transit use.

These broader objectives can be summarized as follows:

- A reduction of average motor vehicle speeds to discourage long-distance road travel in urban areas and promotion of a more compact urban form; traffic calming of main roads is included in this approach.
- Specific land use policies that better integrate transit and land development; the policies are directed at reducing the number, length, and need for motor vehicle trips.
- Strong promotion of walking, cycling, and transit.
- Restrictive measures against private traffic, including parking restrictions, limited major road building, and the direction of funds into transit and nonmotorized modes, as well as taxation policies on fuels and cars, including policies on company cars and road pricing.
- A shift in transportation planning philosophy from a traffic-generation approach of seeking to predict future traffic levels and the roads and parking needed to cope with them, to a traffic-dissolving approach of setting limits on motor vehicle growth and ensuring that transportation/land use policies and practices are aimed at minimizing the need for more motor vehicle facilities.

A good example of a broader traffic-calming policy in action is the Dutch national policy from 1982 that openly promotes transit, walking, and cycling. It states that:

Henceforth other functions will be given priority over motor traffic [and] the car's dominance should be diminished by deliberately increasing travel times, by creating a less dense network of main roads, and by reducing speeds.[22]

NOTES

1 See Transport and Environment Studies (TEST). 1989. *Quality Streets – How Traditional Urban Centers Benefit from Traffic Calming.* London: TEST.

2 Minister of Transport. 1963. *Traffic in Towns.* London: HMSO.

3 Standing Advisory Committee on Trunk Road Assessment (SACTRA). 1994. *Trunk Roads and the Generation of Traffic.* London: Department of Transport, United Kingdom; Department of Environment. 1994. *Planning Policy Guidance 13: Transport.* Whitehall, London: Department of Environment and Department of Transport.

4 Gehl, J. and Gemzøe, L. 1996. *Public Spaces, Public Life.* City of Copenhagen.

5 More detail on the evolution of traffic calming may be found in Hass-Klau, C. 1990. *The Theory and Practice of Traffic Calming: Can Britain Learn from the German Experience?* Discussion Paper 10, Rees Jeffreys Road Fund. Oxford: Transportation Studies Unit, Oxford University; Tolley, R. 1990. *Calming Traffic in Residential Areas.* Wales, UK: Brefi Press; Newman, P. and Kenworthy, J. 1991. *Towards a More Sustainable Canberra: An Assessment of Canberra's Transport, Energy and Land Use.* Institute for Science and Technology Policy, Murdoch University.

6 Hass-Klau. *The Theory and Practice of Traffic Calming.*

7 Hass-Klau, C. 1990. *The Pedestrian and City Traffic.* London: Belhaven Press.

8 Danish Road Data Laboratory. 1987. *Consequence Evaluation of Environmentally Adapted Through Road in Vinderup.* Report 52, Danish Road Data Laboratory, Danish Roads Directorate, Herlev, Copenhagen; *Consequence Evaluation of Environmentally Adapted Through Road in Skærbæk.* Report 63, Danish Road Data Laboratory, Danish Roads Directorate, Herlev, Copenhagen.

9 E.g. Tolley. 1990. *Calming Traffic in Residential Areas.*

10 Hass-Klau, C. 1990. *An Illustrated Guide to Traffic Calming: The Future Way of Managing Traffic.* London: Friends of the Earth.

11 Hass-Klau, C. (ed.) 1986. New Ways of Managing Traffic. *Built Environment*, 12 (1 and 2).

12 Pharoah, T. and Russell, J. 1989. *Traffic Calming: Policy Evaluation in Three European Countries.* Occasional Paper 2/89, Department of Planning, Housing and Development. London: South Bank Polytechnic.

13 Hass-Klau. 1990. *The Theory and Practice of Traffic Calming.*

14 Gehl and Gemzøe. 1996. *Public Spaces, Public Life.*

15 Goodwin, P.B. 1997. Solving Congestion. Inaugural Lecture for the Professorship of Transport Policy, University College, London, 23 October.

16 Appleyard, D. 1981. *Livable Streets.* Berkeley: University of California Press.

17 Roberts, J. 1988. Where's Downtown? 'It Went Three Years Ago.' *Town and Country Planning*, May, pp. 139–141.

18 Hass-Klau, C. 1993. Impact of Pedestrianization and Traffic Calming on Retailing: A Review of the Evidence from Germany and the UK. *Transportation Policy*, 1 (1), pp. 21–31.

19 Center for Livable Communities. 1998. Benefits of Traffic Calming Realized Across the Country. *Livable Places Update.* March.

20 Friends of the Earth. 1997. *Less Traffic, More Jobs: Direct Employment Impacts of Developing a Sustainable Transport System in the United Kingdom.* London: Friends of the Earth.

21 Hass-Klau. 1990. *The Theory and Practice of Traffic Calming.*

22 Ministry of Transport and Public Works. 1982. *From Local Traffic to Pleasurable Living.* The Hague: Ministry of Transport and Public Works, The Netherlands.

"Bicycling Renaissance in North America?"

from *Transportation Research A* (1999)

John Pucher, Charles Komanoff, and Paul Shimek

Editors' Introduction

Despite attempts to develop new devices such as the Segway scooter as an alternative to the automobile, the tried-and-true solution for short-distance personal mobility in many parts of the world has been the bicycle. Simple, cheap, pollution-free, and easy to maintain, the bike has been used widely in nations ranging from China to Cuba. At rush hour waves of cyclists pass down the streets of European cities such as Copenhagen or Amsterdam, as well as those of countless cities in the developing world. Many nations have also sought to promote cycling as a convenient way for public transit patrons to reach transit stations.

While both bicycling and walking seem ideal transportation modes for a sustainable city, a key question is how these can be encouraged within low-density, automobile-dependent communities such as in North America and Australia. While there are signs that cycling is on the increase, much bicycle use in these places is for recreation. What would it take for the bicycle to become a more all-round transportation alternative? In this selection leading transportation researchers John Pucher, Charles Komanoff, and Paul Shimek explore this question. Pucher is a professor in the Department of Urban Planning at Rutgers University in New Jersey, where he has written widely on transportation topics. Komanoff is president of Komanoff Energy Associates in New York City, and a leading consultant on issues of energy use, transportation pricing, and alternative transportation modes. Shimek is a researcher at the Volpe National Transportation Systems Center in Cambridge, Massachusetts.

Other useful materials on bicycle use include an article by Pucher and Lewis Dijkstra, "Making Walking and Cycling Safer: Lessons from Europe," published in *Transportation Quarterly*, 54(3), 2000; *Pedestrian and Bicycle Planning: A Guide to Best Practices*, by Todd Litman *et al.*, available from the Victoria Transportation Policy Institute at www.vtpi.org; and material from the Association of Pedestrian and Bicycle Professionals at www.apbp.org.

Several northern European countries have been enjoying a bicycling boom. Over the past two decades, cycling has increased significantly in Denmark, Germany, Switzerland, and The Netherlands.[1] The number of bicycle trips has grown substantially in these countries, and in many cities cycling's share of travel has risen as well. In Germany, for example, bicycling's modal share for urban trips rose by half between 1972 and 1995, from 8 per cent to 12 per cent.[2] Currently, the bicycle's share of local trips is 30 per cent in The Netherlands, 20 per cent in Denmark, 12 per cent in Germany, and 10 per cent in Switzerland – over ten times higher than in the United States.[3]

All these European countries have very high standards of living, and all have experienced rising incomes, growing auto ownership, and rapid suburbanization. Yet bicycling is thriving in this environment, primarily due to long-term commitments to enhance the safety, speed, and convenience of bicycling while making driving more difficult and expensive. These policies were adopted by democratic political systems, partly to mitigate the social and environmental harm of excessive auto use in cities, but also to accommodate rising demands for mobility within the physical constraints of congested urban roads, high-density cities, and limited land supply for parking.

Many groups have been advocating increased bicycling in the United States, not just for recreation but also for commuting and other utilitarian purposes. The League of American Bicyclists, the Bicycle Federation of America, and bicycling groups in virtually every state and many cities coordinate bicycling events, offer training courses, and lobby for cycling facilities and cycling-friendly roads and traffic policies. Many environmental organizations, community activists, and urban planners support cycling because it is an energy-efficient and non-polluting transport mode, and some transport planners view space-efficient cycling as a way to reduce roadway congestion. Aside from the cost of travel time, cycling is also cheaper than any mode except walking and thus affordable to even the poor. Moreover, the public costs of bicycling are modest, especially compared to motorized transport. Finally, fitness experts and health professionals advocate cycling for its cardiovascular benefits.

In recognition of the benefits of bicycling, and in response to strong public pressure, public policies in the United States have become more supportive of bicycling, especially since passage of the Intermodal Surface Transportation Efficiency Act (ISTEA) in 1991. The successor to ISTEA, the 1998 Transportation Equity Act for the Twenty-first Century (TEA21), continues this trend. . . . [T]he decade of the 1990s has witnessed a dramatic increase in funding of bicycling facilities in the US, with the focus so far on investments in new bike paths. Most states and many cities now have programs to facilitate bicycling, including bicycle coordinators in state departments of transportation. Traffic policies and roadway design in some locales are gradually becoming more conducive to bicycling. Unfortunately, little has been done to educate motorists about cyclists' rights and to enforce traffic laws that allow cycling on most streets and roadways. . . .

[. . .]

Do the growing interest in bicycling and the accompanying policy shifts suggest that America may be poised for a bicycling renaissance? Some bicycling advocates and trade publications already claim a massive boom in bicycling in the United States in recent years.[4] While cycling has certainly increased, sightings of a boom are open to question. This article uses a variety of sources to assess the actual extent of growth in cycling in the United States over the past two decades. . . . On the basis of our seven North American cities, and using information from European experience, we conclude by assessing the effectiveness of alternative policies to promote cycling. . . .

[. . .]

FACTORS AFFECTING CYCLING IN NORTH AMERICA

Cycling has increased in North America over the past two decades, both in the aggregate and for seven case study cities. While the increases are encouraging, the share of total trips by bike in the US still stands at only about 1 per cent, far lower than in most European countries[5]. . . .

Although climate and topography affect cycling levels, the case studies show that they do not explain differences in cycling rates among North American cities. A more important deterrent is the low-density sprawl of most American metropolitan areas, which increases average travel distances and renders utilitarian cycling less feasible. This factor alone may explain the higher cycling levels in Canadian cities, which are more than twice as dense as American cities.[6] European cities are denser still, leading to average trip lengths only about half those in the US.[7]

Nevertheless, even in the United States, a considerable percentage of urban trips are within cycling distance. According to the NPTS, 28 per cent of trips by all modes are one mile or shorter, and another 20 per cent are one to three miles. Of course, some of those short trips are links of longer trip chains that are less readily bikeable. Nevertheless, the high percentage of short trips suggests great potential for increased bicycling, even in the low-density, sprawled cities of the US.

Why, then, does bicycling in the United States remain at low levels? Here we summarize eight key factors.

Public attitude and cultural differences

Is bicycling for transportation considered a normal thing to do? In The Netherlands and Denmark, it is usual for young and old, rich and poor, and students and executives alike to bicycle for many different purposes. In the United States, most cycling is for recreation, and most cycle commuters are men. Even though a majority of Americans own a bicycle, cycling is considered a "fringe mode" in the US,[8] befitting its 0.9 per cent share of total trips. Utilitarian cycling is even less mainstream, with the bicycle used for only 0.3 per cent of all work trips in 1995, according to the NPTS.

Culture, custom, and habit are important. While the other factors listed below help explain which forms of travel behavior become widespread and thus considered "normal," countries with unbroken traditions of utilitarian cycling have an easier time maintaining that tradition. Where cycling is viewed as normal, people consider doing it when it is convenient, and they have access to the necessary equipment and knowledge. Similarly, motorists exhibit more respect toward cyclists, partly because they are more likely to cycle themselves or know others who do. In general, where there are few bicyclists, cycling is considered abnormal, and this climate tends to be self-perpetuating.

Public image

There is no single image of bicycling in America, but a multiplicity of perceptions dependent upon the type of cycling and the context in which it is viewed. Recreational cycling has a youthful, vigorous image since it is associated with sport and fitness; some car ads even feature recreational cyclists. Bicycling as a whole also has a positive environmental image, thanks to zero air pollution,

negligible noise, and minimal energy use. In cities, where the vast majority of utilitarian cycling takes place, cyclists suffer from a renegade image associated with disobedience of traffic laws, and a pervasive sense of cyclists as an alien presence on roads intended for cars. Indeed, the various images of cycling are so heavily determined in relation to automobiles that utilitarian cyclists are variously seen as too poor to own a car, "anti-auto," eccentric, or deviant. The perceptions of cycling as lying outside the mainstream of American life discourage bicycle use.

City size and density

Small, compact cities are more amenable to cycling since more destinations are accessible within a short bike ride, motor traffic volumes are lower, and there are less likely to be obstacles such as expressways and bridges. Indeed, to our knowledge, no very large city (1 million or larger) in either Europe or North America has bike use exceeding 10 per cent of trips. Europe has many more small, dense cities where biking is convenient for reaching many destinations.

Cost of car use and public transport

The cost, speed, and convenience of alternative modes have a crucial impact on modal choice. In the US, the low user-cost of autos is crucial in discouraging virtually all other modes, even walking. Low gasoline taxes, few road tolls, and ubiquitous free parking make auto use almost irresistible in the United States. At negligible marginal user costs, car use becomes a habit even for short trips that could be walked or cycled.[9] Not only are road tolls, taxes and fees far higher in Europe, but the extensive availability of transit makes car ownership less essential, thus reducing the number of car owners and increasing the tendency to use bicycles for many utilitarian trips.

Income

Rising incomes make car ownership and use more affordable. Every econometric analysis of the relationship between income and auto ownership finds a very high positive correlation. This helps explain why university students are more likely to bicycle, and suggests that the bike share of travel should decrease over time as countries get richer and an ever-larger share of the population can afford cars. This generalization does not always hold, however. Although Denmark, The Netherlands, and Germany are among the wealthiest countries in the world, they have very high bike modal shares.

Climate

Cycling levels are obviously affected by climate. Three case study cities with relatively high levels of cycling (Davis, San Francisco, and Seattle) enjoy mild winters and, in the case of the first two, little rain. The extreme heat and humidity of summers in the southern United States clearly discourage cycling there. Yet the effect of climate on cycling may be exaggerated. In spite of mostly cloudy days and frequent rain and drizzle, northern Europe has the highest cycling levels, far higher than in southern Europe, where it is drier, sunnier, and warmer.

Danger

The possibility of accidental injury and death is a major obstacle to bicycling. Making cycling as safe or safer than driving will require behavioral changes by both drivers and bike riders, as well as development of more cycle-appropriate infrastructure. While several European countries have national cycle training programs and more strictly enforce traffic rules for both drivers and cyclists, efforts at such behavior modification have been far less extensive and less successful in the US. Moreover, in the United States the elevated risks of cycling appear to be magnified by cultural attitudes that attribute cycling accidents to the supposedly intrinsic perils of bicycles. In contrast, motorist casualties are not ordinarily associated with the idea that driving is dangerous.[10] From there it is a short step to blaming cyclists for their own peril, an attitude that permeates the reactions of everyone from police and courts to the cyclist's own

family and friends and contributes to cyclists' marginal status. Thus, measures to reduce the statistical frequency of cycling accidents may need to be coupled with efforts to change public understanding of the nature of road dangers – a difficult task at best.

Cycling infrastructure

Unquestionably, separate bike lanes and paths for cyclists, together with better parking facilities, make cycling more attractive to noncyclists. However, we are not aware of any rigorous statistical studies of their actual impact on increasing cycling levels; to some extent, such facilities may be a response to increased cycling instead of its cause. Nevertheless, every European city with high cycling levels has an extensive route system, including separate bike paths and lanes as well as general street use in traffic-calmed neighborhoods.

STEPS TO INCREASE CYCLING IN NORTH AMERICA

Following are seven proposals for making cycling more widespread in the US and Canada.

Increase cost of auto use

Probably the most effective way to increase bicycling in North America would be to discourage auto use and increase its marginal cost, particularly for short auto trips that are both underpriced and most amenable to cycling. A sizeable increase in the price or inconvenience of driving would encourage people to seek other ways to travel and begin loosening the automobile's domination of daily transportation. Unfortunately, this approach is politically difficult. Indeed, the new federal transportation legislation (TEA21) fixes the federal gasoline tax at the same low level (approximately two cents per liter) for the next six years, and recently taxes on auto ownership have been rolled back in several states. A more promising approach may be restructuring road taxes and auto insurance to shift lump-sum charges into marginal use fees, thus providing positive incentives to shorten trips and make greater use of non-auto modes.[11] Blocking highway expansion also increases the time cost to drive and can make cycling more attractive, although it could also work against cycling by fomenting "rat-running" (driver use of local streets) and "road rage."

Clarify cyclists' legal rights

To a great extent, cyclists in the United States and Canada operate outside the prevailing system of traffic governance. As we have noted, many motorists and even police are not cognizant of cyclists' right to use ordinary roads, and there is scant appreciation of the vulnerability cyclists feel when autos impinge too closely. In contrast, many northwestern European cities actively promote cycling through conferences, fairs, and school programs, and their traffic rules, policing, licensing, and judicial systems uphold cyclists' rights far more than do their North American counterparts. However difficult it may be, establishing motorists' accountability for their actions toward cyclists is crucial to improving bicycling safety and encouraging cycling. A key first step, noted in the Toronto case study, would be to establish as a principle of law that cyclists have precedence over motor vehicles where both are vying for the same road space and neither clearly has right of way over the other. With their preferential right of way established in law, cyclists might improve their adherence to traffic laws, leading in turn to greater consideration from motorists in a reinforcing process of mutual respect.

Expand bicycle facilities

As discussed earlier, separate facilities (bike paths and lanes) are not a panacea for making cycling easier and safer. Nevertheless, rail trails and mixed-use greenway paths have increased recreational bicycling, and strategically located cut-through paths (as in Davis) can reduce trip times and thus encourage utilitarian cycling as well. The most successful bicycling programs examined in this article – in Davis, Madison and Seattle – include separate facilities in their overall strategy. Moreover, in every European country with at least 10 per cent bike modal split, separate cycling facilities (and traffic-calmed neighborhood streets) are integral parts of the bike route system. Separate paths and lanes are especially important for those unable or unwilling to do battle with cars for space on streets. Training courses may help, but they do not eliminate the inherent danger of cycling on the same right of way with motor vehicles, particularly for those whose mental or physical conditions limit their capacity to safely negotiate heavy traffic. The slowed reflexes, frailty, and deteriorating hearing and eyesight of many elderly make them especially vulnerable, while limited experience, incomplete judgment, and unpredictable movements put children at special risk on streets. And regardless of age, many people prefer to avoid the anxiety and tension of cycling in mixed traffic, aside from safety hazards. Bicycling should not be reserved for those who are trained, fit, and daring enough to navigate busy traffic on city streets.

Make all roads bikeable

More than other countries, the United States must rely heavily on the general road network for bicycling. Some cities have bike lanes and paths that link up to some extent, but none has a complete route network approaching the dense network of bike paths and lanes in virtually every Dutch, Danish and German city and throughout the countryside, with official route designations, signage and maps. Even Davis and Seattle, with their impressive cycleways, must also rely on the general road system. Thus, a fundamental strategy to make America bikeable must be to improve roads through wider curbside lanes and shoulders, drain grate replacement, pothole patching, clear lane striping, and bike-activated traffic signals, while punishing motorist behavior that infringes upon cyclists' legal right of way. Seattle's efforts to improve the road infrastructure are a good model, but no US jurisdiction has taken real steps to inculcate motorist responsibility for cyclist safety.

Hold special promotions

Bike-to-work weeks and employer-based promotions appear to have been helpful in inducing North

Americans to try – and then continue – cycling for transportation. Similarly, large-scale rides ranging from recreational and charity events to San Francisco's monthly Critical Mass rides help build cyclist confidence and provide mutual support and enthusiasm for cycling. In some cases such rides have also focused public attention on the needs of cyclists and helped force a shift toward more cycling-friendly public policies.

Link cycling to wellness

Numerous studies have documented the health benefits of regular exercise, and physical inactivity has come to be seen as a major cause of premature death in industrial societies, second only to tobacco. Cycling, potentially an ideal, low-cost way of getting that activity, has been linked in the public mind to risktaking and danger, in part by health-based helmet promotions that implicitly link cycling to danger. The British Medical Association's finding that cardiovascular-related gains to longevity from cycling far outweigh collision risks, though widely reported in Europe, is little known in North America.[12] New programs from the California Department of Health Services and the US Centers for Disease Control and Prevention seek to integrate routine physical activity into people's travel, work, leisure, and family life by making physical environments more amenable to walking and bicycling.[13] Holistic and pro-active efforts by the health community could boost cycling by casting it as a prudent, healthful choice.

Broaden and intensify political action

As emphasized by transportation researcher Martin Wachs,[14] political action is essential to bring about changes in public policy to encourage more and safer cycling. Bicyclists in many parts of the United States are already well-organized, and have learned to wield political clout to obtain funding for cycling facilities. Cyclists have won pro-bicycling provisions in ISTEA and TEA21 that portend major expansions and improvements to systems of bike paths, lanes, and parking. TEA21 also encourages better roadway design, which provides

an important basis for making more roads bikeable. Nevertheless, it remains to be seen how effectively cycling groups can pressure state highway departments to carry out the federal mandates. Similarly, cycling groups will have to continue to exert pressure at the local level to maintain and improve existing elements of the cycling infrastructure, such as bridge access, against the threat of prohibitions or banishment to substandard facilities. Cyclists will also need to open up another front: inducing police and courts to enforce the rights of bicyclists to use city roads and to curb driving privileges of motorists who violate those rights.

PROSPECTS FOR BICYCLING IN NORTH AMERICA

With the right set of public policies, bicycling in the United States could increase dramatically. As noted by both Wachs and Gordon and Richardson, to date there has not been sufficient political support to pass and implement those policies. So far, only the easiest no-conflict measures have been implemented; most new bike paths and lanes in the United States do not directly compete with auto use. By contrast, many European cities have implemented policies that sharply restrict auto use in favor of walking and bicycling, especially in city centers.[15] German, Dutch, and Danish cities give cyclists priority on certain streets and intersections and routinely employ "advanced" green lights and traffic-calmed streets. Some one-way streets have been made two-way for bicyclists, and cyclists are exempted from many turn restrictions for cars. Some European cities have dedicated car parking space to bike lanes or bike parking, not just to enable cycling but to discourage auto use. Enacting such measures has taken concerted political pressure, even in cities where 20 per cent of the populace cycles regularly. Such auto-restrictive initiatives do not yet appear politically feasible in America. Too many Americans drive cars (and would feel hurt by such measures), and too few Americans presently bicycle (and feel they would benefit enough to fight for such measures).

It is possible to imagine a *deus ex machina* giving a strong boost to cycling in America – perhaps an oil shock, or a cultural or style change toward bikes and away from cars, or ascendancy of a charismatic politician closely identified with cycling. But the more likely scenario is slow, painstaking progress: modest extensions and improvements in separate bicycle facilities, even more modest improvements in roadway design, and isolated instances of effective enforcement of cyclist rights to use public roads. Those measures may produce significant growth in bicycling in those cities that implement them. But overall, they will not produce a bicycling boom, unless the visible success of cycling enhancements in one or two major cities attracts imitators elsewhere.

NOTES

1 Dutch Ministry of Transport. 1995. *Cities Make Room for Cyclists: Examples from Towns in the Netherlands, Denmark, Germany, and Switzerland.* The Hague: Ministry of Transport, Public Works, and Water Management, The Netherlands; Tolley, R. (ed.). 1997. *The Greening of Urban Transport.* Chichester, Sussex: John Wiley & Sons; Zegeer, C. (ed.). 1994. *FHWA Study Tour for Pedestrian and Bicyclist Safety in England, Germany, and The Netherlands.* Washington, DC: US Department of Transportation; Pucher, J. 1997. Bicycling Boom in Germany: A Revival Engineered by Public Policy. *Transportation Quarterly*, 51(4), pp. 31–46.

2 Pucher. 1997. Bicycling Boom in Germany.

3 Ibid.

4 Sani, M. 1997. Could Better Roads Boost Commuting? *Bicycle Retailer and Industry News*, 1 December.

5 Pucher. 1997. Bicycling Boom in Germany.

6 Schimek, P. 1996. Automobile and Public Transit Use in the United States and Canada: Comparison of Postwar Trends. *Transportation Research Record*, 1521, pp. 3–11; Pucher, J. 1994. Canadian Public Transport: Recent Developments and Comparisons with the United States. *Transportation Quarterly*, 48(1), pp. 65–78.

7 Pucher, J. 1995. Urban Passenger Transport in the United States and Europe: A Comparative Analysis of Public Policies. *Transport Reviews*, 15(2), pp. 89–107, and 15(3), pp. 261–277.

8 Gordon, P. and Richardson, H. 1998. Bicycling in the United States: A Fringe Mode? *Transportation Quarterly*, 52(1), pp. 9–11.

9 Pucher. 1995. Urban Passenger Transport.

10 Komanoff, C. 1997. Restoring Cycling Habitat. *Bicycle Forum*, 45, pp. 6–13.

11 Litman, T., Komanoff, C. and Howell, J. 1998. *Road Relief – Tax and Pricing Shifts for a Fairer, Cleaner, and Less Congested Transportation System in Washington State.* Olympia, WA: Energy Outreach Center.

12 British Medical Association. 1992. *Cycling Toward Health and Safety.* London: Oxford University Press.

13 Seeley, Anne (California Department of Health Services, Active Communities Coordinator). 1998. Personal communication.

14 Wachs, M. 1998. Creating Political Pressure for Cycling. *Transportation Quarterly*, 52(1), pp. 6–8.

15 Pucher. 1997. Bicycling Boom in Germany.

URBAN ECOLOGY
AND RESTORATION

"City and Nature"

from *The Granite Garden: Urban Nature and Human Design* (1984)

Anne Whiston Spirn

Editors' Introduction

Although landscape architects and park designers have long sought to bring nature into cities, this need was often ignored by developers and the nascent city planning profession in the nineteenth and twentieth centuries. Engineers and developers filled in or paved over streams, wetlands, and shorelines to make way for urban expansion. Highways or railroad lines cut many cities off from their waterfronts. Hills were leveled and native vegetation removed. Landowners platted lots and built roads without considering the implications for wildlife, native plant species, or human recreation. With the advent of central heating, electric lighting, air-conditioning, long-distance food transport, and huge dams and pipelines bringing water from hundreds of miles away, urban residents became well insulated from nature in all its forms, and even from the limitations of climate and local geography.

To be sure, historically urban elites have at times created parks for the benefit of city residents. Central Park in New York City is one of the most famous examples. Often these bits of urban nature have been designed in a pastoral English landscape tradition or more manicured continental style. In European cities, estates belonging to royalty or the nobility have sometimes been turned into public green spaces, as well as lands once occupied by city walls or defensive fortifications, while city squares, cemeteries, the occasional botanical garden, and the remnant "commons" of former grazing land at the center of many New England towns provided green oases within American metropolises. On the suburban fringe twentieth-century developers at times sought to create garden suburbs emulating English country estates. But these amenities did not fundamentally alter the fact that as cities and suburbs grew, their residents were increasingly living in a manufactured world with very little connection to natural ecosystems.

Only with the environmental revolution of the 1960s did activists and policy-makers come to think more systematically about integrating urban development with the natural world, as well as protecting human beings from some of the worst abuses of urban environments. Efforts to restore damaged natural systems within cities gained speed in the 1980s and 1990s, and new fields such as landscape ecology provided conceptual tools for thinking about how reconstructed ecosystems might function. Communities experimented with watershed planning, citizens groups worked to restore creeks and rivers, and use of native, climate-appropriate species soared within landscape architecture.

One of the classic pieces first calling attention to systematic relationships between nature and cities was Anne Whiston Spirn's book *The Granite Garden* (New York: Basic Books, 1984). While McHarg had focused on the interaction of new suburban or regional development with natural landscapes, Spirn looked at nature within densely built cities themselves. A professor of architecture at the University of Pennsylvania, she analyzed the role of different natural entities such as soil, water, wind, and light within urban landscapes, and argued that the city should be seen as part of nature, not as something existing outside of it. If nature is welcomed into the city, in her view, a delightful urban environment can be created; if nature is ignored, disaster may result. Michael Hough, a landscape architect at the University of Toronto, took a very similar approach in his books *City Form and Natural Processes: Toward an Urban Vernacular* (New York: Van Nostrand Reinhold, 1984) and *Cities and Natural Process* (New York: Routledge, 1995). As with McHarg's writings, Spirn's eloquent, passionate style inspired many in environmental planning and landscape architecture professions who have since worked out specific ways to implement her philosophy.

▦ ▦ ▦ ▦ ▦ ▦

Nature pervades the city, forging bonds between the city and the air, earth, water, and living organisms within and around it. In themselves, the forces of nature are neither benign nor hostile to humankind. Acknowledged and harnessed, they represent a powerful resource for shaping a beneficial urban habitat; ignored or subverted, they magnify problems that have plagued cities for centuries, such as floods and landslides, poisoned air and water. Unfortunately, cities have mostly neglected and rarely exploited the natural forces within them.

More is known about urban nature today than ever before; over the past two decades, natural scientists have amassed an impressive body of knowledge about nature in the city. Yet little of this information has been applied directly to molding the form of the city – the shape of its buildings and parks, the course of its roads, and the pattern of the whole. A small fraction of that knowledge has been employed in establishing regulations to improve environmental quality, but these have commonly been perceived as restrictive and punitive, rather than as posing opportunities for new urban forms. Regulations have also proven vulnerable to shifts in public policy, at the mercy of the political concerns of the moment, whereas the physical form of the city endures through generation after generation of politicians.

[. . .]

The city is a granite garden, composed of many smaller gardens, set in a garden world. Parts of the granite garden are cultivated intensively, but the greater part is unrecognized and neglected. To the idle eye, trees and parks are the sole remnants of nature in the city. But nature in the city is far more than trees and gardens, and weeds in sidewalk cracks and vacant lots. It is the air we breathe, the earth we stand on, the water we drink and excrete, and the organisms with which we share our habitat. Nature in the city is the powerful force that can shake the earth and cause it to slide, heave, or crumple. It is a broad flash of exposed rock strata on a hillside, the overgrown outcrops in an abandoned quarry, the millions of organisms cemented in fossiliferous limestone of a downtown building. It is rain and the rushing sound of underground rivers buried in storm sewers. It is water from a faucet, delivered by pipes from some outlying river or reservoir, then used and washed away into the sewer, returned to the waters of river and sea. Nature in the city is an evening breeze, a corkscrew eddy swirling down the face of a building, the sun and the sky. Nature in the city is dogs and cats, rats in the basement, pigeons on the sidewalks, raccoons in culverts, and falcons crouched on skyscrapers. It is the consequence of a complex interaction between the multiple purposes and activities of human beings and other living creatures and of the natural processes that govern the transfer of energy, the movement of air, the erosion of the earth, and the hydrologic cycle. The city is part of nature.

Nature is a continuum, with wilderness at one pole and the city at the other. The same natural processes operate in the wilderness and in the city. Air, however contaminated, is always a mixture of gasses and suspended particles. Paving and building stone are composed of rock, and they affect heat gain and water runoff just as exposed rock surfaces do anywhere. Plants, whether exotic or native, invariably seek a combination of light, water, and air to survive. The city is neither wholly natural nor wholly contrived. It is not "unnatural" but, rather, a transformation of "wild" nature by humankind to serve its own needs, just as agricultural fields are managed for food production and forests for timber. Scarcely a spot on the earth, however remote, is free from the impact of human activity. The human needs and the environmental issues that arise from them are thousands of years old, as old as the oldest city, repeated in every generation, in cities on every continent.

The realization that nature is ubiquitous, a whole that embraces the city, has powerful implications for how the city is built and maintained and for the health, safety, and welfare of every resident. Unfortunately, tradition has set the city against nature, and nature against the city. The belief that the city is an entity apart from nature and even antithetical to it has dominated the way in which the city is perceived and continues to affect how it is built. This attitude has aggravated and even created many of the city's environmental problems: poisoned air and water; depleted or irretrievable resources; more frequent and more destructive floods; increased energy demands and higher construction and maintenance costs than existed prior to urbanization; and, in many cities, a pervasive ugliness. Modern urban problems are no different, in essence, from those that plagued ancient cities, except in degree, in the toxicity and persistence of new contaminants, and in the extent of the earth that is now urbanized. As cities grow, these issues have become more pressing. Yet they continue to be treated as isolated phenomena, rather than as related phenomena arising from common human activities, exacerbated by a disregard for the processes of nature. Nature has been seen as a superficial embellishment, as a luxury, rather than as an essential force that permeates the city. Even those who have sought to introduce nature to the city in the form of parks and gardens have frequently viewed the city as something foreign to nature, have seen themselves as bringing a piece of nature to the city.

To seize the opportunities inherent in the city's natural environment, to see beyond short-term costs and benefits, to perceive the consequences of the myriad, seemingly unrelated actions that make up daily city life, and to coordinate thousands of incremental improvements, a fresh attitude to the city and the molding of its form is necessary. The city must be recognized as part of nature and designed accordingly. The city, the suburbs, and the countryside must be viewed as a single, evolving system within nature, as must every individual park and building within that larger whole. The social value of nature must be recognized and its power harnessed, rather than resisted. Nature in the city must be cultivated, like a garden, rather than ignored or subdued.

T
W
O

"Land Development and Endangered Species: Emerging Conflicts"

from *Habitat Conservation Planning* (1994)

Timothy Beatley

Editors' Introduction

Although the science of ecology has been developing since the late nineteenth century, only in the 1970s did a knowledge of the importance of habitat and biodiversity come to be applied to landscapes in and around metropolitan areas. One source of inspiration was the field of landscape ecology, pioneered by Richard Forman and others, which developed a language for describing landscapes in terms of "patches" of habitat, "edge" environments, "corridors" of wildlife movement, and "mosaics" of these features. Also important were new public movements to restrain urban growth, to restore nature within urban areas, and to manage watersheds so as to enhance wildlife habitat and reduce flooding through preservation of natural floodplains.

Timothy Beatley here describes the emergence of concern about biodiversity, as well as the issues behind one of the main strategies to preserve species in or near urban areas: habitat conservation plans. Although controversial because they allow some urban expansion to go forward, these plans have frequently been used since the 1980s in an attempt to balance nature with development, or at least to preserve key elements of natural ecological habitat and function when urbanization cannot be avoided altogether. Beatley is a professor of environmental planning at the University of Virginia. His other writings include *Ethical Land Use: Principles of Policy and Planning* (Baltimore, MD: Johns Hopkins University Press, 1994) and *The Ecology of Place* (Washington, DC: Island Press, 1997).

Public awareness and concern about the extinction of species have undoubtedly increased in recent years. Environmental groups like the World Wildlife Fund and the Audubon Society have been quite successful in elevating concern about the anthropogenic impacts on our great storehouse of flora and fauna. The loss of biological diversity, or "biodiversity," has been added prominently to the list of major environmental problems facing the planet. Even ten years ago the term *biodiversity*

would have had little meaning even to many environmentalists or conservationists, and still less to the average person on the street. The writings of such scientists as Paul Erhlich and Edward O. Wilson have done much recently to popularize the concerns about the loss of biodiversity.[1]

Yet citizens and public officials in this country tend to see the biodiversity problem, if they see one at all, as primarily occurring in other countries. Species are facing extinction, in the minds of many, primarily as a result of tropical deforestation in countries such as Brazil and of illegal poaching in Africa and elsewhere. While these are in fact major threats to global biodiversity, there is sometimes a tendency to de-emphasize threats to biodiversity in this country, or in our own backyards. . . .

It is useful and entirely appropriate to place the US problem in the context of the larger global problem. Globally, species and habitat are threatened by numerous activities, including destroying habitat, over-harvesting/over-exploiting, and invasive species disturbing habitat. In recent years habitat loss has become the primary threat to biodiversity as the extent of human settlements continues to grow.

In many parts of the globe this is clearly the direct result of dramatic rises in population levels and the attendant demands placed on the land to feed and shelter these populations. There has been an incredible rise in the global human population from a little over 1 billion at the turn of the century to around 5.4 billion today. A recent United Nations' report predicts that global population levels may rise as high as 12 billion before stabilizing.[2]

Estimating current rates of global extinction, and predicting future rates, are tenuous at best. There is little certainty about the total number of species on Earth, but estimates put the number between 10 and 30 million. Wilson has estimated that if current rates of deforestation continue, extinction rates may exceed the loss of 17,000 species per year.[3] Others have concluded that as much as 25 per cent of our existing species may become extinct by the beginning of the next century. While the predictions vary there is general agreement that the rates are very high and a large segment of the world biota stock is at risk.[4]

In the United States, the causes of habitat loss are more complex than simple population growth. Clearly population levels have risen substantially here, as well. However, compared with those in other nations around the world, the amounts of land and space per capita are quite large in the United States. The problem, it seems, in recent years has been the inefficient and wasteful nature of our land usage. The dominance of the automobile, the impact of federal subsidies provided for home ownership, major federal investments in a national highway system, and equal neglect of mass transit, among other factors, have led to the sprawling land-intensive patterns of development common in the United States.

[. . .]

The conflicts between species protection and urban growth and development appear all around us, and virtually in every part of the country. Not surprisingly, conflicts are more frequent where the number of rare and endangered species are greatest and where population and development pressures are most severe. Much of the conflict, then, has focused on high-diversity and high-growth states like California, Texas, and Florida. . . . [But] there is diversity in every state and some degree of development and changes as well – thus, the potential for species/development conflicts.

Furthermore, environmental degradation in this country has gradually whittled away at these biological resources. The number of endangered or threatened species listed on the Endangered Species Act continues to climb and is currently well in excess of 700. As well, thousands of additional species have been classified as candidates for listing and could appear at some point in the future. The trends in biodiversity loss appear to move entirely in one direction – species become listed and remain on the list because they rarely recover sufficiently to be removed from it. Notable exceptions are the American Bald Eagle (in the continental United States) and the American alligator.

These increasing conflicts typically pit environmentalists and supporters of biodiversity conservation against developers and supporters of community development and growth. In addition, there are typically a variety of different stakeholder groups involved in these conflicts and in the preparation of habitat conservation plans, and all have varying perspectives and points of view on the issue, which may or may not fall on this conservation/development continuum.

JUSTIFICATION/RATIONALE FOR PROTECTING ENDANGERED SPECIES

Protection of biodiversity and endangered species often conflicts with demands for housing, economic development, and other social and individual objectives. To many these conflicts suggest that those advocating biodiversity must put forth good reasons why such societal sacrifices should be made. Why should we be concerned with the loss of biodiversity in the first place?

There are a number of arguments for protecting endangered species, and more broadly, biodiversity, and they range from utilitarian and instrumental views to views which support protection based on the intrinsic value and inherent worth of other forms of life. Before proceeding to a discussion of habitat conservation plans, it will be useful to briefly review these different positions.

Many have argued that species and biodiversity should be protected by humans because they produce, or will produce, numerous direct benefits for human society. These benefits may be medicinal in nature, for example, in that the globe's existing stock of flora and fauna represents an immense and largely untapped pharmaceutical storehouse. A large portion of commercial pharmaceutical products are derived directly from wild plants and animals.[5] These medicinal benefits are illustrated well by recent discoveries of the importance of the yew tree, indigenous to the forests of the northwest United States. The bark of the yew tree has been found to be a highly effective treatment for certain types of cancer. It has been estimated, however, that only 5 per cent of all plant species have been examined for their potential pharmaceutical benefits. . . .

Protecting biodiversity also holds the potential for numerous other economic and commercial benefits, such as the discovery of new disease-resistant crops or crops that may adjust better to changing climatic conditions (e.g., the buffalo gourd, which requires little water). As another example, a plant native to Central Africa (the kenof) is currently thought to be a much cheaper and less environmentally harmful source of pulp and paper fibers than trees.[6]

Perhaps more fundamentally, conservation of biodiversity is essential to protecting the viability of the larger ecosystem upon which all species depend. Endangered and threatened species are important indicators of how healthy and sustainable our planet really is. The loss of the Least Bell's Vireo or other songbirds may have little direct impact on people, yet may be indicative of the occurrence of broader environmental degradation as well as being a harbinger of more severe environmental calamities to come.

[. . .]

Biodiversity is also important in a deeper emotional sense. It seems that humans do value the existence and qualities of other forms of life as is seen in the names of automobiles and other product lines, the images in advertising and business affairs, and the animal symbols representing important societal and governmental institutions. The loss of each species diminishes our lives in important ways. The prospect of an increasingly empty planet in terms of the number and diversity of species is a depressing one. Species extinction represents innumerable lost opportunities for human enrichment.

While these arguments are convincing in their own right, is the existence of a species justified only if it holds some instrumental value or benefit to humans? This attitude, many writers and ethicists believe, epitomizes humans' arrogance as a species (what some have called "speciesism") by failing to perceive the intrinsic value of other forms of life.

[. . .]

CENTRAL POLICY QUESTIONS

For each habitat conservation plan (HCP) experience, there are a number of specific technical and policy questions which must be addressed. One of the more central of these is what the level of habitat protection must or should be. While certain standards are specified by the Endangered Species Act (ESA) and the US Fish and Wildlife Service (USFWS), there are necessarily differences of opinion about what is required. Should only the bare minimum be protected, or should much larger habitats be set aside to ensure species survival? What types of recovery actions are necessary to ensure long-term survival and recovery, and what level of habitat rehabilitation is needed? Despite the desire for clear and definitive scientific answers to these questions, HCPs . . . illustrate the judgmental and speculative responses to many of these important questions.

Questions also arise about the best strategy for maximizing conservation dollars. Must habitat be protected where the destructive pressures are most evident (i.e., areas subject to urban growth) or should conservation efforts be focused in areas where greater amounts of habitat can be protected for less cost, and where the long-term ecological viability of the habitat is perhaps more secure?

There is as well in the HCP experiences the common tension between protecting the habitat for a single species, and protecting the integrity of larger systems of which that species may be a part. Reminiscent of the debate over the Northern Spotted Owl is the increasing belief that larger ecosystem integrity is more important, and that efforts should be made to protect habitat for multiple species of concern, not just a single species. . . .

The HCP experience also vividly illustrates a common policy dilemma in many other environmental areas – namely the question of who should bear the burden for conservation efforts. Whether it's the loss of logging jobs in the habitat of the Northern Spotted owl, or the diminution of land value under wetlands regulations, distribution of conservation program costs is an important policy question. While each HCP committee has approached it somewhat differently, they all illustrate the ability to put together funding packages which distribute costs over a number of different sources, including development mitigation fees, federal and state conservation funding, and local bond referenda. Determining the precise package is another major topic in HCP deliberations and inevitably the result of a mixture of compromise and political reality.

NOTES

1 For instance, see Edward O. Wilson and Francis M. Peter (eds). 1988. *Biodiversity*. Washington, DC: National Academy Press; Paul Ehrlich and Anne Ehrlich. 1981. *Extinction: The Causes and Consequences of the Disappearance of Species*. New York: Balantine Books.

2 For a discussion of global population trends, see Paul Ehrlich *et al.* 1990. *The Population Explosion*. New York: Simon and Schuster.

3 "The Current State of Biological Diversity." In Wilson and Peter (eds), *Biodiversity*.

4 See Walter V. Reid and Kenton R. Miller. 1989. *Keeping Options Alive: The Scientific Basis for Conserving Biodiversity*. Washington, DC: World Resources Institute.

5 See, e.g., Norman Myers. 1979. *The Sinking Ark: A New Look at the Problem of Disappearing Species*. New York: Pergamon Press.

6 See Jane E. Brody. 1988. Scientists Eye Ancient Plant as Better Source of Pulp for Paper. *New York Times*, 10 December.

TWO

"What Is Restoration?"

from *Restoring Streams in Cities* (1998)

Ann L. Riley

Editors' Introduction

The early environmental movement in the late nineteenth and early twentieth centuries focused on "conservation" or "preservation" of natural lands, resources, and species. In contrast, many urban environmental groups after about 1980 came to focus on restoring previously damaged urban ecosystems. "Restoration" has thus become a catchword of the urban sustainability agenda. Restoration activities may take many forms, but often focus on cleaning up contaminated lands (often known as "brownfield" sites), replanting native vegetation, and restoring streams, wetlands, or other watershed elements.

In this selection, stream restoration pioneer Ann L. Riley discusses some main issues surrounding urban environmental restoration, especially in the context of waterways. She describes what restoration is and what it is not, and gives examples in the context of creek restoration, a movement particularly active in the western United States. Examples include San Luis Creek through the center of San Luis Obispo, California, portions of Strawberry Creek in Berkeley, California, and the Guadalupe River in San Jose. Restoration of native species and habitats is a closely related movement, as is xeriscaping (use of drought-tolerant plants) in arid or semi-arid cities and towns. Other writings on the subject of restoration, use of native species, and permaculture (a philosophy of basing landscape design and "permanent agriculture" on sustainable natural systems) include *Design for Human Ecosystems: Landscape, Land Use, and Natural Resources*, by John Lyle (Washington, DC: Island Press, 1999), *The Ecological City: Preserving and Restoring Urban Biodiversity*, edited by Rutherford H. Platt, Rowan A. Rowntree, and Pamela C. Muick (Amherst, MA: University of Massachusetts Press, 1994), *Sustainable Landscape Construction: A Guide to Green Building Outdoors*, by J. William Thompson and Kim Sorvig (Washington, DC: Island Press, 2000), and *Permaculture: A Practical Guide for a Sustainable Future*, by Bill Mollison (Washington, DC: Island Press, 1990).

The Society for Ecological Restoration defines ecological restoration as "the process of intentionally altering a site to establish a defined indigenous, historical ecosystem. The goal of this process is to emulate the structure, function, diversity, and dynamics of the specified ecosystem."

[Another] interesting definition that adds more of a human and social component is "the process of intentionally compensating for damage by humans to the biodiversity and dynamics of indigenous ecosystems by working with and sustaining natural regenerative processes in ways which lead to the reestablishment of sustainable and healthy relationships between nature and culture."

Using these definitions, the first problem the restorationist needs to address is what historical and indigenous (native to the location) conditions to restore to. In some circumstances it may be most practical to restore a waterway to its condition during a particular period of history, such as when it became formally integrated into the urban landscape as a 1930s Works Progress Administration (WPA) city park. The restoration project could include, for example, restoration of a creek's native vegetation and historical WPA rock work if the rock does not harm the waterway. Perhaps the history of a waterway from the late 1800s to the present has been as a degraded, polluted industrial channel. You may want to use records or maps from before this era to determine restoration goals.

It may be institutionally or ecologically impossible to restore a waterway to a landscape representing conditions before European settlers transformed the landscape to something else. For example, when we select objectives to restore the Chicago River, we cannot return it to a shallow, far-spreading prairie wetland as it was before its lowlands were dredged by humans for use as a shipping channel. Our options at this point are to use a riverine model to guide restoration attempts for the channel and to encourage, to the extent possible, the return of some of the pre-European-settlement prairie wetland species.

Restoration, particularly in urban settings, can require complicated compromises and trade-offs in establishing objectives based on the natural and human-built history that has shaped current land uses and ecological systems. A good practice is to refer to local experts who know the regional landscape well to see if any remnant natural rivers, streams, waterways, or wetlands can provide a restoration model for your degraded waterway.

Both ecological and human settlement needs will be met if you strive to create a landscape that is more self-sustaining than existing conditions. This means that the waterway is changed so that it is in greater balance. For a river or stream, this balanced condition usually means that it is not *excessively* eroding or depositing sediment. (Erosion and deposition are natural to streams; we intervene only when we establish that excessive conditions exist.) It also means that it has biologically diverse aquatic life and does not experience extremes in temperature, nutrients, algae growth, or other chemical parameters. If the natural physical features of the waterway are returned, it will not need as much intervention to correct for erosion, sedimentation, or pollution problems.

The physical features of rivers and streams include the streamside trees and shrubs, the channel and its width and depth, pools, riffles, and meanders. The river also includes its floodplain and may feature terraces, which are old, abandoned floodplains located above the current ones. These physical "structures" perform functions in the river ecosystem, including the transport of water and of sediment, the storage and conveyance of floodwaters, and the creation of terrestrial plant communities and wildlife habitat and aquatic habitat. Finally, stream dynamics include the transport of sediment; conveyance of water; formation of channels, floodplains, and terraces; and the interrelationships among these features and the land uses and vegetation in the watershed. Restoration attempts to return these structures, functions, and dynamics to the extent that it is possible given the constraints of our modern developed landscape.

Sometimes it helps to define what restoration is *not* as a way to clarify its objectives. Fisheries restoration is *not* a fish hatchery, where fish are raised at great expense in captivity and released or sometimes driven to rivers, streams, or lakes for release. Most rivers or lakes with stocked fish cannot support the life cycle of those fish. Consequently, the fish must continually be restocked. Fisheries restoration is reintroducing to a river, creek, or lake wild genetic stock that can maintain a self-sustaining population of fish that are genetically adapted to surviving in natural conditions. Restoration in that case means re-creating spawning and rearing habitats; removing barriers to migration; and restoring shelter, favorable temperatures, and water quality for the species that evolved in those conditions and therefore will survive in them on their own.

Restoration is *not* landscaping. Landscaping at its best has been a means to create new environments that provide sanctuary, adventure, symbolism, recreation, environment, and perhaps sustenance. Landscaping is also done to mitigate for a land-use change such as the building of a freeway; the construction of offices, parking lots, and housing developments; and the construction of water projects. Landscape professionals often use planting designs to screen structures, compensating for noise or lost shade or to cover up what we do not want to see. While those are all legitimate undertakings, they are not restoration. Stream restoration is also not the creation of a "native garden" with water running through it.

Planting trees and shrubs along a stream channelization project is *not* restoration – *even* if native species of plants are used. Planting that is done as an add-on to a flood-control channel, or to try to mitigate some of the lost values of the original river for wildlife or aesthetics, but does not function as a part of a natural riparian system, is landscaping. In such cases, we have not restored; we have only tried to mitigate or compensate for the project's environmental damages. However, if the vegetation functions as a component of a stream environment – if it helps slow the velocity of the water, strengthen stream banks, create vortexes to scour pools, shade the channel to prevent invasion of choking rushes and reeds, or re-create habitat for the species of birds, fish, and mammals that once used the site – then it *is* restoration.

[. . .]

Restoration can be knowing when not to act. Nature is resilient and often adjusts to changes in the watershed. A critical part of a restorationist's role is to know when to allow nature to make adjustments on its own. A variety of human changes might destabilize a stream, including the building of a dam, regulation of stream flows, diversion of water, urban development, fires or timber harvest, culverting, and channel relocation. Natural disasters, such as floods, tornadoes, earthquakes, and hurricanes, may also destabilize the stream's equilibrium. The stream will react to those changes, and its natural adjustments may or may not have unwanted consequences. A restorationist can give local residents insight into the merits and costs of intervention. In many situations, a stream will find a new equilibrium without intervention. In other situations, a stream will defy attempts to manipulate it by blowing out, eroding, or bypassing carefully designed bank protection projects. Sometimes native plant species will return naturally, coming back more quickly and vigorously as volunteers than we can replant them. The uncertainty of these natural changes underscores the importance of consulting with local geomorphologists, hydrologists, and other professionals knowledgable about local stream dynamics. There is a significant history of misdirected and make-work projects on streams that may do more harm than good to the correction of imbalances in channels and watersheds.

[. . .]

We are entering a new era of government engineering programs in which public works projects are going to be designed to accommodate a wider range of values and objectives. The concept of multiobjective floodplain management has gained wide acceptance in the past decade in the river engineering and management professions. This concept states that it is of greatest community benefit to manage river floodplains and flood-prone areas for a range of objectives including flood-damage reduction, protection of wildlife habitat, protection or improvement of water quality, ecological restoration, erosion control, provision of recreation, etc. This contrasts with the many older, singleobjective public works projects for flood or erosion control.

[. . .]

Innovations are now being tried in the design of flood-control projects to avoid environmental impacts and performance and maintenance problems. River meanders are being kept, and floodplains are being restored to both better store and better convey large volumes of water. Revegetation systems . . . are replacing concrete, riprap, and sheet piling on stream banks, waterfronts, and lakesides. Restoration methods are providing an exciting alternative to old methods because they can often solve the important engineering problems of lowering property damages *and* provide environmental benefits. They attempt to return to the stream its structure (riparian forests, meanders, pools, riffles, and other physical features), its functions (instream habitat, flood storage, environmental balance, wildlife habitat), and its dynamics (which determine its shape, dimensions, and meander). By doing this, restoration can reduce excessive erosion, return fish habitat, help the stream recover from pollution, and even reduce flood damages. It becomes a win–win solution.

ENERGY AND MATERIALS USE

"The Metabolism of Cities"

from *Creating Sustainable Cities* (1999)

Herbert Girardet

Editors' Introduction

The flow of natural resources into cities and wastes out of them represents one of the largest challenges to urban sustainability. Many argue that cities must "close the resource loop" by recycling, reusing, re-manufacturing, and otherwise diverting materials from their usual destination in landfills and incinerators. Reducing consumption in the first place is perhaps even more important. Likewise, more efficient urban uses of energy can be sought to reduce dependence on nonrenewable fossil fuels, and renewable energy sources developed such as wind power (the world's fastest growing alternative energy source), solar power, geothermal energy, biomass conversion (the burning of organic materials for energy), and co-generation (the use of waste energy or steam from one industrial process for heat or power).

Herbert Girardet has written eloquently on urban resource flows particularly as they affect the city of London. He is a leading member of sustainable city efforts there, and the author of *The Gaia Atlas of Cities* (New York: Anchor Books, 1993) and other books. His analysis follows in the footsteps of other "appropriate technology" advocates including the great environmental philosopher E.F. Schumacher, author of *Small is Beautiful: Economics as if People Mattered* (New York: Harper & Row, 1973), and energy guru Amory Lovins, author of *Soft Energy Paths: Toward a Durable Peace* (San Francisco: Friends of the Earth, 1977).

The growing understanding developed by the natural sciences of the way ecosystems function has a major contribution to make to solving the problems of urban sustainability. Cities, like other assemblies of organisms, have a definable metabolism, consisting of the flow of resources and products through the urban system for the benefit of urban populations. Given the vast scale of urbanization cities would be well advised to model themselves on the functioning of natural ecosystems, such as forests, to assure their long-term viability. Nature's own ecosystems have an essentially *circular* metabolism in which every output which is discharged by an organism also becomes an input which renews and sustains the continuity of the whole living environment of which it is a part. The whole web of life hangs together in a 'chain of mutual benefit', through the flow of nutrients that pass from one organism to another.

The metabolism of most modern cities, in contrast, is essentially *linear*, with resources being 'pumped' through the urban system without much concern about their origin or about the destination of wastes, resulting in the discharge of vast amounts of waste products incompatible with natural systems. In urban management, inputs and outputs are considered as largely unconnected. Food is imported into cities, consumed, and discharged as sewage into rivers and coastal waters. Raw materials are extracted from nature, combined and processed into

1 *Inputs* (tons per year)

Total tons of fuel, oil equivalent	20,000,000
Oxygen	40,000,000
Water	1,002,000,000
Food	2,400,000
Timber	1,200,000
Paper	2,200,000
Plastics	2,100,000
Glass	360,000
Cement	1,940,000
Bricks, blocks, sand, and tarmac	6,000,000
Metals (total)	1,200,000

2 *Wastes*

CO_2	60,000,000
SO_2	400,000
NO_x	280,000
Wet, digested sewage sludge	7,500,000
Industrial and demolition wastes	11,400,000
Household, civic, and commercial wastes	3,900,000

Table 1 The metabolism of Greater London (population 7,000,000)

Source: Compiled by Herbert Girardet (1995 and 1996), sources available from author.

consumer goods that ultimately end up as rubbish which can't be beneficially reabsorbed into the natural world. More often than not, wastes end up in some landfill site where organic materials are mixed indiscriminately with metals, plastics, glass, and poisonous residues.

This linear model of urban production, consumption, and disposal is unsustainable and undermines the overall ecological viability of urban systems, for it has the tendency to disrupt natural cycles. In the future, cities need to function quite differently. On a predominantly urban planet, cities will need to adopt circular metabolic systems to assure their own long-term viability and that of the rural environments on whose sustained productivity they depend. To improve the urban metabolism, and to reduce the ecological footprint of cities, the application of ecological systems thinking needs to become prominent on the urban agenda. Outputs will also need to be inputs into the production system, with routine recycling of paper, metals, plastic and glass, and the conversion of organic materials,

including sewage, into compost, returning plant nutrients back to the farmland that feeds the cities.

The *local* effects of the resource use of cities also need to be better understood. Urban systems accumulate vast quantities of materials. Vienna, for instance, with 1.6 million inhabitants, every day increases its actual weight by some 25,000 tons.[1] Much of this is relatively inert materials, such as concrete and tarmac which are part of the built fabric of the city. Other materials, such as lead, cadmium metals, nitrates, phosphates, or chlorinated hydrocarbons, build up and leach into the local environment in small, even minute quantities, with discernible environmental effects: they accumulate in water and in the soil over time, with potential consequences for the health of present and future inhabitants. The water table under large parts of London, for instance, has become unusable for drinking water because of accumulations of toxins over the last 200 years. Much of its soil is polluted by the accumulation of heavy metals during the last 50 years.

The critical question today, as humanity moves to 'full scale' urbanization, is whether living standards in our cities can be maintained whilst curbing their local and global environmental impacts. To answer this question, it helps to draw up balance sheets quantifying the environmental impacts of urbanization. We now need figures to compare the resource use by different cities. It is becoming apparent that similar-sized cities supply their needs with a greatly varying throughput of resources, and local pollution levels. The critical point is that cities and their people could massively reduce their throughput of resources, maintaining a good standard of living creating much needed local jobs in the process. I shall now discuss aspects of urban use of resources and energy in more detail.

WATER AND SEWAGE

Our cities consume vast amounts of water: in the UK, typically, some 400 liters per person per day. In the US the figure is as high as 600 litres. In older cities, such as London, water has to be pumped in from elsewhere because it is exceedingly costly to clean it to drinking water standards. Cities externalize the problem. The abstraction of river water, often many miles away from cities, has caused the destruction of river habitats and fisheries; today, many rivers' supply are a pale shadow of their former selves.

Water supplied to households, even if supplied from outside cities, goes through various treatment processes. River water, a source of supply in most countries, has to be cleaned of impurities, including pesticides, phosphates and nitrates from farming. The water is percolated through sand and charcoal filter beds before it is pumped into a city's network of water pipes. Chlorination, which is commonplace, disinfects drinking water, but its unpleasant taste causes many people to switch to bottled drinking water instead. This does not make financial sense, since bottled water, at up to 60p per litre, is often more expensive than the petrol we put in the tanks of our cars. Neither does it make sense environmentally, with vast quantities of bottled water being trucked in from hundreds or even thousands of miles away at great energy cost. It would be desirable to ensure that the quality of urban water could be high enough for it to be commonly used for drinking once again.

Unfortunately, a major function of urban water supply is as a carrier for household and commercial sewage. For this and other reasons, urban sewage systems are of an important issue in the quest for urban sustainability. Their main purpose is to collect human faeces and to separate it from people, to help prevent outbreaks of diseases such as cholera or typhoid. As a result, vast quantities of sewage are flushed away into rivers and coastal waters downstream from population centers. Coastal waters the world over are enriched both with human sewage and toxic effluents, as well as the run-off of mineral fertilizer and pesticides applied to the farmland feeding cities. The fertility taken from farms in the form of crops used to feed city people is not returned to the land. This open loop is not sustainable.

Whilst it is clear that cities need to have efficient sewerage systems, we need to redefine their purpose. Instead of building disposal systems we should construct recycling facilities in which sewage can be treated so that the main output is fertilizers suitable for farms, orchards, and market gardens. It has been too readily forgotten that sewage contains an abundance of valuable nutrients such as nitrates, potash, and phosphates.

Returning these from cities to the land is an essential aspect of sustainable urban development.

A variety of new sewerage systems have been developed for this purpose using several new technologies: membrane systems that separate sewage from any contaminants; so-called 'living machines' that purify sewage by biological methods; and drying technology which converts sewage into granules that can be used as fertilizer. These technologies can be used in combination with each other, making sewerage facilities into efficient fertilizer factories. These sorts of systems are now beginning to be used in cities all over the world.

In Bristol, the water and sewage company Wessex Water now dries and granulates all of the city's sewage. The annual sewage output of 600,000 people is turned into 10,000 tons of fertilizer granules. Most of it is currently used to revitalize the bleak slag heaps around former mining towns such as Merthyr Tydfil in South East Wales. In contrast, Thames Water in London is currently constructing incinerators for burning the sewage sludge produced by 4 million Londoners. This is a decision of historic short-sightedness given that phosphates – only available from North Africa and Russia – are likely to be in short supply within decades. Crops for feeding cities cannot be grown without phosphates.

There is an acknowledged problem with the contamination of sewage with heavy metals and chlorinated hydrocarbons. For this reason, there is growing concern about using sewage-derived fertilizer on farmland. However, the reduction of the use of lead in vehicle fuel and the de-industrialization of our cities is reducing this problem, lessening the load of contaminants that are flushed into sewage pipes. Also, more stringent environmental legislation is further reducing contamination of sewage. The quest for greater urban sustainability will certainly lead to a significant rethink on how we design sewerage systems. The aim should be to build systems to intercept the nutrients contained in sewage whilst assuring that it can be turned into safe fertilizers for the farmland feeding cities.

SOLID WASTE

Solid waste is the most visible output by cities. In recent decades there has been a substantial increase in solid waste produced per head, and the waste mix has become ever more complex. Today's 'garbologists' see a vast difference between early and late twentieth century rubbish dumps. The former contain objects such as horse shoes, enamelled saucepans, pottery fragments and leather straps. The latter contain food wastes, plastic bags and containers, disposable nappies, mattresses, newspapers, magazines and transistor radios. But garbologists will also find plenty of discarded building materials, and crushed canisters containing various undefined, sometimes highly poisonous, liquids.

Urban wastes used to be dumped primarily in holes in the ground. Much of London's waste, for instance, is dumped in a few huge tips, such as at Mucking on the Thames in Essex. Household waste, as well as commercial and industrial waste, is taken here by central London, and 'co-disposed' in pits lined with clay. The compacted rubbish is, eventually, sealed with a top layer of clay, which is then covered with soil and seeded with grass. Inside the dump, methane gas from the rotting waste is now intercepted in plastic pipes and used to run small power stations. However, their output is quite insignificant. Mucking receives the rubbish of some 2 million people, but its methane-powered generators supply electricity to just 30,000 people.[2]

More and more cities, including London, are seeing growing resistance from people in adjoining counties on receiving urban wastes; all the environmental implications of fleets of rubbish trucks, potential groundwater contamination, and stench in the vicinity of waste dumps are a growing concern. As the unwillingness to receive rubbish grows, other waste disposal options are urgently required. Dumping ever growing mounds of waste outside the cities where they originate is a waste of both space and resources that be used more beneficially.

We need to think again about the ways in which urban waste management systems work. Many cities all over the world have chosen incineration as the most convenient route for 'modern' waste management. Incineration has the advantages of reducing waste materials to a small percentage of their original volume. Energy recovery can be an added bonus. But incineration is certainly not the main option for solving urban waste problems. The release of dioxins and other poisonous gases from the smokestacks of waste incinerators has given them a bad name. There have been great improvements in incineration and pollution control techniques, but only those wastes that cannot be recycled should be considered for incineration.

Recently, new objections to incinerators have been voiced in the United States because research has shown conclusively that incinerators compare badly with recycling in terms of energy conservation. Because of the high energy content of many manufactured products that end up in the rubbish bin, recycling paper, plastics, rubber and textiles is three to six [times] more energy-efficient than incineration. These are very significant figures, given that the energy and resource efficiency of urban systems is regarded as critical for future urban sustainability.[3] Many European cities are now deciding against investing in new incinerators, and in favor of a combination of recycling and composting facilities instead, with minimal incineration for waste products that cannot be further recycled.

It has been said that recycling is a red herring because it is so difficult to match the supply of materials to be recycled with regular demand for recycled products. But experiences in many European cities indicate that market incentives can make recycling economically advantageous and that the right policy signals and incentives at national and local level can transform prospects. Whilst not all waste materials can be recycled, much can be done to move in this direction. As concern grows about the continuing viability of the environments on which cities depend, the reuse and recycling of solid wastes is likely to become the rule rather than the exception. Deliberately constructing 'chains of use' that mimic natural ecosystems will be an important step forward for both industrial and urban ecology.

Some modern cities have already made this a top priority. Cities across Europe are installing waste recycling and composting equipment. In German towns and cities, for instance, dozens of new composting plants are being constructed. In Sweden, Gothenburg has taken matters even further by setting up an ambitious program for developing 'eco-cycles', minimizing the leakage of toxic substances into the local environment by helping companies develop advanced non-polluting production processes.[4] Vienna also has an

BOX 1 How Cairo recycles its waste

Cities in the developing world usually make highly efficient use of resources, particularly if people are supported in their recycling activities. Cairo and Manila actively encourage recycling and composting of wastes. There are a growing number of cities that are actually moving towards being zero-waste systems. Cairo, with 15 million people one of the world's largest cities, reuses and recycles most of its solid waste. Much of it is handled by a community of Coptic Christians called the Zaballeen. With the active support of the city authorities, the Zaballeen were able to acquire recycling and composting equipment. Metals and plastics are remanufactured into new products. Waste-paper is reprocessed into new paper and cardboard. Rags are shredded and made into sacks and other products. Organic matter is composted and returned to the surrounding farmland as fertilizer.

The Zaballeen Environment and Development Programme has enabled the 10,000-strong community to substantially increase its income from its recycling and remanufacturing activities. In that way, social and environmental problems affecting Cairo are tackled simultaneously. Had the waste management of the city been given over to a conventional waste management company, thousands of waste collectors would have been out of work. By helping the Zaballeen with appropriate technology, they were able to improve on their traditional waste-handling methods, while Cairo could avoid putting vast waste dumps on the periphery of the city.

In the case of solid waste management, cities in the North have much to learn from the ingenuity of waste recycling in the South. Meanwhile, cities in the South could greatly benefit from the transfer of improved recycling technologies now available in the North.

impressive track record, currently recycling 43 per cent of its domestic wastes.[5] This sort of figure is common to a growing number of European and American cities.

Most European cities exceed the household waste recycling performance of cities in the UK.[6] In some British cities, such as Bath and Leicester, where recycling has advanced a great deal, the benefits for people and the local environment are clearly apparent. The UK landfill tax, introduced in 1996, has increased recycling throughout the UK, helping to achieve the government target of 25 per cent household waste recycling by 2000. This taxation should be extended to approximate a recycling rate of 50 per cent, which is the target in other European countries.

In London, where currently only 7 per cent of household waste is recycled, a proposal by the London Planning Advisory Council is intended to bring recycling up to unprecedented levels. By 2000, every London home would have a recycling box with separate compartments instead of conventional dustbins. Progressively more and more municipal waste would be recycled, establishing new reprocessing industries and creating 1,500 new jobs.[7] Early in the new century, this figure would increase further. Meanwhile, the composting of organic wastes is advancing well, with 'timber stations' that compost shredded branches of pruned trees and leaf litter being established in various locations.

Throughout the developing world, too, cities have made it their business to encourage recycling and composting of wastes.[8] Cairo, Manila, and Calcutta are interesting cases in point.

ENERGY

Looking down on the Earth from space at night, astronauts see an illuminated planet – vast city clusters lit up by millions of light bulbs as well as the flares of oil wells and refineries. Fossil fuels have made us what we are today – an urban–industrial species. Without the power stations they supply and the vehicles they power, our urban lifestyles and our astonishing physical mobility would not have developed.

World-wide, fossil fuel use in the last 50 years has gone up nearly five times, from 1.715 billion tons of oil equivalent in 1950 to well over 8 billion tons today. Fossil fuels provide some 85 per cent of the world's commercial energy, of which oil currently amounts to around 40 per cent. The bulk of the world's energy consumption is *within* cities, and much of the rest is used for producing and transporting goods and people *to and from* cities. This realization is crucial for developing strategies for sustainable use of energy, particularly in the context of global warming.

Energy use is something most of us take for granted. As we switch on electric or gas appliances, we are hardly aware of the refinery, gas field, or the power station that supplies us. And despite publicity about acid rain and climate change, we rarely reflect on the impacts of our energy use on the environment because they are not experienced directly, except when we inhale exhaust fumes on a busy street.

Yet reducing urban energy consumption could make a major contribution to solving the world's air pollution problems. At the 1997 Kyoto conference on climate change, the industrialized nations agreed to cut CO_2 emissions by 5 per cent by 2010, but a world-wide cut of some 60 per cent is needed to actually *halt* global warming. As indicated in the Introduction, large cities and high levels of energy consumption are closely connected, particularly where routine use of motor cars, urban sprawl and air travel define urban lifestyles. Yet the potential exists for cities to be efficient users of energy.

London's 7 million people, for instance, use 20 million tons of oil equivalent per year (two supertankers a week), and discharge some 60 million tons of carbon dioxide. All in all, the per capita energy consumption of Londoners is amongst the highest in Europe. The city's electricity supply system, relying on remote power stations and long distance transmission lines, is no more than 30 to 35 per cent efficient. The know-how exists to bring down London's energy use by between 30 and 50 per cent without affecting living standards, and with the potential of creating tens of thousands of jobs in the coming decades. Significant energy conservation can be achieved by a combination of *energy efficiency* and by more efficient *energy supply systems*.

In the UK, national planning regulations have already substantially improved the energy efficiency of homes, but much more can be done.

At the domestic level, two out of three low-income families lack even the most basic insulation in their homes. Eight million families cannot afford the warmth they need in the winter months. Treating cold-related illnesses costs the National Health Service over £1 billion per year.[9] Only one in twelve domestic properties in Britain have the level of energy efficiency currently required by law.[10] Yet energy efficiency's advantages are impressive:

- reduced fuel bills for everyone;
- benefits to the trade balance through curbing the need for imports;
- the creation of new jobs in the energy efficiency industry;
- the preservation of fossil fuel reserves;
- the alleviation of environmental problems, such as air pollution and global warming, contributed to by energy generation.

There are many examples, particularly from Scandinavia, of how energy efficiency combined with efficient supply systems can dramatically reduce the energy dependence of cities. There is no doubt that the energy supply systems in many cities of the world can be vastly improved. Take electricity: most cities are supplied by power stations located a long way away, fired mainly by coal, with electricity being transferred along high-voltage power lines. On average, these stations are only 34 per cent efficient. Modern gas-fired stations are slightly better, at 40–50 per cent efficiency. Combined heat and power (CHP) stations, in contrast, are about 80 per cent efficient, because instead of wasting heat from combustion, they capture and distribute it through district heating systems.[11]

CHP systems are a very significant technology indeed. They can be fueled by a wide variety of sources – gas, geothermal energy, or even wood chips. CHP systems provide heat and chilled water, as well as electricity to urban buildings and factories. They are now commonplace in many European cities. In Denmark 40 per cent of electricity is produced by CHP; in Finland 34 per cent, and in Holland 30 per cent.

Helsinki has taken the development of CHP further than most cities. Waste heat from local coal-fired power stations is used to heat 90 per cent of its buildings and homes. Its overall level of energy efficiency of 68 per cent was achieved because its compact land use patterns made district heating a viable option. The compactness of the city also made the development of a highly effective public transport system economically viable.

In the development of CHP, the UK has been off to a slow start. Small-scale systems are being installed in some office blocks, schools, hospitals, and hotels, improving their energy efficiency considerably. All have the same high level of efficiency as large-scale systems.

The challenge for national governments and local authorities in the developed world is to put in place new energy policies, particularly to improve urban energy efficiency. The scenario includes the creation of municipally owned and operated energy systems. In some cities, such as Vienna and Stockholm, energy systems are operated by the 'city works,' which also supply water and run the transport and waste management systems. The synergies possible between these services are much harder to achieve in cities where privatization of services is the norm. It appears that the largest improvements in power distribution and consumption are realized by cities with a municipality-owned electricity company, such as Toronto and Amsterdam.[12]

The UK is just seeing the first schemes where greenhouse cultivation is being combined with CHP, utilizing their hot water and waste CO_2 to enhance crop growth for year-round cultivation.[13] Policies for encouraging CHP could thus also be used for enhancing urban agriculture, bringing producers closer to their markets instead of flying and trucking in vegetables from long distances. Once again, local job creation would result.

In addition to CHP, other significant new energy technologies are becoming available for use in cities. These include heat pumps, fuel cells, solar hot water systems and photovoltaic (PV) modules. In the near future, enormous reductions in fossil fuel use can be achieved by the use of PV systems, a technology particularly suited to cities. In the late 1990s there are only a few thousand buildings around the world using electricity from solar panels on their roofs or facades. Solar electricity could meet some of a building's requirements, with the rest of the power coming from the grid.

According to calculations by the oil company BP, London could supply most of its current summer electricity consumption from photovoltaic modules on the roofs and walls of its buildings. While this technology is still expensive, large scale automated production will dramatically reduce unit costs. And the only maintenance they require is cleaning once or twice a year.

Currently, solar energy is about eight times more expensive than conventional, but it is expected to be competitive as early as 2010 as the technology develops and the market grows. Major development programs have been announced in Japan, the USA, the Netherlands and the European Union to stimulate market growth. The technical potential for the generation of electricity from building integrated solar systems is very large indeed and could contribute significantly to the building energy requirements, even in a northerly climate like that in the UK. Of course, not all buildings will be suitable for the installation of a solar roof or facade, and adoption will be more rapid in countries with the highest sunshine level.

BOX 2 Solar energy in Saarbrücken, Germany

Saarbrücken, a city of 190,000 people, has a major investment program in solar energy. Since 1986 US$1.7 million has been spent on solar heating, PV systems, and other renewable energy sources. The state offers a 50 per cent subsidy for technical assistance, and the local savings bank offers residential energy users favorable lending terms for the installations. The local energy utility owns the PV array, but the inhabitants of each house benefit from the solar electricity supply. In addition to domestic systems, there are also municipal PV installations, incorporated into highway noise barriers. The solar initiative has the support of the entire community because it is helping to lay the foundation for a sustainable future. A former coal-mining center, Saarbrücken has now become a center for the development of urban applications of solar energy systems.

Experimental solar buildings are springing up all over Europe. The new German government, elected in 1998, has a national program for installing 100,000 PV modules. PV programs in Japan and the USA are on a similar scale. In the UK, experimental systems have proved to be very promising. The Photovoltaics Centre at Newcastle University, a 1960s building recently clad with PV panels, has proved to be a great success. In Doxford near Newcastle, Europe's largest solar-powered office building was completed in 1998.

In the new millennium, building designers will routinely incorporate this technology when designing a new building or refurbishing an existing one. In the meantime, to get experience with the technology, governments and urban authorities should vigorously encourage the installation of PV modules in our cities, enhancing the capacity to install PV systems. Every city should have buildings to test the potential of PV and to develop the local know-how.

Another energy technology of great promise, *fuel cells*, is fast coming of age. Fuel cells convert hydrogen, natural gas, or methanol into electricity by a chemical process without involving combustion. Fuel cells, like photovoltaic cells, have taken a long time to become commercially viable. Their development is now accelerating as the world searches for practical ways to produce cleaner electricity. Several companies have made great strides in making fuel cells competitive, in a variety of applications: from running generators and power stations, to buses, trucks and cars. Large-scale commercial production of fuel cells will be getting under way early in the new millennium. The combination of photovoltaic cells and fuel cells is a particularly compelling option. Electric energy from PV cells could split water into oxygen and hydrogen, and the latter could be stored and then used to run fuel cell power stations, or generators for individual buildings.

It is plausible that even large cities, whose genesis depended on the routine use of fossil fuels in the first place, may be able to make significant use of renewable energy in the future. To make their energy systems more sustainable, cities will require a combination of energy-efficient systems such as CHP with heat pumps, fuel cells and photovoltaic modules, and the efficient use of energy. Regulating the energy industry to improve generating efficiency, reduce discharge of waste gases and to adopt renewables will profoundly reduce the environmental impact of urban energy systems.

NOTES

1 Prof. Paul Brunner, TU, Vienna, personal communication.
2 Western Riverside Authority. 1991–2. *Annual Report*.
3 Worldwatch Institute. 1994. *Worldwatch Paper 121*. Washington, DC.
4 International Council on Local Government Initiatives. 1996. *The Local Agenda 21 Planning Guide*. Toronto.
5 Dr. Gerhard Gilnreiner, Vienna, personal communication.
6 Prof. Gerhard Vogel, Vienna, personal communication.
7 *Evening Standard*. 30 December 1996. London.
8 *Warmer Bulletin*. Summer 1995. London.
9 National Energy Action. 1997. Newcastle.
10 Energy Savings Trust. 1992. *Meeting the Challenge to Safeguard Our Future*. London.
11 Combined Heating and Power Association. 1998. London.
12 Nijkamp, Peter and Adriaan Perrels. 1994. *Sustainable Cities in Europe*. London: Earthscan.
13 Energy Savings Trust. 1992. *Meeting the Challenge to Safeguard Our Future*. London.

"Waste as a Resource"

from *Regenerative Design for Sustainable Development* (1994)

John Tillman Lyle

Editors' Introduction

One leading designer of sustainable resource systems was the late John Tillman Lyle, who taught at Cal Poly Pomona and founded the Center for Regenerative Studies there. Here, he discusses an aspect of industrial civilization that keenly illustrates its unsustainability – and that is widely ignored in daily life – the disposal of wastes. An architect and landscape architect, Lyle uses Mumford's terminology in labeling the current era "Paleotechic"; that is, one based on primitive, exploitative technologies rather than regenerative ones. Much of Lyle's professional work involved creating "regenerative" systems such as sewage treatment marshes that use natural processes to improve the environment while processing human wastes, in contrast to "mechanical" or technology-based systems that try to intervene in or override natural systems.

Two other pioneers of ecological waste treatment are John and Nancy Todd, founders of the New Alchemy Institute on Cape Cod, who emphasize the creation of "living machines" that can process wastes. Such projects are described in their book *From Eco-Cities to Living Machines: Principles of Ecological Design* (Berkeley, CA: North Atlantic Books, 1993). Bay Area architect Sim Van der Ryn has also created such devices, which frequently consist of a series of clear tanks with water hyacinths and other waste-processing plants within a greenhouse-type environment. Van der Ryn's books include *Sustainable Communities* (San Francisco: Sierra Club Books, 1986; with Peter Calthorpe), and *Ecological Design* (Washington, DC: Island Press, 1996; with Stuart Cowan).

Much of our difficulty with waste is embedded in the word itself. Waste is defined as material considered worthless and thrown away after use. In this sense it is a human invention, essential to the one-way flows of the throughput system; this definition depends on the assumption that energy and materials, having once served our immediate purposes, can simply cease to exist in any functional sense.

The laws of thermodynamics tell us otherwise. Energy continuously degrades and materials change form and state, but they are not destroyed and they do not disappear. In the functional order of natural ecosystems, materials are always reused. Natural processes have evolved a number of ways of accomplishing this on various time scales. In quantities matched to the evolved capacities of the landscape, materials are reintroduced after use into the processes of assimilation, filtration, storage, and production to continue their roles in nature's cycles. When the chemical composition of the waste materials is such that nature has not evolved a means of reprocessing them, or when their quantities are beyond the processing capacity of the landscape, then the sink side of the flow equation develops a serious problem of pollution or overload.

Throughout the Paleotechnic period and especially during the last half of the twentieth century, both overloading of sinks and introducing unassimilable materials into them have become regular, ongoing occurrences. There are several reasons for this. The populations of cities have multiplied, increasing the concentrations of people and thus of their wastes. Especially in the industrial nations, increasing levels of consumption have meant increasing amounts of waste produced per person. In the United States each person produces 50,000 pounds of waste each year and almost 20,000 gallons of sewage.

The preferred method in the United States for dealing with solid waste (or trash) in recent decades has been to bury it in landfills, which is the municipal equivalent of sweeping dirt under the rug; in Europe, incineration has been more common. Both have serious difficulties.

Historically, municipal landfills have been responsible for a great deal of soil and groundwater pollution. About 20 per cent of the sites in the EPA's Superfund cleanup programs are municipal landfills. However, most of these are old sites. The technology of landfilling has improved considerably over the past few decades, but the improvements have been mostly palliative. Sealing the bottoms of landfills has reduced the chances of wastes getting into groundwater, at least temporarily. The heavy plastic liners now being used are expected to last at least 30 years. Decomposition processes, by contrast, are likely to go on for hundreds of years. Thus we might expect a plague of leaking landfills in the twenty-first century.

Effective drainage systems on the surfaces of finished landfills have reduced infiltration of water into the buried trash and thus the danger of chemicals being leached through the trash levels and through or around the sealed bottom. Problems with rodents, birds, odors, and blowing debris have been reduced to manageable levels by sanitary filling, that is, by covering each day's trash deposits with a layer of soil. Somewhat more regenerative in character are the methane collection systems that draw off the gases escaping from decomposition processes. These consist of networks of pipes buried in the layers of trash, which collect the methane as it is generated and convey it to boilers where it is burned to make steam. In most cases the steam is then used to generate electricity, which is fed into the electrical grid.

Such technological improvements have rendered landfilling a relatively harmless means for dealing with nontoxic, nonhazardous wastes. With methane collection, a small portion of the energy embedded in trash is returned for reuse. Even with these improvements, however, landfills remain a degenerative way of dealing with waste; they are a means for wasting waste. Material discarded and consigned to a landfill is effectively removed from the realm of human use. The materials, many of them nonrenewable, and the energy embedded within them, are no longer in the economy. Most of these materials are in fact reusable by some means.

Landfills do not facilitate nature's recycling processes either. Decomposition of buried trash is extremely slow. A number of researchers have dug up materials long ago deposited in landfills and found them hardly changed since the day they were covered over. Finding 30-year-old newspapers that are still readable and food items like hot dogs that are still recognizable is fairly common. Certain microorganisms are needed to decompose these materials and make their components available for reorganizing into new forms. The dry, anaerobic conditions inside a landfill provide a poor environment for them. Thus within a landfill, nature's continuous regeneration is slowed virtually to a standstill. In terms of ecological function, landfills provide not assimilation but storage.

Trash burning, which is commonly practiced in many cities in Europe and in a few in the United States, has appealing short-term advantages. It comes closer than any other technology to simply making waste disappear, and this makes it attractive within the Paleotechnic ethos. A second advantage is that, like methane collection, burning can recover some of the energy imbedded in waste materials by using the heat to generate electricity. The disadvantages of burning, on the other hand, are numerous. Even with the best pollution control equipment, incinerators release into the air considerable volumes of carbon monoxide, sulfur and nitrogen dioxides, dioxin (which is extremely toxic even in minute quantities), and numerous metals, including lead and mercury. The fluidized bed gasifier being tested by the Southern California Edison Company may be a pollution-free incineration technology, but we will not know that for sure for some time.

With burning, groundwater contamination also remains a problem. Incineration does not make the trash entirely disappear. At least 25 per cent of the original weight and 10 per cent of the original volume remain in the form of an ash residue that still must be disposed of, usually by landfilling. This ash still contains in concentrated form a considerable amount of the dangerous materials that were in the trash before burning, especially metals actually released in the burning process.

Moreover, incinerators are costly devices. For each ton of burning capacity per day, the cost has been estimated at $100,000 to $150,000, which is several times the capital cost of materials recovery systems. Furthermore, incinerators, like landfills, fail to make use of the potential utility that remains in a great deal of the material considered as waste.

For some public officials and even for some citizens' groups who opposed landfills and incinerators in their own environs, exporting trash also has the advantage of seeming disappearance. In Los Angeles there have been several proposals to ship it to the desert. For a great many people, the austere, sparsely vegetated, and almost unpopulated landscape of the desert is a wasteland and thus a suitable place for urban refuse. This has made it a likely location for facilities that would not be acceptable in cities, such as coal-fired generating stations. It sometimes happens that desert towns with economies as sparse as their landscape are quite willing to accept the exported urban pollution for a price. The effects on the fragile natural systems of the desert, however, can be considerably more damaging than they are in the more resilient urban environs.

In eastern US cities, where lack of landfill space has been an even more pressing problem than in western cities, exporting garbage has been an even more enticing solution. Philadelphia exports trash to eastern Ohio and northern Virginia. The famous garbage barge *Mobro* dramatized both the problem and some of the difficulties with the export solution: It wandered about the world for 55 days in 1988 searching for a place to dump its cargo of urban refuse from New York and eventually ended the journey where it began. Eastern cities nevertheless routinely propose exporting trash to nonindustrial countries in South America and Africa. Whether intentionally or not, they provide a fitting metaphor for the relationship between the industrial and nonindustrial worlds.

Since Thomas Crapper invented the water closet in the nineteenth century, the prevailing means for dealing with human excrement in the United States and Europe has been to mix it with water in that device and then convey the mixture through underground pipes to the nearest sizable body of water, usually a river or bay. Along the way a highly mechanized sewage treatment plant separates the solids from the liquids. That is, after conventional treatment, the sewage is still there, though in different form.

As long as populations are small and dispersed in relation to the volume of water, rivers and bays can assimilate and dilute the nutrients and other materials in the sewage. However, at some point, as cities grow, the volume of nutrients and other materials becomes too great for the assimilative capacities of the water bodies, resulting in an excessive buildup of nutrients and other pollutants. This has happened in waters around most of the world's cities. Since the 1960s, Congress has tried repeatedly to solve the water pollution problem in the United States through a series of Water Quality Control acts. It is now illegal to dump treated sewage water into a body of water if the quality is lower than what is already there. Nevertheless, in 1991, 19 years after amendments to the Clean Water Act established this requirement, over 2000 beaches were reportedly closed along US coasts due to sewage pollution.[1]

In response to this requirement, many sewage plants have added a secondary level of treatment involving mechanical and biological devices to break down the organic solids still remaining in the water after primary screening. Some have also added various forms of advanced treatment to remove other specific materials. Nevertheless, despite hundreds of billions of dollars spent on treatment plants, pollution problems persist in the rivers and bays where cities dump their sewage, especially the larger ones. . . .

PRACTICES AND TECHNOLOGIES FOR MATERIALS REGENERATION

Among the most serious difficulties with waste management in the industrial nations is the immense quantities of materials to be dealt with. The industrial economies' high emphasis on productivity necessarily results in large volumes of waste. This is the essence of the throughput system. Given that the capacity of any environment – land or water – to assimilate waste is limited, large quantities create a basic conflict.

Regenerative design applies the [strategy of] letting nature do the work to increase the assimilative capacity of land and water. At the same time, thoughtful design can make more land available by multiple functions; that is, land used for processing wastes can often be used also for other purposes. However, these strategies can accomplish only so much. Regenerative goals make it clear that the volume of waste to be processed should be limited by the capacity of the environment to assimilate it.

Waste in other cultures

The success of industrial sewage systems in achieving their overriding goal, the control of disease vectors, has made it easy to assume that the western industrial way is the only way. In fact, there are numerous radically different ways of dealing with sewage in operation in other cultures. In China the excrement of a household, called night soil, is still left near the door in buckets to be collected during the night in some areas. In these places collectors take it to nearby farms where it is used as fertilizer. While this system does not provide the same protection from the spread of disease that western pipe-and-water systems do, it does have the advantage of returning nutrients to the soil. Whatever its shortcomings, it is regenerative. For obvious reasons, however, farmers find it objectionable and many have turned to other sources of fertilizer in recent years. In some parts of Japan, the night soil is stored in tanks in many cities and periodically collected by trucks that pump it from the holding tanks into their own tanks. This somewhat more sanitary process also returns the nutrients to the soil in nearby agricultural fields.[2]

Solid-waste practices are not so varied simply because few societies outside of contemporary western industrial societies have produced enough material goods to present serious problems in their disposal. Solid-waste issues were born of the one-way flow system and the extraordinary effectiveness of industrial technology in speeding the flow and thus the quantity of material collecting in the sinks at flow's end. In addition to reducing the volume of material by various means, regenerative practices can return these materials to the processes of natural and human ecosystems by two fundamentally different means: reuse and environmental reassimilation. There are two kinds of reuse: direct reuse and mechanical recycling.

Direct reuse

In industrial societies, the low cost of material goods often causes them to be discarded long before their usefulness is exhausted. In the reuse of these goods there is enormous potential for slowing the flow and thus for reducing both resource use and waste. Varied means for accomplishing this have appeared spontaneously, most of them operating outside the mainstream of the market economy. Among the examples are garage sales, flea markets, swap meets, junkyards, and thrift stores. In some third-world cities like Mexico City, small entrepreneurs have made a business of collecting usable items after they have been dumped in landfills and then reselling them. In cities like Los Angeles, complex underground economies have developed among ethnic minorities, dealing in secondhand goods, often on the basis of barter or credit. Thrift stores, usually operated for particular charities, also serve to keep reusable goods in circulation. Many of these do especially brisk business in clothing, furniture, and children's toys.

The sales volume of all of these combined is still minuscule in comparison with that of mainstream retailers. Nevertheless, they demonstrate that a reuse marketing network does function effectively even with no institutional incentives. Should social forces move away from one-way flows – should materials become suddenly less available or more expensive, for example, or the cost of disposal abruptly increase – then the importance of the reuse markets could become much greater. Each recession foreshadows such a trend, when thrift store and secondhand sales increase while those of conventional retailers decrease. A trend toward more durable, longer-lasting goods might also give impetus to this market.

Besides personal items, there are a great many materials commonly used on a large scale in industrial societies that might be reused to a far greater extent than they presently are. Prominent among these are building materials and containers, especially metal and plastic food and beverage containers.

Mechanical recycling

As compared to reuse, mechanical recycling requires the reshaping or remanufacturing of an old material into new form and thus involves energy use. Most of the items in the typical waste stream of an industrial society not suitable for reuse are suitable for recycling, through either mechanical or biological processes.

The composition of household trash varies considerably by region, by season, and even by district within a single city. The National Solid Wastes Management Association estimates the composition of household trash on a national basis as follows:

Paper	40%
Food	17%
Yard waste	13%
Glass	9%
Metals	9%
Wood	3%
Misc. organics	3%
Plastics	2%
Rubber and leather	2%
Textiles	2%

While all of these categories are recyclable by some means, the value of recycling is greater for some than others. Recycling metals is especially important because of the nonrenewable materials and the energy used in manufacturing them. Making a can from recycled aluminum requires only about one-third the energy needed to make one from new aluminum. The ratio for steel is about two-thirds. The energy ratio for glass is roughly the same

as for steel. However, the raw materials for glass are far more common than for aluminum or steel.

While the recycling of metals and glass became common practice in the 1980s and is increasing with growing government incentives, plastic recycling is still difficult and limited in its effectiveness. Most plastics are recycled into products far less valuable than those from which the material came. Low-grade packing and building materials are common uses.

The recycling potential of paper is limited by the fact that fibers are weakened with each remanufacture. Thus, with each recycling paper becomes weaker, lower in quality, and eventually loses its usefulness entirely. At that point, it can be biologically recycled, or composted.

As recycling becomes more common, the processes for it are increasingly superimposed on existing community structures which so far have evolved with no concern for such matters. However, to be truly effective, recycling will have to become an integral part of the community. Recycling centers, composting sites, and separation facilities can become important activity nodes. Building design can also facilitate recycling. The renovated Audubon Society Headquarters ... features chutes running vertically through its nine stories for carrying used materials to a recycling center in the lower basement. There are separate chutes for aluminum, glass, organic materials, paper, and plastics. The Society's goal is to recycle 79 per cent of the materials that enter the building.

As they become integral parts of our culture, recycling processes affect the built environment in myriad ways. With recycling, materials often become far more diverse in their uses than the specific functions of their first generation. Consider automobile tires for example. The best second generation function of tires is retreading and reuse. They may be retreaded a second and perhaps a third and fourth time as well, but eventually they become too worn for their original purpose. Then they can spread out into the environment. Without changing form, tires can become bumpers on boat docks or loading docks or crash barriers or swings on playgrounds. They are sometimes sunk in the ocean as artificial reefs or to support growth of crustaceans such as mussels. By cutting them up, it is possible to shape paving blocks, stair treads, or, commonly in nonindustrial countries, soles for sandals. Ground into crumbs, tire rubber makes an excellent roadbed or mulch for agricultural or sports fields. Tire crumbs can also be mixed with asphalt to make RUMAC, which is quieter, more resilient, holds more heat, and lasts twice as long as asphalt alone. Finally, tires can be decomposed by pyrolysis to yield fuel oil. While it is true that the economy has difficulty absorbing some of these products and tires continue piling up as waste, it seems almost certain that repeated and diverse recycling pathways will soon bring dramatic change to both the economy and the landscape.

Biological reassimilation

Biological reassimilation differs from reuse in that it follows the first strategy, letting nature do the work, by drawing on natural processes of decomposition to reintegrate materials into the landscape. Filtration and reassimilation of materials depend on the decomposing activity of countless bacteria and other microbes working unseen in our environment. Though we are hardly aware of them most of the time, these microbes account for most of the earth's biological activity. We can make use of their efforts in a number of ways, three of which are especially important: composting, natural sewage treatment, and bioremediation. These technologies are extremely significant for the future, likely to develop in effectiveness and sophistication.

Composting

The composting process biologically decomposes organic material under controlled conditions. The product is a loosely structured soil-like material that can be handled, stored, and applied to the land as a beneficial soil amendment without adversely affecting the environment. In some areas, compost can provide at least a partial antidote to prevalent conditions of soil degradation.

Important to the utility of compost is its very low, virtually nonexistent health risk. In the composting process, the activity of bacteria in decomposition causes the material to heat up. The high temperature kills pathogens and insect larvae in the mixture as well as most weed seeds.[3]

Most organic materials can be composted, though some decompose faster than others. A mass of organic material can be organized for composting in any number of ways. Piles and rows are common forms. For small operations on the scale of a backyard, it is common practice to load the organic material into a bin with openings in the sides for air movement. Since composting is an aerobic process involving a community of bacteria, fungi, and other microorganisms, providing for air movement is essential. There are several means for accomplishing this, ranging from periodically turning small piles and bins by hand, to force-air devices, to large machines that move between the rows of large composting operations, turning the material mechanically. Composting can also be accomplished in large containers which provide optimum conditions and also control odors.

Under optimum conditions the composting process generally takes from three weeks to two months, though the period can be as short as 12 days. . . .

Aquatic sewage treatment

Since the 1960s, researchers have developed an array of treatment systems that use the capacities of both plants and microorganisms to process sewage without using the elaborate, energy-intensive, often unreliable mechanical devices used in industrial sewage plants. That is, they let nature do the work of sewage treatment. Essentially, these systems replicate and intensify the processes of nature in organic recycling. They simultaneously filter the water and assimilate the solids into living organisms. They use water as a medium and treat sewage as it comes from the conventional collection systems.

The essential point concerning natural treatment systems is that they are landscapes in their basic character and operation, while the conventional treatment systems of the industrial period are basically machines. This is a fundamental difference with far-reaching implications that transcend the technological distinction to involve the design of environment, the shaping of cities, and even the character of societies. . . .

Before entering a natural treatment process, the solids are separated out of the water by settling or screening. The treatment systems fall into three general types: aquacultural ponds, wetlands, and rootzone beds, all of which have certain characteristics in common. The sewage water travels slowly among the roots and stems of aquatic plants, which take up some nutrients and other materials from the water in the process of supporting their own growth. However, the bulk of the work is done by bacteria and other microorganisms living on the roots and stems. Plants and microorganisms are capable of taking almost any materials out of the water, including nutrients, metals, and pathogens. The degree of treatment depends on the time; given enough time, natural systems can produce water suitable for human consumption from the densest raw sewage.

In natural treatment systems the limiting factor is usually biological oxygen demand, or the oxygen content of the water. The oxygenating activity of aquatic plants is important in maintaining oxygen levels as many of the plants take in oxygen through their leaves and release it through their roots. The oxygen-rich environment at the roots supports a rich microbial community that is very effective in the treatment process: bacteria, fungi, filter feeders, detritivores, and their predators....

Aquatic treatment concepts are not limited to municipal or industrial facilities but can be applied in myriad ways at any scale in any situation where purer water or a richer aquatic environment is desirable. Water in a polluted stream, for example, might be diverted through a series of wetlands for treatment and then returned to the stream. A series of ponds and wetlands located at strategic points in a city's drainage system could treat urban runoff to a level suitable for return to natural waterways.

Aquatic treatment systems in all their forms are among the best and clearest examples of regenerative technologies. They are in themselves complex ecosystems which naturally do the work that human society needs done. They can treat water to any level of quality. They do not need the concrete and steel structures or the machinery, the pumps and pipes of conventional industrial treatment. They can work at any scale and are not subject to the breakdowns that plague mechanized treatment systems. They do not use fossil fuels or pollute the air. Finally, they cost far less than do mechanized systems.

The main disadvantage of aquatic systems is that they occupy more land than mechanized systems. In urban areas this can present serious problems. However, in this respect too, aquatic systems are quintessentially regenerative in that they incorporate all of the processing capabilities of the landscape: assimilation, filtering, storage, and production. They are also integrally related to other life-support processes. Besides supplying clean water for any number of uses, they can provide biomass for energy conversion, fertilizer for food production, and food for animals, both domestic and wild.

BOX 1 Arcata marsh treatment system

Arcata is a town of about 15,000 people located on the northern California coast. For decades its sewage was piped into an oxidation pond for a few days and then moved on to the Pacific Ocean. The federal Water Pollution Control Act of 1972 rendered this treatment system clearly inadequate. Some form of secondary treatment would have to be added, and the conventional technologies were very expensive. So in the late 1970s, Arcata developed a small pilot wetland system to function between the oxidation pond and ocean release. The results showed that the wetland could provide advanced treatment with capital and maintenance costs considerably lower than those of a mechanized secondary treatment system. On the basis of those results, the city developed its present system.

This system begins with primary settling after which 2 to 3 million gallons of sewage move into three oxidation ponds each day and an equal amount moves out. After that, a 5.3-acre intermediate marsh, planted mostly with the hardstem bulrush (*Scirpus acutus*), reduces suspended solids. Mosquito fish control mosquito populations. Chlorination and dechlorination follow the intermediate marsh; then the water moves into the 154-acre Arcata Marsh and Wildlife Sanctuary and from there into Humboldt Bay.

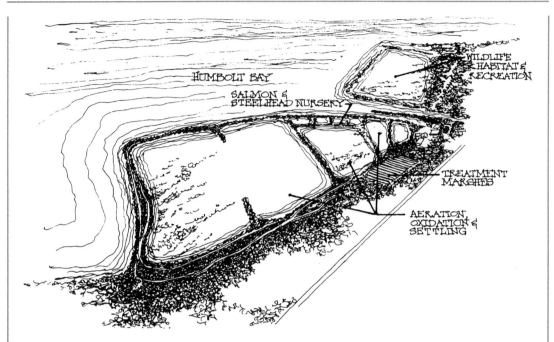

Figure 1. Arcata sewage treatment marsh.

In this marsh, cattails grow along with bulrushes, and duckweed covers much of the surface. Analyses required by the state to maintain the facility's permit to discharge into the bay show that the water moving into the bay meets the standards for secondary treatment. Studies have shown that refinements in the design of the marsh could improve the treatment to the level of standards for advanced treatment.

As important as sewage treatment in the Arcata Marsh is the combination of other functions that it supports. Wildlife habitat is varied and extensive. This is due partly to the inherently diverse character of land–water interactions in a wetland and to the way in which this inherent quality was further augmented by design. In its final grading the levels of the marsh bottom were varied to provide conditions for a range of marsh plants that grow at different levels of submergence. These plants in turn provide conditions for a range of microbes, mollusks, fish, and birds. Over 220 bird species have been recorded here, many of them migratory species that stop over on their journey along the Pacific flyway.

Researchers have also established a fish hatchery in the lagoon. Salmon fry are imprinted to return here as adults to lay their own eggs.

The recreational value of the marsh has been proven as well. Over 100,000 people visit the area each year. Most of these are from the city of Arcata, since the marsh has become a local focal point, but many are tourists. A significant increase in the number of tourists visiting Arcata has been documented by the city.

NOTES

1 Stammer, Larry B. 1992. Sewage Forced Closure of 2000 Beaches in 1991. *Los Angeles Times*, 24 July.

2 Pradt, Louis A. 1971. Some Recent Developments in Night Soil Treatment. *Water Resources*, 5.

3 Hornick, S.B. *et al.* 1979. *Use of Sewage Sludge Compost for Soil Improvement and Plant Growth.* US Department of Agriculture ARM-NE-6.

ENVIRONMENTAL JUSTICE
AND SOCIAL EQUITY

"People-of-Color Environmentalism"

from *Dumping in Dixie: Race, Class, and Environmental Quality* (1990)

Robert Bullard

Editors' Introduction

Sustainability goals are often presented in terms of the "three Es" – environment, economy, and equity – which in a sustainable society would all be enhanced rather than undermined over the long term. Of these, equity has been by far the least represented within public policy debates. There are relatively few well-organized groups advocating on behalf of low-income or otherwise disadvantaged communities. Even the environmental movement, with its relatively progressive middle-class constituency, developed with little consideration of the equity implications of its issues.

The link between social justice and environmental issues in the USA was developed beginning in the 1980s in large part by working-class communities fighting against the location of garbage incinerators, land fills, and toxic chemical hazards near their neighborhoods. African-American and Latino activists also criticized mainstream environmental groups for their lack of diversity, and demanded changes in federal regulation to produce more equitable public participation within environmental decision-making. At the same time, Third World activists were calling attention to the inequitable impacts of development policies internationally – a separate but parallel set of equity debates. The environmental injustices suffered by disenfranchised communities in North America, in other words, came to be seen by many activists as similar to the condition of less well-off groups worldwide.

Atlanta sociology professor Robert D. Bullard has been at the forefront of chronicling and defining the environmental justice movement in the USA. Here he discusses the roots of the movement, links with gender issues, and prospects for future organizing. Other leading writings on the subject of environmental justice include *Sprawl City: Race, Politics, and Planning in Atlanta*, edited by Bullard, Glenn S. Johnson, and Angel O. Torres (Washington, DC: Island Press, 2000), *Environmental Injustices, Political Struggles: Race, Class, and the Environment*, edited by David E. Camacho (Durham, NC: Duke University Press, 1993), and *Just Sustainabilities: Development in an Unequal World*, edited by Julian Agyeman, Robert D. Bullard, and Bob Evans (Cambridge, MA: MIT Press, 2003).

It is time for people to stop asking the question "Do minorities care about the environment?" The evidence is clear and irrefutable that white middleclass communities do not have a monopoly on environmental concern, nor are they the only groups moved to action when confronted with the threat of pollution. Although a "concern-and-action gap" may still exist between people of color and whites, communities of color are no longer being bullied into submission by industrial polluters and government regulators.[1]

Clearly, a "new" form of environmentalism has taken root in America and in communities of color. Since the late 1970s, a new grassroots social movement has emerged around the toxics threat. Citizens mobilized around the anti-waste theme. These social activists acquired new skills in areas where they had little or no prior experience. They soon became resident "experts" on toxics issues. . . . However, they did not limit their attacks to well-publicized toxic-contamination issues but sought remedial actions on problems like housing, transportation, air quality, and even economic development – issues the traditional environmental agenda had largely ignored.

Environmental justice embraces the principle that all people and communities are entitled to equal protection of environmental, health, employment, housing, transportation, and civil rights laws. Activists even convinced the EPA to develop a definition of environmental justice. The EPA defines environmental justice as:

> The fair treatment and meaningful involvement of all people regardless of race, color, national origin, or income with respect to the development, implementation, and enforcement of environmental laws, regulations and policies. Fair treatment means that no group of people, including racial, ethnic, or socio-economic group should bear a disproportionate share of the negative environmental consequences resulting from industrial, municipal, and commercial operations or the execution of federal, state, local, and tribal programs and policies.[2]

A major paradigm shift occurred in the 1990s. This shift created a new framework and a new leadership. Women led much of this grassroots leadership. The impetus behind this change included grassroots activism, redefinition of environmentalism as a "right," research documenting disparities, national conferences and symposia, emphasis on pollution and disease prevention, government initiatives, interpretation of existing laws and mandates, and grassroots alliances and coalitions.

Environmentalism has been too narrowly defined. Concern has been incorrectly equated with check writing, dues paying, and membership in environmental organizations. These biases have no doubt contributed to the misunderstanding of the grassroots environmental justice movement in people-of-color communities. People-of-color activists in this new movement focused their attention on the notion of deprivation. For example, when people of color compare their environmental quality with that of the larger society, a sense of deprivation and unequal treatment, unequal protection, and unequal enforcement emerges. Once again, institutional racism and discriminatory land-use policies and practices of government – at all levels – influence the creation and perpetuation of racially separate and unequal residential areas for people of color and whites. Too often the disparities result in groups fighting another form of institutional discrimination.[3]

All communities are not created equal. Institutional barriers have locked millions of people of color in polluted neighborhoods and hazardous, low-paying jobs, making it difficult for them to "vote with their feet" and escape these health-threatening environments. Whether in the ghetto or barrio, on the reservation, or in rural "poverty pockets," environmental injustice is making some people sick. Government has been slow to take these concerns as legitimate environmental and health problems. Mainstream environmentalists have also been slow in recognizing these grassroots activists as "real" environmentalists.[4]

The environmental justice movement is an extension of the social justice movement. Environmental justice advocates should not have to apologize for this historical fact. Environmentalists may be concerned about clean air but may have opposing views on public transportation, highway construction, industrial-facility siting, or the construction of low-income housing in white, middle-class suburban neighborhoods. On the other hand, environmental justice advocates also want clean air. People of color have come to understand that environmentalists are no more enlightened than nonenvironmentalists when it comes to issues of justice and social equity. But then, why should they be more enlightened? After all, we are all products of socialization and reflect the various biases and prejudices of this process. It is not surprising that mainstream environmental organizations have not been active on issues that disproportionately impact people of color, as in the case of toxics, workplace hazards, rural and urban housing needs, and the myriad of problems resulting from discriminatory zoning and strains in the urban, industrial complex. Yet people of color are the ones accused of being ill-informed, unconcerned, and inactive on environmental issues.

Environmental decision-making operates at the juncture of science, economics, politics, and ethics. It has been an uphill battle to try to convince some government and industry officials and some environmentalists that unequal protection, disparate impact, and environmental racism exist. Nevertheless, grassroots activists have continued to argue and in many instances have won their case. Working together, community stakeholders can assist government decision-makers in identifying "at-risk" populations, toxic "hot spots," research gaps, and action plans to correct existing imbalances and prevent future threats.[5] In order to accomplish their mission in an era of dwindling resources, environmental policymakers are increasingly turning to strategies that incorporate a *community-empowerment* approach. For example, community environmental protection (CEP) is being touted by the EPA as a "new" way of doing business.

Strengthening grassroots community groups can build a supportive social environment for decision-making. Residents and government authorities (local, state, and federal), often working together through creative partnerships with grassroots community groups, universities, nonprofit agencies, and other institutions, can begin solving environmental and health problems and design strategies to prevent future problems in low-income areas and communities of color. But the US Environmental Protection Agency and other governmental agencies cannot resolve all environmental problems alone. Communities also need to be in the position to assist in their own struggle for clean, safe, healthy, livable, and sustainable communities.

THE RIGHT TO BREATHE CLEAN AIR

Before the federal government stepped in, issues related to air pollution were handled primarily by states and local governments. Because states and local governments did such a poor job, the federal government established national clean-air standards.

Congress enacted the Clean Air Act (CAA) in 1970 and mandated the EPA to carry out this law. Subsequent amendments (1977 and 1990) were made to the CAA that form the current federal program. The CAA was a response to states' unwillingness to protect air quality. Many states used their lax enforcement of environmental laws as lures for business and economic development.[6]

Transportation policies are also implicated in urban air-pollution problems. Automobile-choked highways create health-threatening air pollution.[7] Freeways are the lifeline for suburban commuters, and millions of central-city residents are dependent on public transportation as their primary mode of travel.[8] Are people of color concerned about air quality and transportation? The answer is yes. The air-quality impacts of transportation are especially significant to people of color, who are more likely than whites to live in urban areas with reduced air quality. . . .

Asthma is an emerging epidemic in the United States. The annual age-adjusted death rate from asthma increased by 40 per cent between 1982 and 1991, from 1.34 to 1.88 per 100,000 population,[9] with the highest rates being consistently reported among blacks between the ages of 15 and 24 years during the period 1980–1993.[10] Poverty and minority status are important risk factors for asthma mortality. Children are at special risk from ozone.[11] Children also represent a considerable share of the asthma burden, that affliction being the most common chronic disease of childhood. Asthma affects almost 5 million children under 18 years of age. . . .

The public health community has insufficient information to explain the magnitude of some of the air pollution-related health problems. However, they do know that people suffering from asthma are particularly sensitive to the effects of carbon monoxide, sulfur dioxides, particulate matter, ozone, and nitrogen oxides.[12] Ground-level ozone may exacerbate health problems such as asthma, nasal congestion, throat irritation, respiratory-tract inflammation, reduced resistance to infection, changes in cell function, loss of lung elasticity, chest pains, lung scarring, formation of lesions within the lungs, and premature aging of lung tissues.[13]

African Americans, for example, have significantly higher prevalence of asthma than the general population.[14] A 1996 report from the federal Centers for Disease Control and Prevention shows hospitalization and death rates from asthma increasing for individuals 25 years old or younger.[15] The greatest increases occurred among African Americans. African Americans are two to six times more likely than whites to die from asthma.[16] Similarly, the hospitalization rate for African Americans is 3.4 times the rate for whites. . . . Air pollution, for many environmental justice advocates, translates into poor health, loss of wages, and diminished quality of life.

THE THREAT OF ECONOMIC EXTORTION

Why were people-of-color organizations late in challenging the environmental imbalance that exists in the United States? People-of-color organizations and their leaders have not been as sensitive to the environmental threats as they have been to problems in education, housing, jobs, drugs, and, more recently, the AIDS epidemic. In some cases, they have operated out of misguided fear and speculation that environmental justice will erode hard-fought civil rights gains or thwart economic development in urban core neighborhoods. There is no evidence that environmental justice or the application of Title VI of the Civil Rights Act of 1964 has hurt business or "brownfields" (abandoned properties that may or may not be contaminated) redevelopment opportunities in communities of color.[17] On the other hand, we do not have to speculate about the harm inflicted on the residents from racial red-lining by banks and insurance companies and the targeting of communities of color for polluting industries and locally unwanted land uses, or LULUS. The harm is real and measurable.

Grassroots groups in communities of color are beginning to take a stand against threatened plant closure and job loss as a trade-off for environmental risks. These threats are tantamount to economic extortion. This extortion has lost some of its appeal, especially in those areas where the economic incentives (jobs, taxes, monetary contributions, etc.) flow outside of the host community. People can hardly be extorted over economic benefits they never receive from the local polluting industry. There is a huge difference between the promise of a job and a real job. People will tell you, "You can't eat promises." Because of the potential

to exacerbate existing environmental inequities, community leaders are now questioning the underlying assumptions behind so-called trade-offs as applied in poor areas.

In their push to become acceptable and credible, many mainstream environmental organizations adopted a corporate model in their structure, demeanor, and outlook. This metamorphosis has had a down side. These corporate-like environmental organizations have alienated many grassroots leaders and community organizers from the larger movements. The environmental justice movement – with its egalitarian worldview and social justice agenda – offers an alternative to the more staid traditional environmental groups.

Local community groups may be turned off by the idea of sitting around a table with a waste-disposal giant, a government regulator, and an environmentalist to negotiate the siting of a toxic-waste incinerator in their community. The lines become blurred in terms of the parties representing the interests of the community and those of business. Negotiations of this type fuel residents' perception of an "unholy trinity," where the battle lines are drawn along an "us-versus-them" power arrangement. Moreover, overdependence on and blind acceptance of risk-assessment analysis and "the best available technology" for policy setting serves to intimidate, confuse, and overwhelm individuals at the grassroots level.

Talk of risk compensation for a host community raises a series of moral dilemmas, especially where environmental imbalances already exist. Should risks be borne by a smaller group to spare the larger groups? Past discriminatory facility-siting practices should not guide future policy decisions. Having one polluting facility makes it easier to site another in the same general area. The "one more won't make a difference" logic often becomes the dominant framework for decision-making. Any saturation policy derived from past siting practices perpetuates equity impacts and environmental injustice. Facility siting becomes a ritual for selecting "victims for sacrifice."

MOBILIZING THE GRASS ROOTS

It is unlikely that the environmental justice movement will ever gain unanimous support in communities of color. Few social movements can count on total support and involvement of their constituent groups. All social movements have "free riders," individuals who benefit from the efforts of a few. Some people shake the trees, while others pick up the apples. People-of-color environmentalism has been and will probably remain wedded to a social-action and social-justice framework. The issues raised by environmental justice advocates challenge the very core of privilege in our society. Some people make money and profit off the misery of poisoning others. Some communities are spared environmental assaults because of industrial-siting practices of concentrating locally unwanted land uses in communities with little or no political power and limited resources. After all, American society has yet to achieve a race-neutral state where race- and ethnic-based organizations are no longer needed.

Although the color barrier has been breached in most professional groups around the country, blacks still find it useful to have their own organizations. The predominately black National Bar Association (NBA), National Medical Association (NMA), National Association of Black Social Workers (NABSW), Association of Black Psychologists (ABP), and Association of Black Sociologists (ABS) are examples of race-based professional organizations that will probably be around for some time in the new millennium.

Grassroots environmental organizations have the advantage of being closer to the people they serve and the problems they address. Future growth in the environmental movement is likely to come from the bottom up, with grassroots environmental groups linking up with social-justice groups for expanded spheres of influence and focus.

Communities of color do not have a long track record in challenging government decisions and private industries that threaten the environment and health of their residents. Many of the organizations and institutions were formed as a reaction to racism and dealt primarily with social-justice issues. Groups such as the NAACP, Urban League, Southern Christian Leadership Conference, and Commission for Racial Justice operate at the multi-state level and have affiliates in cities across the nation. With the exception of Reverend Joseph Lowery of the Southern Christian Leadership

Conference, Benjamin R. Chavis Jr. of the United Church of Christ's Commission for Racial Justice, and Reverend Jessie Jackson of the National Rainbow Coalition, few national black civil rights leaders and organizations embraced an ideology that linked environmental disparities with racism.[18] It was not until the 1980s that national civil rights organizations began to make such links. This linking of institutional racism with the structure of resource allocation (clean environments) has led people-of-color social-action groups to adopt environmental justice as a civil rights issue, an issue well worth "taking to the street."

NIMBYism [not-in-my-backyard politics] has operated to insulate many white communities from the localized environmental impacts of waste facilities while providing them the benefits of waste disposal. NIMBYism, like white racism, creates and perpetuates privileges for whites at the expense of people of color. Citizens see the siting and unequal protection question as an all-out war. Those communities that can mobilize political influence improve their chance of "winning" this war. Because people of color remain underrepresented in elected and appointed offices, they must, most often, rely on indirect representation, usually through white officials who may or may not understand the nature and severity of the community problem. Citizen redress often becomes a political issue. Often the only science involved in the government response and decision-making is political science.

Who are the frontline leaders in this quest for environmental justice? The war against environmental racism and environmental injustice has been waged largely by people of color who are indigenous to the communities. People-of-color grassroots community groups receive some moral support from outside groups, but few experts are down in the trenches fighting alongside the warriors. On the other hand, it was the mothers and grandmothers, ministers from the churches, and the activist leaders from community-based organizations, civic clubs, neighborhood associations, and parents' groups who mobilized against the toxics threat. Few of these leaders may identify themselves as environmentalists or see their struggle solely as an environmental problem. Their struggles embrace larger issues of equity, social justice, and resource distribution. Environmental justice activists question the fairness of the decision-making process and the outcome.

Many environmental justice disputes revolve around siting issues, involving government or private industry. Proposals for future sites are more likely to attract environmentalists' support than are existing sites. It is much easier to get outside assistance in fighting a noxious facility that is on paper than one that is in operation. Again, plant closure means economic dislocation. Because communities of color are burdened with a greater share of existing facilities – many of which have been in operation for decades – it is an uphill battle of convincing outside environmental groups to support efforts to close such facilities.

It makes a lot of sense for the organized environmental movement in the United States to broaden its base to include people-of-color, low-income, and working-class individuals and issues. Why diversify? People of color now form a potent voting bloc. Diversification makes good economic and political sense for the long-range survival of the environmental movement. However, it is not about selfishness or "quota filling." Diversification can go a long way in enhancing the national environmental movement's worldwide credibility and legitimacy in dealing with global environmental and development issues, especially in Third World nations.[19]

NOTES

1 See Bullard, R.D. 1996. *Unequal Protection: Environmental Justice and Communities of Color.* San Francisco: Sierra Club Books, pp. 3–22; Bryant, B. 1995. *Environmental Justice: Issues, Policies, and Solutions.* Washington, DC: Island Press, pp. 8–34.

2 US Environmental Protection Agency. 1998. *Guidance for Incorporating Environmental Justice Concerns in EPA's NEPA Compliance Analysis.* Washington, DC: EPA.

3 Bullard, R.D. 1999. Dismantling Environmental Racism in the USA. *Local Environment*, 4, pp. 5–19.

4 Westra, Laura and Wenz, Peter S. 1995. *Faces of Environmental Racism: Confronting Issues of Global Justice.* Lanham, MD: Rowman and Littlefield.

5 Bullard, R.D. 1999. Leveling the Playing Field Through Environmental Justice. *Vermont Law Review*, 23, pp. 453–478; Collin, Robert W. and Robin M. 1998. The Role of Communities in Environmental Decisions: Communities Speaking for Themselves. *Journal of Environmental Law and Litigation*, 13, pp. 37–89.

6 Reitze, Arnold W. Jr. 1991. A Century of Air Pollution Control Law: What Worked; What Failed; What Might Work. *Environmental Law*, 21, pp. 15–49.

7 Davis, Sid. 1997. Race and the Politics of Transportation in Atlanta. In R.D. Bullard and G.S. Johnson (eds). *Just Transportation: Dismantling Race and Class Barriers to Mobility.* Gabriola Island, DC: New Society Publishers, pp. 84–96; Environmental Justice Resource Center. 1999. *Sprawl Atlanta: Social Equity Dimensions of Uneven Growth and Development.* Atlanta, GA: Report prepared for the Turner Foundation, Clark Atlanta University.

8 For an in-depth discussion of transportation investments and social equity issues, see Bullard and Johnson (eds). *Just Transportation.*

9 Centers for Disease Control. 1995. Asthma – United States, 1982–1992. *Morbidity and Mortality Weekly Report*, 43, pp. 952–955.

10 Centers for Disease Control. 1996. Asthma Mortality and Hospitalization Among Children and Young Adults – United States, 1980–1993. *Morbidity and Mortality Weekly Report*, 45, pp. 350–353.

11 Pribitkin, Anna E. 1994. The Need for Revision of Ozone Standards: Why Has the EPA Failed to Respond? *Temple Environmental Law & Technology Journal*, 13, p. 104.

12 See Mann, Eric. 1991. *L.A.'s Lethal Air: New Strategies for Policy, Organizing, and Action.* Los Angeles: Labor/Community Strategy Center.

13 US Environmental Protection Agency. 1996. *Review of National Ambient Air Quality Standards for Ozone, Assessment of Scientific and Technical Information.* Research Triangle Park, NC: OAQPS staff paper, EPA; American Lung Association. 1995. *Out of Breath: Populations-at-Risk to Alternative Ozone Levels.* Washington, DC: American Lung Association.

14 See Mak, H.P., Abbey, H., and Talamo, R.C. 1983. Prevalence of Asthma and Health Service Utilization of Asthmatic Children in an

Inner City. *Journal of Allergy and Clinical Immunology*, 70, pp. 367–372; Goldstein, I.F. and Weinstein, A.L. 1986. Air Pollution and Asthma: Effects of Exposure to Short-Term Sulfur Dioxide Peaks. *Environmental Research*, 40, pp. 332–345; Schwartz, J., Gold, D., Dockey, D.W., Weiss, S.T., and Speizer, F.E. 1990. Predictors of Asthma and Persistent Wheeze in a National Sample of Children in the United States. *American Review of Respiratory Disease*, 142, pp. 555–562; Crain, F., Weiss, K., Bijur, J. *et al.*, 1994, An Estimate of the Prevalence of Asthma and Wheezing Among Inner-city Children. *Pediatrics*, 94, pp. 356–362.

15 Centers for Disease Control and Prevention. 1996. Asthma Mortality and Hospitalization Among Children and Young Adults – United States, 1980–1993. *Morbidity and Mortality Weekly Report*, 45.

16 Centers for Disease Control. 1992. Asthma – United States, 1980–1990. *Morbidity and Mortality Weekly Report*, 39.

17 US Environmental Protection Agency. 1999. *Brownfields Title VI Case Studies: Summary Report*. Washington, DC: Office of Solid Waste and Emergency Response.

18 United Church of Christ Commission for Racial Justice. 1998. *From Plantation to Plant: Report of the Emergency National Commission on Environmental Justice in St. James Parish, Louisiana*. Cleveland: United Church of Christ.

19 See Bullard, R.D. 1993. *Confronting Environmental Racism: Voices from the Grassroots*. Boston: South End Press, ch. 1.

T W O

"Domesticating Urban Space"

from *Redesigning the American Dream: The Future of Housing, Work, and Family Life* (1984)

Dolores Hayden

Editors' Introduction

Just as current patterns of urban development have disproportionate impacts on communities of color, so they also profoundly affect women, children, the elderly, and other less powerful groups within society. Developing socially and environmentally healthy cities means recognizing and addressing these impacts. Although planners and designers have become more sensitive to the needs of women and other groups in recent years, much remains to be done to make cities friendly to all types of people.

In this selection from her book *Redesigning the American Dream: The Future of Housing, Work, and Family Life* (New York: Norton, 1984), urban historian Dolores Hayden examines how women have been systematically excluded or made to feel unsafe or uncomfortable in urban environments. Other similar writings include the essays in Catherine R. Stimpson's 1980 volume *Women and the American City* (Chicago: University of Chicago Press, 1981), and those in a Third World-oriented book edited by Rosi Braidotti and others, entitled *Women, the Environment and Sustainable Development: Towards a Theoretical Synthesis* (London: Zed Books, 1994). On a more detailed level, Clare Cooper Marcus and Wendy Sarkissian's *Housing as if People Mattered* (Berkeley: University of California Press, 1986) provides design guidelines for how medium-density housing can meet the needs of all residents, especially women, children, and the elderly. Gerda R. Wekerle and Carolyn Whitzman examine similar issues in the design of public spaces in their work *Safe Cities: Guidelines for Planning, Design, and Management* (New York: Van Nostrand Reinhold, 1995).

The phrase "A woman's place is in the home" has defined much housing policy and urban design in American society. The query "What's a nice girl like you doing in a place like this?" has reflected the prevailing attitude toward women in public, urban space. Both phrases have their roots in a Victorian model of private and public life. The first involves the patriarchal home as haven; the second defines the Victorian male double standard of sexual morality. Both are implicit rather than explicit principles of urban planning; neither will be found stated in large type in textbooks on land use. Both attitudes are linked to a set of nineteenth-century beliefs about female passivity and propriety in the domestic setting ("woman's sphere") versus male combativeness and aggression in the public setting ("man's world").

When nineteenth-century men (and women) argued that the good woman was at home in the kitchen with her husband, they implied that no decent woman was out in city streets, going places where men went. Thus, it was "unladylike" for a woman to earn her own living. Because the working woman was no *one* urban man's property (her father or her husband had failed to keep her at home), she was *every* urban man's property. She was the potential victim of harassment in the factory, in the office, on the street, in restaurants, and in places of amusement such as theaters or parks.

While the numbers of employed women and women in active public life have increased, many of these spatial stereotypes and patterns of behavior remain. Just as the haven houses hobble employed women, so the double standard harasses them when they are alone or with their children.[1] Men do not escape the problem. As husbands and fathers they share the stresses of the isolated houses and the violent streets they and their wives and children must negotiate. But rarely do men attribute the problems of housing and the city to the Victorian patriarchal views that reserved urban, public life for men only.[2]

THE FREEDOM OF THE CITY

At a theoretical level, women have never explicitly demanded or enjoyed the *droit de la ville*, the medieval right to the freedom of the city that distinguished urban citizens from feudal serfs.[3] The existing literature on urban history and theory conveys this in dozens of titles like *City of Man* or *The Fall of Public Man*. The experiences of women in urban space are absent from the content of many academic studies as well as from the titles.[4] The implicit assumption has been that either respectable women (and children) had the same urban experiences as men when they were with

Plate 1. Public space on the male double standard is shown in J.N. Hyde's *Running the Gauntlet*, New York City, 1874. A respectable woman walks down the street with ten men ogling as if she were a prostitute. Her body is tense and corseted, her eyes averted; the men lounge and stare boldly, providing a graphic example of what Nancy Henley calls "body politics," the dominance of one race, class, or gender over another through positioning of bodies in space.

Plate 2. Inhospitable environments. Older women wait for a bus while the car culture passes them by; a young woman waits for a bus while "Kim" advertises the double standard.

men, or else that women (and children) had no urban experiences, since their place was in the home or other segregated spaces. The closest that American women of all classes and races have come to challenging this view was in the Progressive Era, when Frances Willard attacked the double standard as expressed in saloons and the wardboss political culture of the late nineteenth century and attempted to domesticate the American city.

Some historians and critics have suggested that women failed to establish lasting power in the public sphere in the Progressive Era because they failed to develop an urban political theory suited to their needs.[5] Critics who complain of women's lack of "ideology" might first examine definitions of the nature of political theory and political activity. As long as the domestic world remains a romantic haven "outside" of public life and the political economy, politically active women can always be sent back to it, and men can justify the exclusion of women and children from their public debates and analyses. Yet the reverse is also true. If women can overcome what Lyn Lofland

has called the "thereness" of women,[6] if they can transcend what Jessie Bernard has defined as "the female world of a segregated place,"[7] new kinds of homes and neighborhoods might become the most powerful base in America for progressive political coalitions on urban issues.

A political program to overcome the "thereness" of women and win all female and male citizens, and their children, access to safe, public urban space requires that the presence of women (and their children) in public space be established as a political right; and that gender stereotypes be eliminated from architecture, urban design, and graphic design in public space. Such a political program would share many common features with Olmsted's attempt to create public space accessible to all in his parks; it would share many goals of the campaign Frances Willard launched in the 1890s for a "homelike world" in America's cities; it would require many new institutions like Jane Addams' Hull-House, a public center for community organizing on the model of a collective home open to all; it would link the campaigns of the anti-flirt club of the 1920s to the "Take Back the Night" marches of the 1970s.

Many professionals in the design fields are ready to support the political struggles necessary to bring domestic standards of amenity and safety into public space. In the recent past, disdainful speculators and politicians have claimed that the men engaged in these causes were not "real men," and the women were "little old ladies in tennis shoes," but on this subject, Lewis Mumford can be very reassuring. Calling for a serious study of resources and settlement design in 1956, Mumford challenged Americans to go beyond the old machismo of previous urban development patterns. "This new adventure," he said, "demands psychological maturity as the boyish heroism of the old adventures did not; for it is an exploration in depth, to fathom all the potential resources of a region . . . and to assess its possibilities for continued enjoyment."[8] The ultimate proof of maturity is the ability to nurture and protect human life, to develop public safety, public mobility, public amenities. Small, commonsense improvements in urban design can be linked to larger ideas about nurturing to help end the split between private life and public life.

RECAPTURING PUBLIC SPACE FOR THE NEEDS OF PARENTS

Anyone who has ever cared for babies or small children, even for a few hours away from a domestic environment, knows how little thought has been spent on making public space accessible to parents. When they begin a family, parents find that having a baby puts great spatial limits on their public, urban life. Although a first priority is to have adequate infant care and day care in residential neighborhoods for parents, it is also necessary to make it easier for adults to move in public space with their children. Other countries do this. In Denmark banks provide children's play areas with small furniture and toys while parents do their banking; in New Zealand, department stores offer day care to customers.[9]

To develop public facilities with the expectation that parents and grandparents with babies and young children will be using them is not a technically difficult task: it requires a commitment to better programming and a little imagination. Baby-changing spaces as a standard feature of both men's rooms and women's rooms would help; well-located seating and small children's furniture in banks, stores, and restaurants suggest that children are expected and welcomed with their fathers or mothers; windows at children's eye levels, as well as adults', are attractive in spaces where children represent a substantial number of users. Such changes in scale to accommodate children add liveliness and diversity to the urban scene. Concern for building materials and interior finishes neither too fragile nor too tough also helps define places that children can use. Play spaces can add a sense of joy for people of all ages, especially when they are organized to incorporate trees and flowers, public art or local landmarks such as ruins of old buildings, traces of past settlement patterns, or artifacts conveying economic history. Public space for children is, at its best, not only warm and educational but also fun. . . .

GREENLIGHTS AND SAFEHOUSES

Many adult Americans are afraid of public urban space. In 1967 the US President's Commission on Law Enforcement and the Administration of Justice concluded that "One-third of a representative sample of Americans say it is unsafe to walk alone at night in their neighborhoods."[10] Another study by a Congressional Committee in 1978 reported that fully one-half of all Americans were afraid to go out at night.[11] Yet another study by sociologists in Chicago states that one-half of all women and 20 per cent of all men were afraid to walk alone in their neighborhoods at night.[12]

Two programs designed to bring a greater sense of security into the lives of citizens are the Greenlight Program developed by the Women's Safety Committee of a group called City Lights in Jamaica Plain, Massachusetts, and the Safehouse Program created by Tenderloin Tenants for Safer Streets in San Francisco. As one reporter put it, "Safehouses, the traditional refuges of intelligence agents, fugitive radicals – and more recently havens for battered wives – now are being established for senior citizens."[13]

Both programs attempt to extend a sense of domestic security into the public realm. Greenlight aims to make women feel more secure from mugging and rape. Any house in the neighborhood showing a green light in the window is a place where a frightened woman can find shelter, a telephone, and emergency counseling. The Safehouse Program, identified by signs showing a peaked roof and a dove, operates in stores, bars, and hotels in a very rough district of San Francisco. All people in distress are welcome to enter and ask for safety, help getting to the hospital, or assistance from the police. The group distributes maps of the area to show the locations of the refuges, and encourages residents to patronize them.

Obviously neither Greenlight nor Safehouse signs can prevent crime. However these two efforts do show how citizens and planners can begin to give public space a more homelike quality.

RECLAIMING ACCESS TO PUBLIC SPACE: BETTER PUBLIC TRANSPORTATION

[. . .]

Most of the United States' expenditures on transportation in the last thirty years have supported freeway construction and car culture rather than projects like the [Washington, DC] Metro. The United States urgently needs better public transit systems. In the meantime, the services of existing public transit – subways, commuter railroads, and buses – must be improved by better social programming and by recognition that minorities, the young, the elderly, and women are most often the citizens without private cars.

Access to the public domain is especially difficult for older women. After age sixty-five, many women reap the results of a lifetime of low earnings, limited mobility, and self-sacrifice. In a study of 82,000 widows in Chicago, Helena Lopata found that over half of them did not go to public places, and over a fifth did not even go visiting. While 82 per cent were not in a position to offer transportation to others, 45 per cent had no one, of any age, to rely on for transportation.[14]

There are many ways that bus schedules and fares can be tailored for older people. Security issues can be worked on for both female passengers and the elderly, who are apprehensive about long waiting periods, especially at night, in deserted bus stops, train stations, and subway stations. Child-oriented and elderly-oriented schedules offer a further bonus: they relieve adults, usually women, from the responsibility of driving the elderly and the young, and at the same time encourage independence and self-reliance.

Good transportation is also a key factor in rape prevention. Recent estimates suggest that one woman in three in the United States will experience an attempted or completed rape in her lifetime.[15] Of course, if most citizens, including politicians and police officers, believe that a woman's place is in the home to begin with, they will not necessarily be blamed for unsafe streets. Instead, they may blame the rape victim for being in urban public space. Several innovative transportation projects, documented by Rebecca Dreis and Gerda Wekerle, meet the demand for greater safety on the streets by adding needed, flexible transportation.[16] In 1973 the Madison, Wisconsin, Women's Transit Authority began a service operating two cars seven nights a week, serving 1,000 women per month, on a fixed route shuttle service plus a flexible service within a four mile radius of the University of Wisconsin campus. Volunteers drive; the university, city and county pay the costs of the vehicles. In Whitehorse, Alaska, the Yukon Women's Minibus Society reached an even broader constituency. Women concerned about access to public space and about security created a system of four minibuses with sixteen seats each that now serve 700 passengers per day. The women's project provided the first bus system for the whole community.[17]

Public transportation not only provides safe access to public space, it can also educate riders about their city and about political struggles to make public space hospitable for everyone. This can range from adult classes on commuter trains, such as those initiated by Michael Young in England, to exhibits in key places – bus shelters and subway stations, and on buses and subways – that can provide essential information about cities, infrastructure, and public safety. Riders are thus united as a constituency for better services.

ADVERTISEMENTS, PORNOGRAPHY, AND PUBLIC SPACE

Americans need to look more consciously at the ways in which the public domain is misused for spatial displays of gender stereotypes. These appear in outdoor advertising, and to a lesser extent in commercial displays, architectural decoration, and public sculpture. While the commercial tone and violence of the American city is often criticized, there is little analysis of the routine way that crude stereotypes appear in public, urban spaces as the staple themes of commercial art. Most Americans are accustomed to seeing giant females in various states of undress smiling and caressing products such as whiskey, food, and records. Male models also sell goods, but they are usually active and clothed – recent ad campaigns aimed at gay men seem to be the first major exception. . . .

This double standard is the result of advertising practices, graphic design, and urban design. Sanctioned by the zoning laws, billboards are approved by the same urban planning boards who will not permit child-care centers or mother-in-law apartments in many residential districts. But the problem with billboards is not only aesthetic degradation. By presenting gender stereotypes in the form of non-verbal body language, fifty feet long and thirty feet high, billboards turn the public space of the city into a stage set for a drama starring enticing women and stern men.

NOTES

1 Saegert, S. 1981. Masculine Cities, Feminine Suburbs: Polarized Ideas, Contradictory Realities. In Stimpson, C. *et al.* (eds). 1980. *Women and the American City*. Chicago: University of Chicago Press; Chatfield-Taylor, A. 1973. Hitting Home. *Architectural Forum*, 138, pp. 58–61.

2 Katznelson, I. 1981. *City Trenches: Urban Politics and the Patterning of Class in the United States*. Chicago: University of Chicago Press. Discusses the split between "home" and "work" as an event of political importance but makes no gender analysis.

3 Lefebvre, H. 1974. *La Production de L'espace*. Paris: Anthropos. Suggests the political importance of this demand for men. On women's exclusion, see de Pizan, C. 1982. *The Book of the City of Ladies*, trans. E.J. Richards. New York: Persea Books.

4 See Gerde Werkerle's review essay in Stimpson *et al.*, 1980, *Women and the American City*,

pp. 188–214. Also see Enjeu, C. and Savé, J. 1974. The City: Off-Limits to Women. *Liberation*, 8, pp. 9–15; Kjaer, B. 1982. A Woman's Place. *Architect's Journal*, 176, p. 87. These last two essays from France and from Denmark (published in England) show that the double standard in public places is an international problem, not simply a result of American housing.

5 Degler, C. 1979. Revolution Without Ideology. *Daedalus*, 93, pp. 653–670, and also his 1979 *At Odds: Women and the Family*. New York: Oxford University Press.

6 Lofland, L. 1975. The 'Thereness' of Women: A Selective Review of Urban Sociology. In Milliman, M. and Kanter, R.M. (eds). *Another Voice: Feminist Perspectives on Social Life and Social Science*. New York: Anchor.

7 Bernard, J. 1980. *The Female World*. New York: Simon and Schuster.

8 Mumford, L. 1956. *The Transformations of Man*. New York: Harper and Row.

9 For a description of one mother's problems in public, see Chesler, P. 1979. *With Child: A Diary of Motherhood*. New York: Thomas Crowell, pp. 149–150, 175.

10 Quoted in Edgerton, R.B. 1978. *Alone Together: Social Order on an Urban Beach*. Los Angeles: University of California Press, p. 4.

11 Ibid., p. 5.

12 Gordon, M., Riger, S., LeBailly, R.K., and Heath, L. "Crime, Women, and the Quality of Urban Life." In Stimpson *et al.*, 1980, *Women and the American City*, pp. 161–169.

13 Hager, P. 1982. Safehouses Ease Fears of Aged Residents of San Francisco. *Los Angeles Times*, 21 November, p. 1.

14 Lopata, H. The Chicago Woman: A Study of Patterns of Mobility and Transportation. In Stimpson *et al.*, 1980, *Women and the American City*, pp. 161–169.

15 Griswold Johnson, A. 1980. On the Prevalence of Rape in the United States. *Signs: Journal of Women in Culture and Society*, 6 (Autumn), pp. 136–146. See also Brownmiller, S. 1975. *Against Our Will: Men, Women, and Rape*. New York: Simon and Schuster, p. 15; Griffen, S. 1971. Rape: The All-American Crime. *Ramparts*, 10 (September), pp. 26–35; Herman, D. The Rape Culture. In Freeman (ed.) 1975, *Women: A*

Feminist Perspective, Palo Alto. CA: Mayfield Pub. and Co., pp. 469–473.

16 Wekerle, G. 1979. Women's Self-Help Projects in the City: Transportation and Housing. Paper presented at conference on Social Practice, UCLA Urban Planning Program, 2 March, p. 5.

17 Ibid., pp. 5–7. On women as providers of urban services, also see Cranz, G. Women and Urban Parks: Their Roles as Users and Suppliers of Park Services. In Keller (ed.). *Building for Women*. 1991, Lexington MA: Lexington Books, pp. 151–171; Women in Urban Parks. In Stimpson *et al.* (eds). *Women and the American City*, pp. 79–95.

ECONOMIC DEVELOPMENT

"The Economic System and Natural Environments"

from the "Introduction" to *Blueprint for a Sustainable Economy* (2000)

David Pearce and Edward B. Barbier

Editors' Introduction

Economics and sustainable development co-exist uneasily. The common denominator for many observers is a realization that current economic theory and practice – the driving forces behind much of the world's development – leave out many important social and environmental factors that represent the costs as well as the benefits of development. Traditional economic tools make it difficult to incorporate externalities such as pollution, resource depletion, and degradation of human living environments. They also have no good way to take long-term costs and benefits into account, assume endless growth in material consumption, inadequately take into account market distortions caused by subsidies and regulations, and in many other ways fail to promote sustainable development.

British authors David Pearce, Edward B. Barbier, Anil Markandya, and other colleagues have been at the forefront of developing the discipline known as environmental economics, which seeks ways to better incorporate ecological concerns into economic analysis. In 1989 they authored an influential book entitled *Blueprint for a Green Economy* (London: Earthscan, 1989), followed by *Blueprint 2: Greening the World Economy* (London: Earthscan, 1991). These books provided popular explanations of a wide range of eco-economic issues. In 2000 Pearce and Barbier updated these resources in a volume entitled *Blueprint for a Sustainable Economy* (London: Earthscan, 2000) from which the following selection is taken. Here they review the challenges facing economists in incorporating environmental factors into their work.

Other writings in the field of environmental economics include Pearce and Jeremy J. Warford's *World Without End: Economics, Environment and Sustainable Development* (New York: Oxford University Press, 1993), and Barry C. and Martha K. Field's *Environmental Economics: An Introduction* (New York: McGraw-Hill, 2002).

A second, somewhat overlapping perspective has been that of ecological economics, which generally has less faith in markets, casts a more skeptical eye on the ability of economics to incorporate environmental and social factors, and seeks new tools and techniques to balance social, environmental, and economic objectives. Useful readings in this field include Robert Costanza *et al.*, *An Introduction to Ecological Economics* (Boca Raton, FL: CRC Press, 1997), *The Local Politics of Global Sustainability*, by Tom Prugh, Robert Costanza, and Herman E. Daly (Washington, DC: Island Press, 2000), and articles in the journal *Ecological Economics*.

The source of most environmental problems lies in the failure of the economic system to take account of the valuable services which natural environments provide for us.... Environmental assets are akin to other capital assets like machinery and roads and the stock of knowledge and skills. Increasing one form of asset while running another one down is likely to be a short-sighted prescription to increasing human well-being. Yet in many ways this has been the history of past economic development.

Environmental or ecological services and functions include:

- the provision of non-renewable natural resources such as coal and oil, bauxite and iron;
- the provision of renewable natural resources such as timber, fish and water;
- the provision of waste sinks to receive and assimilate solid, liquid and gaseous wastes from economic systems;
- the provision of amenity;
- the provision of biogeochemical cycles which help stabilize climates, provide nutrients to living things, and purify water and air; and
- the provision of information in the form of genetic blueprints and behavioural observation.

Economic systems are generally good at providing only the first of these. While some commentators still express concern about the exhaustion of fossil fuel energy and mineral resources, the functioning of market systems can, to a considerable extent, be relied upon to signal when these resources are approaching exhaustion. These signals will include rising prices, which constitute an early warning that we should be moving out of those resources and into others, such as into renewable energy sources. Moreover, the rising prices themselves stimulate new discoveries, substitution of now relatively cheaper resources, and encourage technological change. We can argue about the precise pace at which this transition should happen, but the evidence is that what we might call old-fashioned resource exhaustion is not a major environmental problem. But economic systems seem to go wrong when it comes to providing the rest of the environmental functions listed above. Why?

... One of the central messages of *Blueprint I* was that many environmental resources have no market.

They are not bought and sold. Accordingly, there are no price signals to alert us to their scarcity or to induce discovery, substitution and technological change. The same feature of these missing markets also results in an uneven playing field between environmental conservation and the immediate factors which threaten conservation. To the slash-and-burn farmer there is little benefit in pointing to the many ecological functions served by the forest if he receives no income, in cash or kind, from those services. The fact that the trees act as a store of carbon is of no immediate consequence to the farmer, even though it is a matter of great concern if we believe in the science of global warming, for then the carbon is better stored in the biomass than released to the atmosphere as carbon dioxide through burning the trees.

So it is with a great many environmental assets. The upstream polluter has little incentive to take account of the downstream river user, unless forced to do so. Manufacturers of chlorofluorocarbons had no incentive to be concerned about the effects of these chemicals on the stratospheric ozone layer and hence on excess skin cancers. European farmers have little incentive to take into account the loss of wildlife arising from hedgerow removal or pesticide and fertilizer applications. All of these cases are examples of externalities, uncompensated third-party effects. It is difficult enough to internalize the externality (that is to make the polluter regard the externality as a cost to himself or herself) when the externality results in cash losses to others. It is even harder when the losses show up as non-monetary losses, as for example in the loss of amenity or impaired health.

The missing-market phenomenon therefore biases our so-called economic development decisions against the environment and in favour of economic activity which harms the environment. Ultimately some of the development in question will be worthwhile – this is a matter of comparing costs and benefits. But it would be fairly obvious that much development will be far more doubtful once we recognize that it is at the cost of ecological functions which themselves have economic value. That economic value is simply not realized because there are no markets through which its value can be expressed.

In this central theme we have three of the main features of environmental economics and the way in which environmental economists seek to find solutions to environmental problems.

First, we see that without markets there is a clear bias towards economic activity which, at best, downgrades the environment and, at worst, ignores it altogether. Hence there is a need to establish or create markets where none exist, or to modify markets where they exist but fail adequately to reflect environmental impacts. This conclusion is mitigated by the extent to which those creating degradation may themselves be unaware of the impact they are having and be willing to change their behaviour because they do have wider social and environmental concerns.

Second, we see that these missing or imperfect markets are a cause of environmental degradation. This is a major advance on most casual discussions of the causes of degradation, which tend to focus on the agents of destruction – the farmer, the logging company, the multinational company – rather than on the reasons for their behaviour. To devise a policy we first have to know the underlying causes. In this case, the solution lies in creating markets.

Third, while we could simply catalogue environmental impacts and show these alongside the economic benefits of the economic activity in question, it is far more forceful to put a money value on the environmental damage done. Then, costs and benefits can be compared directly, using the same monetary language that is used to justify economic development. Moreover, we have a substantial and lengthy experience of the former non-monetary approach in the form of environmental impact assessments (EIA). An EIA is a critically important activity, but it is difficult to resist the view that much of it is cosmetic, more designed to say that the environment has been taken into account, than to overrule environmentally damaging developments. Going the one stage further and putting money values on environmental damage puts the environment on the same footing as the economic arguments in favour of development, provided, of course, that the exercise does not remain confined to the paper on which the analysis is written. There have to be incentives based on the monetary value analysis.

Posing the issue in terms of development versus conservation is a little dangerous. It implies that we can have one but not the other and, indeed, those who are opposed to economic growth see it in precisely those terms. We do not agree with that view. The comparison between development and environment has to be one of the types of economic development that can be secured at the same time as minimizing the risks to the environment. . . . The reality is that the world has to accommodate at least 50 per cent more people than it has now and they will need space, food and water, shelter and infrastructure. That cannot be secured without environmental loss. While this may appear to lend support to those who argue against not only economic growth but also population growth, the population growth in question is unavoidable. A further 50 per cent on top of that is avoidable and everything that can be done that is humane and respectful of individual liberties should be done. But the first 50 per cent is demographically unavoidable. The debate has to be about the way in which we foster future economic development to meet the legitimate aspirations of people; that is, it has to be about sustainable development.

"Natural Capitalism"

from *Mother Jones*, March/April 1997

Paul Hawken

Editors' Introduction

Co-founder of the successful Smith & Hawken garden supply business, Paul Hawken has been a leading philosopher of ecological capitalism for several decades. His work includes *The Ecology of Commerce: A Declaration of Sustainability* (New York: HarperCollins, 1993) and *Natural Capitalism: Creating the Next Industrial Revolution* (London: Earthscan, 1999; with Amory and L. Hunter Lovins). Hawken has also been chair of The Natural Step, an influential worldwide organization begun in Norway which works with businesses to develop more sustainable practices.

Hawken argues in this selection from *Mother Jones* magazine that powerful economic mechanisms can eventually be harnessed for restorative rather than exploitative ends. To bring this about, he believes that existing incentives and subsidies must be changed to eliminate those supporting unsustainable ways of doing business, and to encourage alternative economic activity. Doing this, he believes, can bring about a "natural capitalism" that is capable of transforming society. Other writings along this line include Lester Brown's volume *Eco-Economy: Building an Economy for the Earth* (New York: Norton, 2001). An alternative point of view is supplied by the essays in Martin O'Connor's volume *Is Capitalism Sustainable? Political Economy and the Politics of Ecology* (New York: The Guilford Press, 1994), which are generally skeptical about the fundamental ability of capitalism to change in directions that are not exploitative.

Somewhere along the way to free-market capitalism, the United States became the most wasteful society on the planet. Most of us know it. There is the waste we can see: traffic jams, irreparable VCRs, Styrofoam coffee cups, landfills; the waste we can't see: Superfund sites, greenhouse gases, radioactive waste, vagrant chemicals; and the social waste we don't want to think about: homelessness, crime, drug addiction, our forgotten infirm and elderly.

Nationally and globally, we perceive social and environmental decay as distinct and unconnected. In fact, a humbling design flaw deeply embedded in industrial logic links the two problems. Toto, pull back the curtain: The efficient dynamo of industrialism isn't there. Even by its own standards, industrialism is extraordinarily inefficient.

Modern industrialism came into being in a world very different from the one we live in today: fewer people, less material well-being, plentiful natural resources. As a result of the successes of industry and capitalism, these conditions have now reversed. Today, more people are chasing fewer natural resources.

But industry still operates by the same rules, using more resources to make fewer people more productive. The consequence: massive waste – of both resources and people.

Decades from now, we may look back at the end of the twentieth century and ponder why business and society ignored these trends for so long – how one species thought it could flourish while nature ebbed. Historians will show, perhaps, how politics, the media, economics, and commerce created an industrial regime that wasted our social and natural environment and called it growth. As author Bill McKibben put it, "The laws of Congress and the laws of physics have grown increasingly divergent, and the laws of physics are not likely to yield."

The laws we're ignoring determine how life sustains itself. Commerce requires living systems for its welfare – it is emblematic of the times that this even needs to be said. Because of our industrial prowess, we emphasize what people can do but tend to ignore what nature does. Commercial institutions, proud of their achievements, do not see that healthy living systems – clean air and water, healthy soil, stable climates – are integral to a functioning economy. As our living systems deteriorate, traditional forecasting and business economics become the equivalent of house rules on a sinking cruise ship.

One is tempted to say that there is nothing wrong with capitalism except that it has never been tried. Our current industrial system is based on accounting principles that would bankrupt any company.

Conventional economic theories will not guide our future for a simple reason: They have never placed "natural capital" on the balance sheet. When it is included, not as a free amenity or as a putative infinite supply, but as an integral and valuable part of the production process, everything changes. Prices, costs, and what is and isn't economically sound change dramatically.

Industries destroy natural capital because they have historically benefited from doing so. As businesses successfully created more goods and jobs, consumer demand soared, compounding the destruction of natural capital. All that is about to change.

NATURAL CAPITAL

Everyone is familiar with the traditional definition of capital as accumulated wealth in the form of invest-ments, factories, and equipment. "Natural capital," on the other hand, comprises the resources we use, both nonrenewable (oil, coal, metal ore) and renewable (forests, fisheries, grasslands). Although we usually think of renewable resources in terms of desired materials, such as wood, their most important value lies in the services they provide. These services are related to, but distinct from, the resources themselves. They are not pulpwood but forest cover, not food but topsoil. Living systems feed us, protect us, heal us, clean the nest, let us breathe. They are the "income" derived from a healthy environment: clean air and water, climate stabilization, rainfall, ocean productivity, fertile soil, watersheds, and the less-appreciated functions of the environment, such as processing waste – both natural and industrial. *Nature's Services*, a book . . . edited by Stanford University biologist Gretchen C. Daily, identifies trillions of dollars of critical ecosystem services received annually by commerce.

For anyone who doubts the innate value of ecosystem services, the $200 million Biosphere II experiment stands as a reality check. In 1991, eight people entered a sealed, glass-enclosed, 3-acre living system, where they expected to remain alive and healthy for two years. Instead, air quality plummeted, carbon dioxide levels rose, and oxygen had to be pumped in from the outside to keep the inhabitants healthy. Nitrous oxide levels inhibited brain function. Cockroaches flourished while insect pollinators died, vines choked out crops and trees, and nutrients polluted the water so much that the residents had to filter it by hand before they could drink it. Of the original 25 small animal species in Biosphere II, 19 became extinct.

At the end of 17 months, the humans showed signs of oxygen starvation from living at the equivalent of an altitude of 17,500 feet. Of course, design flaws are inherent in any prototype, but the fact remains that $200 million could not maintain a functioning ecosystem for eight people for 17 months. We add eight people to the planet every three seconds.

The lesson of Biosphere II is that there are no man-made substitutes for essential natural services. We have not come up with an economical way to manufacture watersheds, gene pools, topsoil, wetlands, river systems, pollinators, or fisheries. Technological fixes can't solve problems with soil fertility or guarantee clean air, biological diversity, pure water, and climatic stability; nor can they increase the capacity of the environment to absorb 25 billion tons of waste created annually in America alone.

NATURAL CAPITAL AS A LIMITING FACTOR

Until the 1970s, the concept of natural capital was largely irrelevant to business planning, and it still is in most companies. Throughout the industrial era, economists considered manufactured capital – money, factories, etc. – the principal factor in industrial production, and perceived natural capital as a marginal contributor. The exclusion of natural capital from balance sheets was an understandable omission. There was so much of it, it didn't seem worth counting. Not any longer.

Historically, economic development has faced a number of limiting factors, including the availability of labor, energy resources, machinery, and financial capital. The absence or depletion of a limiting factor can prevent a system from growing. If marooned in a snowstorm, you need water, food, and warmth to survive. Having more of one factor cannot compensate for the absence of the other. Drinking more water will not make up for lack of clothing if you are freezing.

In the past, by increasing the limiting factor, industrial societies continued to develop economically. It wasn't always pretty: Slavery "satisfied" labor shortages, as did immigration and high birthrates. Mining companies exploited coal, oil, and gas to meet increased energy demands. The need for labor-saving devices provoked the invention of steam engines, spinning jennies, cotton gins, and telegraphs. Financial capital became universally accessible through central banks, credit, stock exchanges, and currency exchange mechanisms.

Because economies grow and change, new limiting factors occasionally emerge. When they do, massive restructuring occurs. Nothing works as before. Behavior that used to be economically sound becomes unsound, even destructive.

Economist Herman E. Daly cautions that we are facing a historic juncture in which, for the first time, the limits to increased prosperity are not the lack of man-made capital but the lack of natural capital. The limits to increased fish harvests are not boats, but productive fisheries; the limits to irrigation are not pumps or electricity, but viable aquifers; the limits to pulp and lumber production are not sawmills, but plentiful forests.

Like all previous limiting factors, the emergence of natural capital as an economic force will pose a problem for reactionary institutions. For those willing to embrace the challenges of a new era, however, it presents an enormous opportunity.

THE HIGH PRICE OF BAD INFORMATION

The value of natural capital is masked by a financial system that gives us improper information – a classic case of "garbage in, garbage out." Money and prices and markets don't give us exact information about how much our suburbs, freeways, and spandex cost. Instead, *everything else* is giving us accurate information: our beleaguered air and watersheds, our overworked soils, our decimated inner cities. All of these provide information our prices should be giving us but do not.

Let's begin with a startling possibility: The US economy may not be growing at all, and may have ceased growing nearly 25 years ago. Obviously, we are not talking about the gross domestic product (GDP), measured in dollars, which has grown at 2.5 per cent per year since 1973. Despite this growth, there is little evidence of improved lives, better infrastructure, higher real wages, more leisure and family time, and greater economic security.

The logic here is simple, although unorthodox. We don't know if our economy is growing because the indices we rely upon, such as the GDP, don't measure growth. The GDP measures money transactions on the assumption that when a dollar changes hands, economic growth occurs. But there is a world of difference between financial exchanges and growth. Compare an addition to your home to a two-month stay in the hospital for injuries you suffered during a mugging. Say both cost the same. Which is growth? The GDP makes no distinction. Or suppose the president announces he will authorize $10 billion for new prisons to help combat crime. Is the $10 billion growth? Or what if a train overturns next to the Sacramento River and spills 10,000 gallons of atrazine, poisoning all the fish for 30 miles downstream? Money pours into cleanups, hatchery releases, announcements warning people about tainted fish, and lawsuits against the railroad and the chemical company. Growth? Or loss?

Currently, economists count most industrial, environmental, and social waste as GDP, right along with bananas, cars, and Barbie dolls. Growth includes *all* expenditures, regardless of whether society benefits or loses. This includes the cost of emergency room services, prisons, toxic cleanups, homeless shelters, lawsuits, cancer treatments, divorces, and every piece of litter along the side of every highway.

Instead of counting decay as economic growth, we need to subtract decline from revenue to see if we are getting ahead or falling behind. Unfortunately, where economic growth is concerned, the government uses a calculator with no minus sign.

WASTING RESOURCES MEANS WASTING PEOPLE

Industry has always sought to increase the productivity of workers, not resources. And for good reason. Most resource prices have fallen for 200 years – due in no small part to the extraordinary increases in our ability to extract, harvest, ship, mine, and exploit resources. If the competitive advantage goes to the low-cost provider, and resources are cheap, then business will naturally use more and more resources in order to maximize worker productivity.

Such a strategy was eminently sensible when the population was smaller and resources were plentiful. But with respect to meeting the needs of the future, contemporary business economics is pre-Copernican. We cannot heal the country's social wounds or "save" the environment as long as we cling to the outdated industrial assumptions that the summum bonum of commercial enterprise is to use more stuff and fewer people. Our thinking is backward: We shouldn't use more of what we have less of (natural capital) to use less of what we have more of (people). While the need to maintain high labor productivity is critical to income and economic well-being, labor productivity that corrodes society is like burning the furniture to heat the house.

Our pursuit of increased labor productivity at all costs not only depletes the environment, it also depletes labor. Just as overproduction can exhaust

topsoil, overproductivity can exhaust a workforce. The underlying assumption that greater productivity would lead to greater leisure and well-being, while true for many decades, has become a bad joke. In the United States, those who are employed, and presumably becoming more productive, find they are working 100 to 200 hours more per year than 20 years ago. Yet real wages haven't increased for more than 20 years.

In 1994, I asked a roomful of senior executives from Fortune 500 companies the following questions: Do you want to work harder in five years than you do today? Do you know anyone in your office who is a slacker? Do you know any parents in your company who are spending too much time with their kids? The only response was a few embarrassed laughs. Then it was quiet – perhaps numb is a better word.

Meanwhile, people whose jobs have been downsized, re-engineered, or restructured out of existence are being told – as are millions of youths around the world – that we have created an economic system so ingenious that it doesn't need them, except perhaps to do menial service jobs.

In parts of the industrialized world, unemployment and underemployment have risen faster than employment for more than 25 years. Nearly one-third of the world's workers sense that they have no value in the present economic scheme.

Clearly, when 1 billion willing workers can't find a decent job or any employment at all, we need to make fundamental changes. We can't – whether through monetary means, government programs, or charity – create a sense of value and dignity in people's lives when we're simultaneously developing a society that doesn't need them. If people don't feel valued, they will act out society's verdict in sometimes shocking ways. William Strickland, a pioneer in working with inner-city children, once said that "you can't teach algebra to someone who doesn't want to be here." He meant that urban kids don't want to be here at all, alive, anywhere on earth. They try to tell us, but we don't listen. So they engage in increasingly risky behavior – unprotected sex, drugs, violence – until we notice. By that time, their conduct has usually reached criminal proportions – and then we blame the victims, build more jails, and lump the costs into the GDP.

The theologian Matthew Fox has pointed out that we are the only species without full employment.

Yet we doggedly pursue technologies that will make that ever more so. Today we fire people, perfectly capable people, to wring out one more wave of profits. Some of the restructuring is necessary and overdue. But, as physicists Amory Lovins and Ernst von Weizsšcker have repeatedly advised, what we *should* do is fire the unproductive kilowatts, barrels of oil, tons of material, and pulp from old-growth forests – and hire more people to do so.

In fact, reducing resource use creates jobs and lessens the impact we have on the environment. We can grow, use fewer resources, lower taxes, increase per capita spending on the needy, end federal deficits, reduce the size of government, and begin to restore damaged environments, both natural and social.

At this point, you may well be skeptical. The last summary is too hopeful and promises too much. If economic alternatives are this attractive, why aren't we doing them now? A good question. I will try to answer it. But, lest you think these proposals are Pollyannaish, know that my optimism arises from the magnitude of the problem, not from the ease of the solutions. Waste is too expensive; it's cheaper to do the right thing.

RESOURCE PRODUCTIVITY

Economists argue that rational markets make this the most efficient of all possible economies. But that theory works only as long as you use financial efficiency as the sole metric and ignore physics, biology, and common sense. The physics of energy and mass conservation, along with the laws of entropy, are the arbiters of efficiency, not *Forbes* or the Dow Jones or the Federal Reserve. The economic issue is: How much work (value) does society get from its materials and energy? This is a very different question than asking how much return it can get out of its money.

If we already deployed materials or energy efficiently, it would support the contention that a radical increase in resource productivity is unrealistic. But the molecular trail leads to the opposite conclusion. For example, cars are barely 1 per cent efficient in the sense that, for every 100 gallons of gasoline, only one gallon actually moves the passengers. Likewise, only 8 to 10 per cent of the energy used in heating the filament of an incandescent lightbulb actually becomes visible light. (Some describe it as a space heater disguised as a lightbulb.) Modern carpeting remains on the floor for up to 12 years, after which it remains in landfills for as long as 20,000 years or more – less than 0.06 per cent efficiency.

According to Robert Ayres, a leader in studying industrial metabolism, about 94 per cent of the materials extracted for use in manufacturing durable products become waste before the product is even manufactured. More waste is generated in production, and most of that is lost unless the product is reused or recycled. Overall, America's material and energy efficiency is no more than 1 or 2 per cent. In other words, American industry uses as much as 100 times more material and energy than theoretically required to deliver consumer services.

A watershed moment in the study of resource productivity occurred in 1976, when Amory Lovins published his now-famous essay "Energy Strategy: The Road Not Taken?" Lovins' argument was simple: Instead of pursuing a "hard path" demanding a constantly increasing energy supply, he proposed that the real issue was how best to provide the energy's "end use" at the least cost. In other words, consumers are not interested in gigajoules, watts, or Btu, he argued. They want well-illuminated workspaces, hot showers, comfortable homes, effective transport. People want the *service* that energy provides. Lovins pointed out that an intelligent energy system would furnish the service at the lowest cost. As an example, he compared the cost of insulation with that of nuclear power. The policy of building nuclear power plants represented the "supply at any cost" doctrine that still lingers today. He said it made no sense to use expensive power plants to heat homes, and then let that heat escape because the homes lack insulation. Lovins contended that we could make more money by saving energy than by wasting it, and that we'd find more energy in the attics of American homes than in all the oil buried in Alaska. His predictions proved correct, although his proposals remained largely unheeded by the government. Today, the nuclear power industry has become moribund, not because of anti-nuclear protests but because it is uncompetitive.

In 1976, energy experts used to argue about whether the United States could achieve energy savings of 30 per cent. Twenty-one years later, having already obtained savings of more than 30 per cent over 1976 levels – savings worth $180 billion a year – experts now wonder whether we can achieve an additional 50 to 90 per cent. Lovins thinks we might possibly save as much as 99 per cent. That may sound ridiculous, but certainly no more so than the claim that textile workers could use gears and motors to increase their efficiency a hundredfold would have sounded at the beginning of the Industrial Revolution.

The resource productivity revolution is at a similar threshold. State-of-the-shelf technologies – fans, lights, pumps, superefficient windows, motors, and other products with proven track records – combined with intelligent mechanical and building design, could reduce energy consumption in American buildings by 90 per cent. State-of-the-art technologies that are just being introduced could reduce consumption still further. In some cases – wind power, for example – the technologies not only operate more efficiently and pollute less, they also are more labor-intensive. Wind energy requires more labor than coal-generated electricity, but has become competitive with it on a real-cost basis.

The resource revolution is starting to show up in all areas of business. In the forest products industry, clearinghouses now identify hundreds of techniques that can reduce the use of timber and pulpwood by nearly 75 per cent without diminishing the quality of housing, the "services" provided by books and paper, or the convenience of a tissue. In the housing industry, builders can use dozens of local or composite materials, including those made from rice and wheat straw, wastepaper, and earth, instead of studs, plywood, and concrete. The Herman Miller company currently designs furniture that can be reused and remanufactured a number of times; DesignTex, a subsidiary of Steelcase, a leading manufacturer of office furniture, sells fabrics that can be easily composted.

Although a new *hypercar* is now in development, "new urbanist" architects, such as Peter Calthorpe, Andres Duany, Elizabeth Plater-Zyberk, and others, are designing communities that could eliminate 40 to 60 per cent of driving needs. (A recent San Francisco study showed that communities can decrease car use by 30 per cent when they double population density.) Internet-based transactions may render many shopping malls obsolete. Down the road we'll have quantum semiconductors that store vast amounts of information on chips no bigger than a dot; diodes that emit light for 20 years without bulbs; ultrasound washing machines that use no water, heat, or soap; hyperlight materials stronger than steel; deprintable and reprintable paper; biological technologies that reduce or eliminate the need for insecticides and fertilizers; plastics that are both reusable and compostable; piezoelectric polymers that can generate electricity from the heel of your shoe or the force of a wave; and roofs and roads that do double duty as solar energy collectors. Some of these technologies, of course, may turn out to be impractical or have unwanted side effects. Nevertheless, these and thousands more are lining up like salmon to swim upstream toward greater resource productivity.

RESOURCE POLITICS

How can government help speed these entrepreneurial "salmon" along? The most fundamental policy implication is simple to envision, but difficult to execute: We have to revise the tax system to stop subsidizing behaviors we don't want (resource depletion and pollution) and to stop taxing behaviors we do want (income and work). We need to transform, incrementally but firmly, the sticks and carrots that guide business.

Taxes and subsidies are information. Everybody, whether rich or poor, acts on that information every day. Taxes make something more expensive to buy; subsidies artificially lower prices. In the United States, we generally like to subsidize environmental exploitation, cars, big corporations, and technological boondoggles. (We don't like to subsidize clean technologies that will lead to more jobs and innovation because that is supposed to be left to the "market.") Specifically, we subsidize carbon-based energy production, particularly oil and coal; we massively subsidize a transportation system that has led to suburban sprawl and urban decay; we subsidize risky technologies like nuclear fission and pie-in-the-sky weapons systems like Star Wars. (Between 1946 and 1961 the Atomic Energy Commission spent $1 billion to develop a nuclear-powered airplane. But it was such a lemon that the plane could not get off the ground. History's dustbin also includes a nuclear-powered ship, the Savannah, that was retired after the Maritime Administration found she cost $2 million more per year than other ships.)

We subsidize the disposal of waste in all its myriad forms – from landfills, to Superfund clean-ups, to deep-well injection, to storage of nuclear waste. In the process, we encourage an economy where 80 per cent of what we consume gets thrown away after one use.

As for farming, the US government covers all the bases: We subsidize agricultural production, agri-cultural nonproduction, agricultural destruction, and agricultural restoration. We provide price sup-ports to sugarcane growers, and we subsidize the restoration of the Everglades (which sugarcane growers are destroying). We subsidize cattle graz-ing on public lands, and we pay for soil conserva-tion. We subsidize energy costs so that farmers can deplete aquifers to grow alfalfa to feed cows that make milk that we store in warehouses as surplus cheese that does not get to the hungry.

Then there is the money we donate to dying industries: federal insurance provided to flood-plain developers, cheap land leases to ski resorts, deposit insurance given to people who looted US savings and loans, payments to build roads into wilderness areas so that privately held forest pro-duct companies can buy wood at a fraction of replacement cost, and monies to defense suppliers who have provided the Pentagon with billions of dollars in unnecessary inventory and parts.

Those are some of the activities we encourage. What we hinder, apparently, is work and social welfare, since we mainly tax labor and income, thereby discouraging both. In 1994, the federal government raised $1.27 trillion in taxes. Seventy-one per cent of that revenue came from taxes on labor – income taxes and Social Security taxes. Another 10 per cent came from corporate income tax. By taxing labor heavily, we encourage busi-nesses not to employ people.

To create a policy that supports resource pro-ductivity will require a shift away from taxing the social "good" of labor, toward taxing the social "bads" of resource exploitation, pollution, fossil fuels, and waste. This tax shift should be "revenue neutral" – meaning that for every dollar of taxation

added to resources or waste, one dollar would be removed from labor taxes. As the cost of waste and resources increases, business would save money by hiring less-expensive labor to save more-expensive resources. The eventual goal would be to achieve zero taxation on labor and income.

The purpose of this tax shift would be to change *what* is taxed, not *who* is taxed. But no tax shift is uniform, and without adjustments for lower incomes, a shift toward taxing resources would likely be regressive. Therefore, efforts should be made to keep the tax burden on various income groups more or less where it is now. (There are numerous means to accomplish this.) The import-ant element to change is the *purpose* of the tax system because, other than generating revenue, the current tax system has no clear goal. The only incentive provided by the Internal Revenue Code, with its 9,000 sections, is to cheat or to hire tax lawyers.

A shift toward taxing resources would require steady implementation, in order to give business a clear horizon in which to make strategic investments. A time span of 15 to 20 years, for example, should be long enough to permit businesses to continue depreciating current capital investments over their useful lives.

Of course, a tax shift alone will not change the way business operates; a broad array of policy changes on issues of global trade, education, economic development, econometrics (including measures of growth and well-being), and scientific research must accompany it. For the tax shift to succeed, we must also reverse the wrenching breakdown of our democracy, which means addressing campaign finance reform and media concentration.

It is easier, as the saying goes, to ride a horse in the direction it is going. Because the costs of nat-ural capital will inevitably increase, we should start changing the tax system now and get ahead of the curve. Shifting taxes to resources won't – as some in industry will doubtless claim – mean diminishing standards of living. It will mean an explosion of innovation that will create products, techniques, and processes that are far more effect-ive than what they replace.

Some economists will naturally counter that we should let the markets dictate costs and that using taxation to promote particular outcomes is interventionist. But *all* tax systems are interventionist; the question is not whether to intervene but *how* to intervene.

A tax system should integrate cost with price. Currently, we dissociate the two. We know the price of everything but the cost of nothing. Price is what the buyer pays. Cost is what society pays. For example, Americans pay about $1.50 per gallon at the gas pump, but gasoline actually costs up to $7 a gallon when you factor in all the costs. Middle Eastern oil, for instance, costs nearly $100 a barrel: $25 to buy and $75 a barrel for the Pentagon to keep shipping lanes open to tanker traffic. Similarly, a pesticide may be priced at $35 per gallon, but what does it cost society as the pesticide makes its way into wells, rivers, and bloodstreams?

THE FUTURE

In 1750, few could imagine the outcome of industrialization. Today, the prospect of a resource productivity revolution in the next century is equally hard to fathom. But this is what it promises: an economy that uses progressively less material and energy each year and where the quality of consumer services continues to improve; an economy where environmental deterioration stops and gets reversed as we invest in increasing our natural capital; and, finally, a society where we have more useful and worthy work available than people to do it.

A utopian vision? No. The human condition will remain. We will still be improvident and wise, foolish and just. No economic system is a panacea, nor can any create a better person. But as the twentieth century has painfully taught us, a bad system can certainly destroy good people.

Natural capitalism is not about making sudden changes, uprooting institutions, or fomenting upheaval for a new social order. (In fact, these consequences are more likely if we don't address fundamental problems.) Natural capitalism is about making small, critical choices that can tip economic and social factors in positive ways.

Natural capitalism may not guarantee particular outcomes, but it *will* ensure that economic systems more closely mimic biological systems, which have successfully adapted to dynamic changes over millennia. After all, this analogy is at the heart of capitalism, the idea that markets have a power that mimics life and evolution. We should expand this logic, not retract it.

For business, the opportunities are clear and enormous. With the population doubling sometime in the next century, and resource availability per capita dropping by one-half to three-fourths over that same period, which factor in production do you think will go up in value – and which do you think will go down? This basic shift in capital availability is inexorable.

Ironically, organizations like Earth First!, Rainforest Action Network, and Greenpeace have now become the *real* capitalists. By addressing such issues as greenhouse gases, chemical contamination, and the loss of fisheries, wildlife corridors, and primary forests, they are doing more to preserve a viable business future than are all the chambers of commerce put together. While business leaders hotly contest the idea of resource shortages, there are few credible scientists or corporations who argue that we are not losing the living systems that provide us with trillions of dollars of natural capital: our soil, forest cover, aquifers, oceans, grasslands, and rivers. Moreover, these systems are diminishing at a time when the world's population and the demand for services are growing exponentially.

Looking ahead, if living standards and population double over the next 50 years as some predict, and if we assume the developing world shared the same living standard we do, we would have to increase our resource use (and attendant waste) by a factor of 16 in five decades. Publicly, governments, the United Nations, and industries all work toward this end. Privately, no one believes that we can increase industrial throughput by a factor anywhere near 16, considering the earth's limited and now fraying life-support systems.

It is difficult for economists, whose important theories originated during a time of resource abundance, to understand how the decline in ecosystem services is laying the groundwork for the next stage in economic evolution. This next stage, whatever it may be called, is being brought about by powerful and much-delayed feedback from living systems. As we surrender our living systems, social stability, fiscal soundness, and personal health to outmoded economic assumptions, we are hoping that conventional economic growth will save us. But if economic "growth" does save us, it will be anything but conventional.

So why be hopeful? Because the solution is profitable, creative, and eminently possible. Societies may act stupidly for a period of time, but eventually they move to the path of least economic resistance. The loss of natural capital services, lamentable as it is in environmental terms, also affects costs. So far, we have created convoluted economic theories and accounting systems to work around the problem.

You can win a Nobel Prize in economics and travel to the royal palace in Stockholm in a gilded, horse-drawn brougham believing that ancient forests are more valuable in liquidation – as fruit crates and Yellow Pages – than as a going and growing concern. But soon (I would estimate within a few decades), we will realize collectively what each of us already knows individually: It's cheaper to take care of something – a roof, a car, a planet – than to let it decay and try to fix it later.

While there may be no "right" way to value a forest or a river, there is a wrong way, which is to give it no value at all. How do we decide the value of a 700-year-old tree? We need only ask how much it would cost to make a new one. Or a new river, or even a new atmosphere.

Despite the shrill divisiveness of media and politics, Americans remain remarkably consistent in what kind of country they envision for their children and grandchildren. The benefits of resource productivity align almost perfectly with what American voters say they want: better schools, a better environment, safer communities, more economic security, stronger families and family support, freer markets, less regulation, fewer taxes, smaller government, and more local control.

The future belongs to those who understand that doing more with less is compassionate, prosperous, and enduring, and thus more intelligent, even competitive.

"Import Replacement"

from *Going Local: Creating Self-Reliant Communities in a Global Age* (1998)

Michael Shuman

Editors' Introduction

The engine that powers much unsustainable development around the world currently is rapid, export-oriented economic growth. Conventional wisdom holds that such growth is essential to raise standards of living and to enable societies the luxury of funding environmental protection and social welfare programs. But a growing chorus of critics – led by opponents of economic globalization – questions this export-driven model and calls instead for more locally oriented economic development rooted in the needs, resources, and skills of local communities.

Attorney Michael Shuman is the former director of the Institute for Policy Studies in Washington, DC and consults widely on community economics, international development, and citizen diplomacy. In this selection from his book *Going Local: Creating Self-Reliant Communities in a Global Age* (New York: The Free Press, 1998), Shuman argues for an import replacement approach toward economic development, in which communities seek to produce many basic goods and services themselves. The advantages of such a strategy in his view include vastly reduced transportation needs, lower pollution, more local jobs, local ownership of businesses, and retention of capital in the local community. Such import replacement strategies have in fact been a recurrent theme in alternative economic development philosophies since World War II. Jane Jacobs, for example, believed that development of local industries to replace imported goods is behind the rise of most successful cities. Some advocates of local self-sufficiency have even gone so far as to set up local exchange trading systems (LETS) in which community members barter goods and services from one another using paper notes equal to one hour of labor. If long-distance trade is required for certain products, other activists believe that it should be "fair trade" in which producers receive decent wages and workers labor in healthy and safe conditions. However, such strategies are usually downplayed or vigorously opposed by advocates of global free trade, and have yet to catch on widely.

Other writings on this subject include David Morris' *Self-Reliant Cities* (San Francisco: Sierra Club Books, 1982), Jacobs' *Cities and the Wealth of Nations* (New York: Random House, 1984), and *The Case Against the Global Economy and for a Turn Toward the Local* (San Francisco: Sierra Club Books, 1996), edited by Jerry Mander and Edward Goldsmith. More information is also available from the Institute for Local Self-Reliance (www.ilsr.org) and Co-op America (www.coopamerica.org). Information on the Fair Trade movement is available from the Fair Trade Federation (www.fairtradefederation.org) and the Fair Trade Foundation (www.fairtrade.org.uk).

Community self-reliance may be difficult to imagine, but it has been the norm for most of human history. Somewhere around 10,000 BC, *homo sapiens sapiens* pioneered agriculture and used crude tools to plow fields for millet and rice in Southeast Asia.[1] At roughly the same time, counterparts in France were building structures capable of sheltering 400 to 600 people. Three thousand people lived within the walls of the city of Jericho as early as 8,000 BC. Fragmentary records suggest that these early communities were capable of producing enough grains, vegetables, fruits, and livestock to feed their residents. Wood, grasses, stones, and mud provided the essentials for housing and furniture. Animal hides, furs, and fibers were fashioned into clothing. Locally harvested plants and minerals became sources of the first medications. The burning of wood and manure generated warmth and light, and even facilitated basic metallurgy. Mechanical energy from running water and from draft animals helped process raw ingredients like wheat into more usable products like flour. These "simple" communities, the first outposts of human civilization, left an impressive legacy of folklore, music, art, and science.

Few communities in the United States today are this self-reliant. They require oil brought in by truck, coal by rail, natural gas by pipeline, and electricity through expansive power grids. The typical food item travels 1,300 miles before it winds up on the dinner table, and is distributed through supermarkets and chain stores.[2] Wander through your house and you'll probably find more evidence of resources from places thousands of miles away than from your own bioregion: tuna caught in the Gulf of Mexico; beer brewed from water and hops in Germany and bottled in New Jersey; Evian water extracted from southern France. Your closets are filled with shirts and blouses that traveled from silk producers in China to production plants in Malaysia, on to packaging facilities in Guatemala, and finally to the racks at Macy's.

The variety of these well-traveled goods certainly enhances the quality of our lives, but our growing dependence on them carries profound risks. The more essential an item is for our survival, the more dangerous it is to depend on someone outside the community selling it to us. Basic necessities, of course, are difficult to define. But if you were to write down a list of what you need to survive, chances are good you would include what most nations assented to in Article 25 of the Universal Declaration of Human Rights: "Everyone has the right to a standard of living adequate for the health and well-being of himself and his family, including food, clothing, housing, and medical care and necessary social services. . . ." To provide these basics, a community needs farmers, water suppliers, loggers, materials processors, and social-service providers. But is it really possible to structure a viable community around just these economic sectors? Can community-scale businesses meet local citizens' needs cost-effectively?

Nowhere are the dangers of depending on imports of basics more clear than with oil. Over the next decade the United States is expected to import 60 per cent of its oil from foreign suppliers (up from 52 per cent in 1994).[3] Just as OPEC's sudden boycott and price hikes in 1973 and 1979 triggered long queues at gas stations, shortages, inflation, and recessions, our continued dependence threatens economic instability in the years ahead. A 60 per cent import rate means a trade deficit in oil of $100 billion per year.[4] Leaving aside the costs of spills from offshore oil-drilling and air pollution from oil-burning, the transportation of oil across oceans poses huge environmental risks to harbor and coastal ecosystems, as well as to the entire oceanic food chain. Another decade of imports will provide $1 trillion of revenue to Persian Gulf states, much of which will be spent on weapons, wars, and saber-rattling that are hardly in the interests of US foreign policy. America is once again in a position where a handful of governments, most of them undemocratic, can bring the nation's economy to its knees by turning off the spigots.

Dependency on necessities from outside the community means that a remote crisis can reverberate into a local one. A study for the Pentagon concluded that in a single night, without ever leaving Louisiana, a few saboteurs could cut off three-quarters of the natural gas supplies to the eastern United States for more than a year.[5] It also found that low-technology sabotage of any one of the nation's more than 100 power stations could trigger a catastrophic core meltdown, inflicting a Chernobyl-like accident, or worse, on the country. In November 1965, a faulty relay in Canada cascaded into a series of equipment failures that blacked out 30 million electricity customers in the

Northeast for up to 13 hours.[6] On July 13, 1977, the chair of the Consolidated Edison Company in New York assured his customers that he could "guarantee" that a recurrence was extremely unlikely. Three days later, a combination of lightning and equipment failures blacked out nearly 9 million Con Ed customers for up to 25 hours, and set the stage for $121 million worth of looting. Two decades of reflection and planning did not prevent another major collapse of the electric grid on August 10, 1996. A tree falling into a transmission line near the Columbia River in Oregon toppled a series of mechanical dominoes and left millions of consumers in 14 states and two Canadian provinces without power for several hours. Dependence holds a community hostage to mistakes, misdeeds, and misfortunes totally outside its control.

A dependent community also loses the economic benefits of producing necessities for itself. A community that chooses not to generate its own electricity, not to grow its own food, not to process its own lumber, and so on, winds up losing the jobs and income that might have come from these commercial ventures. The more these activities are performed outside the community, the weaker the economic multiplier inside.

Economists are skeptical about the principles of self-reliance and import substitution, because they deprive a community of the benefits of trade. A community focusing inward narrows the range of goods and services available to its citizens, and deprives its businesses of the new machines, new production methods, and competitive forces necessary for progress. Moreover, if local goods and services cost more than those produced outside, going local means losing money, which condemns a community to less consumption and lower investment. Economies that have sought to seal themselves hermetically from the world – like the former Soviet Union, Albania, and Burma – soon crumble from obsolescence.

A better way to think of the goal of import substitution, however, is that it motivates a community to move the most important and valuable types of production back home, not to unplug completely from the national or international economy. And the means to accomplish this need not be tariffs or heavy-handed regulation, but simply smart choices by residents and local officials to buy, invest, and hire locally. Viewed this way, "the theoretical case for emphasizing import substitution is strong," according to Joseph Persky, David Ranney, and Wim Wiewel, three economists at the University of Illinois at Chicago.[7]

Import-substituting growth facilitates the diversification of the local economy and the accumulation of its own capital, skills, and experience. Jane Jacobs, in her 1969 book *The Economy of Cities*, provides two examples of cities that were able to pump up their economies through import substitution:[8] The Los Angeles export industries that nourished the US military during World War II, like aircraft manufacturing, shipbuilding, and petroleum refining, all declined after the war ended in 1945. Yet the number of jobs in the city expanded because of the growth of new import-replacing businesses. Automobile companies in Detroit, for example, opened branch plants to serve customers in southern California. In Chicago, between 1845 and 1855, the city's population grew by nearly a factor of seven, as did the overall economy. The reason again, writes Jacobs, was import substitution:

> [A]t the beginning of the decade Chicago, like any other little Midwestern depot settlement, was importing most kinds of things that every town supplied. But by the end of that decade it was producing a very large range of the common city-made goods of the time and some of the luxuries too – clocks, watches, medicines, many kinds of furniture, stoves, kitchen utensils, many kinds of tools, most building components.[9]

Import substitution actually incorporates some of the export-oriented thinking of mainstream economists. As a community replacing imports grows in size, it naturally attracts more businesses that target national and global markets. Persky, Ranney, and Wiewel summarize the export-led theory of development in the following terms:

> [A]n area that is able to increase significantly its sales of a major export will experience population growth related to the new employment in the basic export sector. This population growth in turn implies that an area may pass thresholds for new products. Now the metropolis will provide some commodities for itself that formerly

it imported. Export growth leads to import substitution.[10]

A community committed to import substitution, however, aims to *minimize* population growth. The goal is to expand the quantity and quality of jobs without drawing new people. University of Montana economist Thomas Michael Power has shown that the states in which jobs grew fastest such as Alaska, Arizona, Nevada, and Utah, also saw family income grow more slowly than the national average.[11] In states like New York and Rhode Island, where employment growth was slowest, income growth was above the national average. The reason for these seemingly paradoxical results is the inflow and outflow of people. Power concludes that "there is no reliable connection between mere quantitative expansion of the local economy and local economic well-being."[12] Expansion of economic activity should be targeted carefully at those people, institutions, and businesses that have a long-term commitment to the community. He winds up recommending a concerted local effort at import substitution:

> The import substitution can be direct, as in the case of energy or of a local bakery replacing imported bread, or it can be indirect. As the variety of goods and services produced locally expands, the richer commercial economy attracts and holds more of the residents' dollars. Local dollars that would have otherwise flowed out of the community, to purchase things that would add variety and quality to residents' lives, stay in the community to purchase local services. Live theater and music, instruction in skills, recreational facilities, and so on attract and hold dollars that otherwise would have flowed out to finance imports.[13]

If import replacement leads to greater exports, is the distinction from export-led development simply a matter of semantics? Hardly. A typical export-led development strategy targets just one or two industries. The World Bank pushed dozens of poor countries in the 1980s to specialize in the production of a few primary commodities like coffee, sugar, cocoa, copper, aluminum, and lumber. The fatal flaw of this approach was that each country's economy became so specialized that it was vulner-

able to the collapse of the price of a targeted commodity. And this is exactly what happened. The nosedive of coffee prices in the late 1980s and early 1990s, for example, destabilized coffee-exporting countries worldwide, from Guatemala to Indonesia. In Rwanda, another country dependent on coffee exports, the consequent economic turmoil set off a chain of events that culminated in the Hutus massacring an estimated million Tutsis. . . .

Can a small community embark upon an import-replacement development strategy? Don't certain industries need a large enough local market to be set up in the first place? No, respond Persky, Ranney, and Wiewel: "[W]e do not observe a well-defined threshold population at which a given industry enters a community. For most industries, we only observe that the share of local demand supplied locally tends to rise with size."[14]

A small community like Eugene, Oregon, couldn't operate a factory to produce cars solely for its own needs, but it *could* build a plant that met the transport demand of the Pacific Northwest region. The key issue is what economists call the optimal *economy of scale*. Whenever the economy of scale of production is large and a plant needs a very high output to operate competitively, it must export to consumers outside the community. And any good or service for which this is true, by definition, can facilitate import substitution in only a limited number of communities. Because the economy of scale of automobile construction is large, it's neither possible nor desirable for every one of America's 36,000 municipalities to manufacture cars.

The prevailing wisdom among economists and businesspeople is that large economies of scale are the rules of thumb for industrial production. Huge factories with global distribution networks are assumed to deliver cheaper and better products than are small factories serving just local markets. This is because certain fixed costs, such as management, machinery, warehouses, marketing, and lawyers, can be spread over more and more units of production. Business consultants like the Boston Company and Wall Street investment houses like Drexel Burnham Lambert have accumulated vast fortunes acquiring, merging, and reorganizing firms to achieve larger economies of scale.

Yet, as the dinosaurs learned, bigger is not always better. Bigger businesses also tend to develop certain diseconomies of scale. The larger the distance between producers and consumers, the harder it is to fine-tune products to the particular tastes of local markets. Local businesses set up to serve the exacting demands of local consumers can be started more quickly, with smaller investments and smaller risks. Once a conscientious process of import replacement begins, a company may be surprised to discover other savings inherent in local production and distribution. Transportation costs go down. So do the costs of marketing.

So do the costs of excessive preservatives and packaging. A study in Europe found that a jar of strawberry yogurt traveled 2,166 miles before reaching the typical German consumer.[15] Most of the mileage wasn't logged by the food; the main ingredients – milk and sugar – came from the surrounding countryside, and the strawberries were grown in Poland. It was attributable to the packaging. Glass for the jar, paper for the label, paste for the paper, aluminum for the top all were produced from sites across Europe.

The bottom line depends on how these economies and diseconomies add up. For basic necessities, economies of scale appear to be shrinking to the point where hundreds or even thousands of US communities could move toward self-reliance. Recent breakthroughs in technology, workforce organization, and resource management are enabling local entrepreneurs to provide food, water, wood, energy, and materials in cost-effective ways. While few communities have deliberately tried to replace all their imports of basic necessities, a number of examples in each of these economic sectors testify to the possibility and promise of doing so.

FOOD INDUSTRIES

One of the saddest stories of the past century in the United States is the destruction of community-based family farming. The number of Americans on a farm today is less than one-fifth what it was in 1920.[16] Those involved in farming have substantially expanded their land holdings. In 1994, 2 per cent of all the farms in the United States were so big that they were responsible for half of all food sales, while three-quarters of the farms were so small that they accounted for only 9 per cent.[17] A closer look at recent developments in agriculture, however, suggests that small-scale systems to grow, process, and market food are becoming not only cost-effective – in both rural and urban areas – but also essential to preserve the genetic integrity of the world's edible plants.

Leaving aside inflation, the value added to the economy by food-growing farms has remained constant throughout this century. What has changed is the explosive growth of the food-marketing business and of corporations providing inputs into farming (seeds, fertilizers, herbicides, pesticides, etc.). In 1910, for every dollar Americans spent for food, 41 cents went to farmers and 59 cents to marketers and input providers; now 9 cents go to farmers, 24 cents to input providers, and 67 cents to marketers.[18]

Economists celebrate these statistics, because fewer farmers can grow more food for American consumers at a lower price. But the costs have been the decimation of once-vibrant rural communities, and increased dependency of urban and suburban communities on far-off sources of nutrition. Another look at the numbers suggests that community-scale production might actually *lower* the cost of food for local consumers. Community-scale agriculture, even if it means higher-cost farming, might be able to bring down the cost of inputs through organic growing methods and the cost of marketing through local distribution.[19] When farmers are involved in distribution, they also have an opportunity to retain more of the "value added."

The economics of small-scale farming actually is improving, though it varies across the country. The valleys of the Mississippi, Missouri, and Ohio Rivers offer richer farming opportunities than do the Mohave desert or Fairbanks, Alaska, But the presence of *some* commercially viable agriculture in every one of America's fifty states (as well as in the less hospitable provinces of Canada) is a reminder that wherever there is land and water, growing one's own food is possible. One out of four fruits and vegetables distributed in today's commercial systems never makes it to the consumer's table, because it spoils during shipment or on the grocery-store shelf.[20] More locally grown produce, in contrast, needs to be thrown away less, and because it's closer to the customer (a typical item travels 200 miles instead of 1,300), transportation costs are lower. Finally, direct relationships between farmers and consumers can greatly reduce or eliminate the costs of packaging, marketing, middlemen, and supermarkets.

A recent study by the United Nations Development Program suggests that even dense urban areas hold enormous potential for cultivation.[21] Some 800 million people in the world who live in cities are engaged in urban agriculture, mainly for their own consumption.[22] In Hong Kong, which has an extraordinary population density, nearly half of all vegetables consumed are grown within the city limits, on 5 to 6 per cent of the city's land. Squatters in Lusaka, Zambia, grow one-third of their food in the city. Residents of Kampala, Uganda, meet 70 per cent of their poultry and egg consumption with local production. Data from the 1980s suggest that the 18 largest cities in China met over 90 per cent of their vegetable needs, and half their meat needs, through urban farming. And Singapore raises 80 per cent of its poultry, and a quarter of its vegetables, within city limits.

A nation's food system is rooted deeply in its history, culture, diet, and land-use policies, and therefore these experiences abroad may not be entirely transferable to the United States. But enough farming is occurring in or near US cities that the opportunities for expanded urban agriculture are certainly worth exploring. More than 30 per cent of the dollar value of US agricultural production originates from farms within major metropolitan areas.[23] Over the past 20 years, New York City has opened a thousand community gardens on public land, and 18 public markets to sell produce grown in them.[24] Boston and Philadelphia have even more gardens per capita. Recognizing the potential of urban farming, several cities such as Chattanooga and Hartford, and states such as Massachusetts and Oregon, have developed comprehensive urban food policies.

Several trends in American cities make the economics of urban agriculture increasingly attractive. First, more land is becoming available for farming. Most US cities are spreading out over larger geographic areas, while their populations are remaining stable or declining. Some 40 per cent of all the land in Detroit is now vacant. The exodus of industry and people from Pittsburgh has left more than 37,000 empty lots. These lots, once cleared of garbage, toxics, and abandoned buildings, have potential for cultivation, along with

parkland, greenhouses, and even rooftops. Second, cities have a growing supply of idle labor. The highest unemployment rates in the country are in urban areas. In principle, much of the work required for municipal farming could be done by unskilled inner-city residents, though some training would be required and psychological barriers to manual labor would need to be overcome. Third, with recent cuts in Aid to Families with Dependent Children, cities will continue to have serious unmet demands for more food. The Food Research and Action Center estimates that nearly half the children under the age of 12 who live in urban households below the poverty line are "either hungry or at-risk of hunger." Developing new, cheap sources of nutrition for these young people is imperative – for their future and for the future of America's cities.

A handful of nonprofits have seized this opportunity in intriguing ways. In South Central Los Angeles, Reverend George Singleton's Hope LAS Horticulture Corps trains young and at-risk gang members to plant flowers, herbs, vegetables, and trees in vacant lots. Sales of the produce grown pay for the project, which combines agricultural coursework with hands on training. Singleton is now working with organizations to spread the program to five other sites in the Los Angeles area, and to the Bronx.

Farms need not be within a city's limits to contribute to its economic well-being. Linking farmers just outside a city with consumers inside it can boost the regional economy. One indication that this is happening is the revival of farmers' markets across the country, in which local growers set up stands (or use the backs of their station wagons) to sell fruits, vegetables, and nuts. This produce often is not as pretty as what's available at the local supermarket; it may well be discolored and irregular. But consumers are attracted to it because it tastes better, holds more nutrition, and often is free of pesticides.

Some clever farmers have organized groups of consumers (and vice versa) into subscribers' clubs. One model of what has become known as community-supported agriculture, or CSA, is that each member family agrees to pay a fee for the growing season, and in exchange the farmer promises to deliver a box of vegetables each week, enough to feed a family. The contents of the box vary throughout the season; one week it might be mostly asparagus and melon, another it might be potatoes and pumpkins. Some CSAs add farm products like eggs, milk, honey, spices, flowers, and firewood. Subscribers can pick up their box of produce at a nearby distribution point or have it delivered to their doorstep.

Some 600 such community-supported agricultural or horticultural operations now exist in 42 of the United States, with 100,000 members.[25] A typical participating farm has about 3 acres (plus grazing land) and serves the needs of 60 to 70 families who each pay about $400.[26] With the subscription fee paid up front, consumers share the risks of failure with the farmer. If consumers were to become co-owners of the land and equipment, effectively shareholders, this model could be seen as a community corporation.

If you can't imagine taking a box of produce chosen and packed by someone else, but still want to support local farmers, you might just try to shop only at community-friendly markets. Retailing food, of course, is another viable business for community corporations, which can set up outdoor farmers' markets or indoor stores, or take over and revitalize failing stores. The Village Retail Services Association (VIRSA) in England has a simple mission: to save small shops endangered by shopping malls and gigantic outlets. It organizes customers to invest in local shops, with the goal of strengthening their buying loyalty and the incentive of advertising by word by mouth. It also works with management to improve the efficiency and quality of operations. Sometimes VIRSA purchases the entire operation and hires new management; and occasionally, where there is very broad local commitment, it will organize the shop to be run entirely by volunteers. In VIRSA's first year it assisted 12 British communities.[27]

How far urban agriculture, CSAs, farmers' markets, and village stores can go in feeding a community is unclear. Even if these innovations can deliver fruits and vegetables cost-effectively, can they also provide grains, meat, and other kinds of food? The answer may only be positive if more and more value-added activity is undertaken by farmers themselves. One recent study found intriguing evidence that small-scale farmers do well supplementing their incomes with homestead chicken production, small dairy herds, and on-farm food processing.[28] Certainly if growing numbers of consumers continue to prefer food grown, raised, processed, packaged, and sold by local farmers, the economics of community production will steadily improve.

NOTES

1 Roberts, J.M. 1976. *History of the World.* London: Penguin, p. 45.
2 Imhoff, Daniel. 1996. Community Supported Agriculture. In Jerry Mander and Edward Goldsmith (eds). *The Case Against the Global Economy.* San Francisco: Sierra Club Books, p. 425.
3 Romm, Joseph J. and Curtis, Charles B. 1996. Mideast Oil Forever? *Atlantic Monthly*, April, p. 57.
4 Ibid., p. 60.
5 Lovins, Amory B. and Hunter, L. 1982. *Brittle Power: Energy Strategy for National Security.* Andover, MA: Brick House Publishers, p. 122; and Lovins, Amory B. and Hunter, L. 1983. The Fragility of Domestic Energy. *Atlantic Monthly*, November, pp. 118–126.
6 Lovins and Hunter, *Brittle Power*, pp. 51–67; Bolden, Tim. 1996. Blackout May Be Caution Sign on Road to Utility Deregulation. *New York Times*, 19 August, p. A14.
7 Persky, Joseph, Ranney, David, and Wievel, Wim. 1993. Import Substitution and Local Economic Development. *Economic Development Quarterly*, February, p. 18.
8 Jacobs, Jane. 1969. *The Economy of Cities.* New York: Vintage, pp. 145–179.
9 Ibid., p. 157.
10 Persky *et al.*, Import Substitution, p. 19.
11 Power, Thomas Michael. 1996. *Environmental Protection and Economic Well-Being.* Armonk, NY: M.E. Sharpe, pp. 155–180.
12 Ibid., p. 155.
13 Ibid., p. 194.
14 Persky *et al.* Import Substitution, p. 19.
15 Douthwaite, Richard. 1996. *Short Circuit.* Devon, UK: Resurgence, pp. 228–229.
16 Alperovitz, Gar and Faux, Jeff. 1984. *Rebuilding America.* New York: Pantheon, p. 196.
17 Lehman, Karen and Krebs, Al. 1996. In Mander and Goldsmith, *The Case Against the Global Economy*, p. 127.

18 Smith, Stewart. 1993. Sustainable Agriculture and Public Policy. *Maine Policy Review*, April, pp. 69–70.

19 Smith, Stewart. 1994. Farming Activities and Family Farms: Getting the Concepts Right. In *Symposium: Agricultural Industrialization and Family Farms: The Role of Federal Policy*. Hearing before the Joint Economic Committee of the United States, 21 October 1992, pp. 57–929. Washington, DC: USGPO, pp. 117–133.

20 Imhoff, supra, p. 429.

21 United Nations Development Programme. 1996. *Urban Agriculture: Food, Jobs and Sustainable Cities*. New York: UNDP.

22 Ibid., pp. 26–27.

23 Ibid., p. 25.

24 Ibid., p. 26.

25 Keen, Elizabeth. 1997. Researcher for CSA of North America. Personal communication, 21 August.

26 Imhoff, supra, pp. 429–430.

27 Douthwaite, supra, pp. 315–318.

28 Integrity Systems Cooperative Co. 1997. *Adding Value to Our Food System: An Economic Analysis of Sustainable Community Food Systems* (monograph). Everson, WA: Integrity Systems Cooperative Co., February.

GREEN ARCHITECTURE
AND BUILDING

"Design, Ecology, Ethics and the Making of Things"

a sermon given at the Cathedral of St. John the Divine, New York City (1993)

William McDonough

Editors' Introduction

Solar energy technologies and passive solar heating of buildings (using the sun's energy to warm interior spaces) were hallmarks of the first wave of modern ecological architecture in the 1960s and 1970s, along with efforts to improve the energy efficiency of new construction. Solar strategies remained on the fringe of new construction, but energy conservation requirements were soon written into many building codes, in part in response to the energy crises of the 1970s. Some pioneering architects also began experimenting with natural or recycled building materials, such as earthen or straw bale walls and remilled lumber, and employed new designs intended to emulate those found in nature.

By the 1990s and early 2000s a much wider variety of sustainable design practices and building materials had begun to enter mainstream construction, encouraged by the emergence of green building standards such as the LEED (Leadership for Energy and Environmental Design) standards first codified by the US Green Building Council in 1998. Even some relatively mainstream buildings began to contain green features. Along the way, many architects rediscovered the vernacular building practices that cultures have used historically to adapt their buildings to local climates, materials, and traditions. Designers also explored ancient systems such as the Chinese *feng shui*, an elaborate set of beliefs about the flow of energy within landscapes and buildings. While green building practices are still far from the mainstream of development, their use is gradually spreading, and there is an increasing number of examples of built projects.

Architect William McDonough has been a leader in the movement to rethink architecture and design in ways that can enhance urban sustainability. Former Dean of the School of Architecture at the University of Virginia, McDonough is the principal of William McDonough & Partners in Charlottesville, Virginia. Among his better-known projects are the headquarters of Environmental Defense in New York City, the Ford Motor Company's renovated River Rouge plant in Dearborn, Michigan, and a green Wal Mart store in Lawrence, Kansas. McDonough was also the author of the Hannover Principles, an influential set of ecological design principles that circulated widely in the early 1990s. His writings include the book *Cradle to Cradle: Remaking the Way We Make Things* (with Michael Braungart; New York: North Point Press, 2002).

Other writings in the field of ecological design include Van der Ryn and Cowan's *Ecological Design* (Washington, DC: Island Press, 1996), Lyle's *Regenerative Design for Sustainable Development* (New York: Wiley, 1994), the Todds' *From Eco-Cities to Living Machines* (Berkeley, CA: North Atlantic Books, 1993), British architects Brenda and Robert Vale's *Green Architecture* (Boston, MA: Little, Brown, 1991) and

The New Autonomous House (London: Thames & Hudson, 2000), and Oberlin professor David Orr's *Ecological Literacy* (Albany: State University of New York Press, 1992), *Earth in Mind* (Washington, DC: Island Press, 1994), and *The Nature of Design* (New York: Oxford University Press, 2002).

■ ■ ■ ■ ■ ■

It is humbling to be an architect in a cathedral because it is a magnificent representation of humankind's highest aspirations. Its dimension is illustrated by the small Christ figure in the western rose window, which is, in fact, human scale. A cathedral is a representation of both our longings and intentions. This morning, here at this important crossing in this great building, I am going to speak about the concept of design itself as the first signal of human intention and will focus on ecology, ethics, and the making of things. I would like to reconsider both our design and our intentions.

When Vincent Scully gave a eulogy for the great architect Louis Kahn, he described a day when both were crossing Red Square, whereupon Scully excitedly turned to Kahn and said, "Isn't it wonderful the way the domes of St. Basil's Cathedral reach up into the sky?" Kahn looked up and down thoughtfully for a moment and said, "Isn't it beautiful the way they come down to the ground?"

If we understand that design leads to the manifestation of human intention and if what we make with our hands is to be sacred and honor the earth that gives us life, then the things we make must not only rise from the ground but return to it, soil to soil, water to water, so everything that is received from the earth can be freely given back without causing harm to any living system. This is ecology. This is good design. It is of this we must now speak.

If we use the study of architecture to inform this discourse, and we go back in history, we will see that architects are always working with two elements, mass and membrane. We have the walls of Jericho, mass, and we have tents, membranes. Ancient peoples practiced the art and wisdom of building with mass, such as an adobe-walled hut, to anticipate the scope and direction of sunshine. They knew how thick a wall needed to be to transfer the heat of the day into the winter night, and how thick it had to be to transfer the coolness into the interior in the summer. They worked well with what we call "capacity" in the walls in terms of storage and thermal lags. They worked with resistance, straw, in the roof to protect from heat loss in the winter and to shield the heat gain in summer from the high sun. These were very sensible buildings within the climate in which they are located.

With respect to membrane, we only have to look at the Bedouin tent to find a design that accomplishes five things at once. In the desert, temperatures often exceed 120 degrees. There is no shade, no air movement. The black Bedouin tent, when pitched, creates a deep shade that brings one's sensible temperature down to 95 degrees. The tent has a very coarse weave, which creates a beautifully illuminated interior, having a million light fixtures. Because of the coarse weave and the black surface, the air inside rises and is drawn through the membrane. So now you have a breeze coming in from outside, and that drops the sensible temperature even lower, down to 90 degrees. You may wonder what happens when it rains, with those holes in the tent. The fibers swell up and the tent gets tight as a drum when wet. And of course, you can roll it up and take it with you. The modern tent pales by comparison to this astonishingly elegant construct.

Throughout history, you find constant experimentation between mass and membrane. This cathedral is a Gothic experiment intregrating great light into massive membrane. The challenge has always been, in a certain level, how to combine light with mass and air. This experiment displayed itself powerfully in modern architecture, which arrived with the advent of inexpensive glass. It was unfortunate that at the same time the large sheet of glass showed up, the era of cheap energy was ushered in, too. And because of that, architects no longer rely upon the sun for heat or illumination. I have spoken to thousands of architects, and when I ask the question, "How many of you know how to find true south?," I rarely get a raised hand.

Our culture has adopted a design stratagem that essentially says that if brute force or massive amounts of energy don't work, you're not using enough of it. We made glass buildings that are more about buildings than they are about people. We've used the glass ironically. The hope that glass would connect us to the outdoors was completely stultified by making the buildings sealed. We have created stress in people because we are meant to be connected with the outdoors, but instead we are trapped. Indoor air quality issues are now becoming very serious. People are sensing how horrifying it can be to be trapped indoors, especially with the thousands upon thousands of chemicals that are being used to make things today.

Le Corbusier said in the early part of this century that a house is a machine for living in. He glorified the steamship, the airplane, the grain elevator. Think about it: a house is a machine for living in. An office is a machine for working in. A cathedral is a machine for praying in. This has become a terrifying prospect, because what has happened is that designers are now designing for the machine and not for people. People talk about solar heating a building, even about solar heating a cathedral. But it isn't the cathedral that is asking to be heated, it is the people. To solar-heat a cathedral, one should heat people's feet, not the air 120 feet above them. We need to listen to biologist John Todd's idea that we need to work with living machines, not machines for living in. The focus should be on people's needs, and we need clean water, safe materials, and durability. And we need to work from current solar income.

There are certain fundamental laws that are inherent to the natural world that we can use as models and mentors for human designs. Ecology comes from the Greek roots Oikos and Logos, "household" and "logical discourse." Thus, it is appropriate, if not imperative, for architects to discourse about the logic of our earth household. To do so, we must first look at our planet and the very processes by which it manifests life, because therein lie the logical principles with which we must work. And we must also consider economy in the true sense of the word. Using the Greek words Oikos and Nomos, we speak of natural law and how we measure and manage the relationships within this household, working with the principles our discourse has revealed to us.

And how do we measure our work under those laws? Does it make sense to measure it by the paper currency that you have in your wallet? Does it make sense to measure it by a grand summation called

GNP? For if we do, we find that the foundering and rupture of the Exxon Valdez tanker was a prosperous event because so much money was spent in Prince William Sound during the clean-up. What then are we really measuring? If we have not put natural resources on the asset side of the ledger, then where are they? Does a forest really become more valuable when it is cut down? Do we really prosper when wild salmon are completely removed from a river?

There are three defining characteristics that we can learn from natural design. The first characteristic is that everything we have to work with is already here – the stones, the clay, the wood, the water, the air. All materials given to us by nature are constantly returned to the earth, without even the concept of waste as we understand it. Everything is cycled constantly with all waste equaling food for other living systems.

The second characteristic is that one thing allowing nature to continually cycle itself through life is energy, and this energy comes from outside the system in the form of perpetual solar income. Not only does nature operate on "current income," it does not mine or extract energy from the past, it does not use its capital reserves, and it does not borrow from the future. It is an extraordinarily complex and efficient system for creating and cycling nutrients, so economical that modern methods of manufacturing pale in comparison to the elegance of natural systems of production.

Finally, the characteristic that sustains this complex and efficient system of metabolism and creation is biodiversity. What prevents living systems from running down and veering into chaos is a miraculously intricate and symbiotic relationship between millions of organisms, no two of which are alike.

As a designer of buildings, things, and systems, I ask myself how to apply these three characteristics of living systems to my work. How do I employ the concept of waste equals food, of current solar income, of protecting biodiversity in design? Before I can even apply these principles, though, we must understand the role of the designer in human affairs.

In thinking about this, I reflect upon a commentary of Emerson's. In the 1830s, when his wife died, he went to Europe on a sailboat and returned in a steamship. He remarked on the return voyage that he missed the "Aeolian connection." If we abstract this, he went over on a solar-powered recyclable vehicle operated by craftspersons, working in the open air, practicing ancient arts. He returned in a steel rust bucket, spilling oil on the water and smoke into the sky, operated by people in a black dungeon shoveling coal into the mouth of a boiler. Both ships are objects of design. Both are manifestations of our human intention.

Peter Senge, a professor at MIT's Sloan School of Management, works with a program called the Learning Laboratory where he studies and discusses how organizations learn. Within that he has a leadership laboratory, and one of the first questions he asks CEOs of companies that attend is, "Who is the leader on a ship crossing the ocean?" He gets obvious answers, such as the captain, the navigator, or the helmsman. But the answer is none of the above. The leader is the designer of the ship because operations on a ship are a consequence of design, which is the result of human intention. Today, we are still designing steamships, machines powered by fossil fuels that have deleterious effects. We need a new design.

I grew up in the Far East, and when I came to this country, I was taken aback when I realized that we were not people with lives in America, but consumers with lifestyles. I wanted to ask someone: when did America stop having people with lives? On television, we are referred to as consumers, not people. But we are people, with lives, and we must make and design things for people. And if I am a consumer, what can I consume? Shoe polish, food, juice, some toothpaste. But actually, very little that is sold to me can actually be consumed. Sooner or later, almost all of it has to be thrown away. I cannot consume a television set. Or a VCR. Or a car. If I presented you with a television set and covered it up and said, "I have this amazing item. What it will do as a service will astonish you. But before I tell you what it does, let me tell you what it is made of and you can tell me if you want it in your house. It contains 4,060 chemicals, many of which are toxic, two hundred of which off-gas into the room when it is turned on. It also contains 18 grams of toxic methyl mercury, has an explosive glass tube, and I urge you to put it at eye-level with your children and encourage them to play with it." Would you want this in your home?

Michael Braungart, an ecological chemist from Hamburg, Germany, has pointed out that we should remove the word "waste" from our vocabulary and start using the word product instead, because if waste is going to equal food, it must also be a product. Braungart suggests we think about three distinct product types:

First, there are consumables, and actually we should be producing more of them. These are products that when eaten, used, or thrown away, literally turn back into dirt, and therefore are food for other living organisms. Consumables should not be placed in landfills, but put on the ground so that they restore the life, health, and fertility of the soil. This means that shampoos should be in bottles made of beets that are biodegradable in your compost pile. It means carpets that break down into carbon dioxide and water. It means furniture made of lignin, potato peels and technical enzymes that looks just like your manufactured furniture of today except it can be safely returned to the earth. It means that all "consumable" goods should be capable of returning to the soil from whence they came.

Second are products of service, also known as durables, such as cars and television sets. They are called products of service because what we want as customers is the service the product provides – food, entertainment, or transportation. To eliminate the concept of waste, products of service would not be sold, but licensed to the end-user. Customers may use them as long as they wish, even sell the license to someone else, but when the end-user is finished with, say, a television, it goes back to Sony, Zenith, or Philips. It is "food" for their system, but not for natural systems. Right now, you can go down the street, dump a TV into the garbage can, and walk away. In the process, we deposit persistent toxins throughout the planet. Why do we give people that responsibility and stress? Products of service must continue beyond their initial product life, be owned by their manufacturers, and be designed for disassembly, remanufacture, and continuous re-use.

The third type of product is called "unmarketables." The question is, why would anyone produce a product that no one would buy? Welcome to the world of nuclear waste, dioxins, and chromium-tanned leather. We are essentially making products or subcomponents of products that no one should buy, or, in many cases, do not realize they are buying. These products must not only cease to be sold, but those already sold should be stored in warehouses when they are finished until we can figure out a safe and non-toxic way to dispose of them.

I will describe a few projects and how these issues are implicit in design directions. I remember when we were hired to design the office for an environmental group. The director said at the end of contract negotiations, "By the way, if anybody in our office gets sick from indoor air quality, we're going to sue you." After wondering if we should even take the job, we decided to go ahead, that it was our job to find the materials that wouldn't make people sick when placed inside a building. And what we found is that those materials weren't there. We had to work with manufacturers to find out what was in their products, and we discovered that the entire system of building construction is essentially toxic. We are still working on the materials side.

For a New York men's clothing store, we arranged for the planting of 1,000 oak trees to replace the two English oaks used to panel the store. We were inspired by a famous story told by Gregory Bateson about New College in Oxford, England. It went something like this. They had a main hall built in the early 1600s with beams forty feet long and two feet thick. A committee was formed to try to find replacement trees because the beams were suffering from dry rot. If you keep in mind that a veneer from an English oak can be worth seven dollars a square foot, the total replacement cost for the oaks was prohibitively expensive. And they didn't have straight forty foot English oaks from mature forests with which to replace the beams. A young faculty member joined the committee and said, "Why don't we ask the College Forester if some of the lands that have been given to Oxford might have enough trees to call upon?" And when they brought in the forester he said, "We've been wondering when you would ask this question. When the present building was constructed 350 years ago, the architects specified that a grove of trees be planted and maintained to replace the beams in the ceiling when they would suffer from dry rot." Bateson's remark was, "That's the way to run a culture." Our question and hope is, "Did they replant them?"

For Warsaw, Poland, we responded to a design competition for a high-rise building. When the client chose our design as the winner after seeing the model, we said, "We're not finished yet. We have to tell you about the building. The base is made from concrete and includes tiny bits of rubble from World War II. It looks like limestone, but the rubble's there for visceral reasons." And he said, "I understand, a phoenix rising." And we said the skin is recycled aluminum, and he said, "That's O.K., that's fine." And we said, "The floor heights are thirteen feet clear so that we can convert the building into housing in the future, when its utility as an office building is no longer. In this way, the building is given a chance to have a long, useful life." And he said, "That's O.K." And we told him that we would have opening windows and that no one would be further than twenty-five feet from a window, and he said that was O.K., too. And finally, we said, "By the way, you have to plant ten square miles of forest to offset the building's effect on climate change." We had calculated the energy costs to build the structure, and the energy cost to run and maintain it, and it worked out that 6,400 acres of new forest would be needed to offset the effects on climate change from the energy requirements. And he said he would get back to us. He called back two days later and said, "You still win. I checked out what it would cost to plant ten square miles of trees in Poland and it turns out it's equivalent to a small part of our advertising budget."

The architects representing a major retail chain called us a year ago and said, "Will you help us build a store in Lawrence, Kansas?" I said that I didn't know if we could work with them. I explained my thoughts on consumers with lifestyles, and we needed to be in the position to discuss their stores' impact on small towns. Click. Three days later we were called back and were told, "We have a question for you that is coming from the top. Are you willing to discuss the fact that people with lives have the right to buy the finest-quality products, even under your own terms, at the lowest possible price?" We said, "Yes." "Then we can talk about the impact on small towns."

We worked with them on the store in Kansas. We converted the building from steel construction, which uses 300,000 BTUs per square foot, to wood construction, which uses 40,000 BTUs, thereby saving thousands of gallons of oil just in the fabrication of the building. We used only wood that came from resources that were protecting biodiversity. In our research we found that the forests of James Madison and Zachary Taylor in Virginia had been put into sustainable forestry and the wood for the beams came from there and other forests managed this way. We also arranged for no CFCs to be used in the store's construction and systems, and initiated significant research and a major new industry in daylighting. We have yet to fulfill our concerns about the bigger questions of products, their distribution and the chain's impact on small towns, with the exception that this store is designed to be converted into housing when its utility as a retail outlet has expired.

For the City of Frankfurt, we are designing a day-care center that can be operated by the children. It contains a greenhouse roof that has multiple functions: it illuminates, heats both air and water, cools, ventilates, and shelters from the rain, just like a Bedouin tent. One problem we were having during the design process was the engineers wanted to completely automate the building, like a machine. The engineers asked, "What happens if the children forget to close the shade and they get too hot?" We told them the children would open a window. "What if they don't open a window?", the engineers wanted to know. And we told them that in that case the children would probably close the shade. And they wanted to know what would happen if the children didn't close the shade. And finally we told them the children would open windows and close shades when they were hot because children are not dead but alive. Recognizing the importance for children to look at the day in the morning and see what the sun is going to do that day and interact with it, we enlisted the help of teachers of Frankfurt to get this one across because the teachers had told us the most important thing was to find something for the children to do. Now the children have ten minutes of activity in the morning and ten minutes of activity when they leave the building, opening and closing the system, and both the children and teachers love the idea. Because of the solar hot-water collectors, we asked that a public laundry be added to the program so that parents could wash clothes while awaiting their children in school. Because of advances in glazing, we are able to create a day-care center that requires no fossil fuels for operating the heating or cooling. Fifty years from now, when fossil fuels will be scarce, there will be hot water for the community, a social center, and the building will have paid back the energy "borrowed" for its construction.

As we become aware of the ethical implications of design, not only with respect to buildings, but in every aspect of human endeavor, they reflect changes in the historical concept of who or what has rights. When you study the history of rights, you begin with the Magna Carta, which was about the rights of white, English, noble males. With the Declaration of Independence, rights were expanded to all landowning white males. Nearly a century later, we moved to the emancipation of slaves, and during the beginnings of this century, to suffrage, giving the right to women to vote. Then the pace picks up with the Civil Rights Act in 1964, and then in 1973, the Endangered Species Act. For the first time, the right of other species and organisms to exist was recognized. We have essentially "declared" that Homo Sapiens are part of the web of life. Thus, if Thomas Jefferson were with us today, he would be calling for a Declaration of Interdependence which recognizes that our ability to pursue wealth, health, and happiness is dependent on other forms of life, that the rights of one species are linked to the rights of others and none should suffer remote tyranny.

This Declaration of Interdependence comes hard on the heels of realizing that the world has become vastly complex, both in its workings and in our ability to perceive and comprehend those complexities. In this complicated world, prior modes of domination have essentially lost their ability to maintain control. The sovereign, whether in the form of a king or nation, no longer seems to reign. Nations have lost control of money to global, computerized trading systems. The sovereign is also losing the ability to deceive and manipulate, as in the case of Chernobyl. While the erstwhile Soviet Republic told the world that Chernobyl was nothing to be concerned about, satellites with ten-meter resolution showed the world that it was something to worry about. And what we saw at the Earth Summit was that the sovereign has lost the ability to lead even on the most elementary level. When Maurice Strong, the chair of the United Nations Conference on the Environment and Development, was asked how many leaders were at the Earth Summit, he said there were over 100 heads of state. Unfortunately, we didn't have any leaders.

When Emerson came back from Europe, he wrote essays for Harvard on Nature. He was trying to understand that if human beings make things and human beings are natural, then are all the things human beings make natural? He determined that Nature was all those things which were immutable. The oceans, the mountains, the sky. Well, we now know that they are mutable. We were operating as if Nature is the Great Mother who never has any problems, is always there for her children, and requires no love in return. When you think about Genesis and the concept of dominion over natural things, we realize that even if we want to get into a discussion of stewardship versus dominion, in the end, the question is, if you have dominion, and perhaps we do have dominion, isn't it implicit that we have stewardship too, because how can you have dominion over something you've killed?

We must face the fact that what we are seeing across the world today is war, a war against life itself. Our present systems of design have created a world that grows far beyond the capacity of the environment to sustain life into the future. The industrial idiom of design, failing to honor the principles of nature, can only violate them, producing waste and harm, regardless of purported intention. If we destroy more forests, burn more garbage, drift-net more fish, burn more coal, bleach more paper, destroy more topsoil, poison more insects, build over more habitats, dam more rivers, produce more toxic and radioactive waste, we are creating a vast industrial machine, not for living in, but for dying in. It is a war, to be sure, a war that only a few more generations can surely survive.

When I was in Jordan, I worked for King Hussein on the master plan for the Jordan Valley. I was walking through a village that had been flattened by tanks and I saw a child's skeleton squashed into the adobe block and was horrified. My Arab host turned to me and said, "Don't you know what war is?" And I said, "I guess I don't." And he said, "War is when they kill your children." So I believe we're at war. But we must stop. To do this, we have to stop designing everyday things for killing, and we have to stop designing killing machines.

We have to recognize that every event and manifestation of nature is "design," that to live within the laws of nature means to express our human intention as an interdependent species, aware and grateful that we are at the mercy of sacred forces larger than ourselves, and that we obey these laws in order to honor the sacred in each other and in all things. We must come to peace with and accept our place in the natural world.

"Principles of Green Architecture"

from *Green Architecture* (1991)

Brenda and Robert Vale

Editors' Introduction

Many sets of principles for an ecological or sustainable architecture have been proposed. Among these are McDonough's Hannover Principles, Van der Ryn and Cowan's Ecological Design Principles, and Lyle's Regenerative Design Principles. Most such manifestos set forth similar themes in slightly different ways – in particular encouraging designers to conserve energy and nonrenewable resources, to take account of local place characteristics, and to work with building users and surrounding communities.

British architects Brenda and Robert Vale here present one of the simplest and most straightforward frameworks for green architecture, contained in their book *Green Architecture: Design for an Energy-Conscious Future* (Boston: Little, Brown, 1991). They illustrate these principles with extensive examples of building designs from Europe, the United Kingdom, and the United States. Like McDonough, the Vales emphasize learning from the vernacular architecture that already embodies generations of people's experience in living with a particular place and climate.

More detailed information on green building principles and implementation is available from many sources, including the American Institute of Architects' Committee on the Environment (www.aia.org/cote), the US Green Building Council (www.usgbc.org), and in the United Kingdom and Europe, Sustainable Homes (www. sustainablehomes.co.uk), the Ecological Design Association (www.edaweb.org), and the European Housing Ecology Network (www.eheneurope.net/). Interested readers may also want to see David Pearson's writings such as *The Natural House Book* (New York: Simon & Schuster, 1989), *The New Natural House Book* (New York: Simon & Schuster, 1998), and *New Organic Architecture* (Berkeley, CA: University of California Press, 2001).

The "green" approach to architecture is not a new approach. It has existed since people first selected a south-facing cave rather than one facing north to achieve comfort in a temperate climate. What is new is the realization that a green approach to the built environment involves a holistic approach to the design of buildings; that all the resources that go into a building, be they materials, fuels or the contribution of the users, need to be considered if a sustainable architecture is to be produced. Many buildings embody at least one of the various identifiable green characteristics. Few as yet embrace the holistic approach. . . .

Because the "green" argument is to suggest that issues are interdependent, and that the ramifications of any decision need to be considered, the notion of separate principles runs counter to a green approach. Inevitably principles will overlap.

Nevertheless, the headings that follow are offered as a series of differing emphases among which a balance needs to he found for a green architecture to emerge.

PRINCIPLE 1: CONSERVING ENERGY

A building should be constructed so as to minimize the need for fossil fuels to run it.

Past societies accepted the necessity of this principle without question. It is only with the recent proliferation of materials and technologies that such a basis for ordinary building has been lost. Whether by the use of materials or the disposition of building elements, buildings modified climate to suit the needs of the users. Moreover, the very idea of community comes from the sheltering of people together, whether to provide maximum areas of shade and cooler air between buildings or to reduce the external surface area of the community as it faced the hostile weather. People constructed their buildings together because of the mutual benefit to be obtained. A policy of cheap energy removed this generator of traditional community as surely as did the automobile.

Recent buildings that have attempted to reduce dependence on fossil fuels have tended to stand alone as separate experiments rather than cluster in patterns that respond to local climate. Consequently, such experiments must be viewed as half-way attempts towards the creation of a green architecture. Many such experiments have also come from the committed individual rather than from the community as a whole, thereby further separating out the single achievement.

[. . .]

PRINCIPLE 2: WORKING WITH CLIMATE

Buildings should be designed to work with climate and natural energy sources. . . .

Building form and the disposition of building elements can alter internal comfort conditions, rather than demonstrating the reduction of fossil fuel demands through the use of insulation in the building fabric. Inevitably there will be some overlap in the two approaches.

In the days before the widespread exploitation of fossil fuels the main source of energy was wood. Firewood still provides about 15 per cent of the world's energy today. As wood became scarce, it seemed natural to many civilizations to make use of the heat of the sun to help reduce the need for wood to provide heat. The ancient Greeks were well aware of the benefits of solar design, and commonly arranged their houses to collect the rays of the winter sun. Greek cities such as Priene, following its relocation to avoid flooding, were laid out on a grid plan with streets running east–west to allow a southerly orientation of the buildings.[1]

The Romans continued the solar design principles that they learned from their contacts with Greek examples, but they were able to make use of window glazing, developed in the first century AD, to increase the heat that could be gained. Growing shortage of wood for fuel made the use of south-facing glazing popular for the villas of the wealthy, and for the public baths.[2]

The tradition of designing with climate to achieve comfort in buildings is not confined to the provision of warmth. In many climates the problem that faces the architect is to cool spaces in order to achieve comfortable conditions. The conventional modern solution, the provision of air conditioning systems, is no more than a crude process of opposing climate with energy, which was foolish when energy was cheap and pollution not considered, but is now verging on the insane.

[. . .]

PRINCIPLE 3: MINIMIZING NEW RESOURCES

A building should be designed so as to minimize the use of new resources and, at the end of its useful life, to form the resources for other architecture.

Although this principle, like the others discussed, is directed towards new buildings, it acknowledges that immense resources are already a part of the existing built environment, and that the rehabilitation and upgrading of the existing building stock for minimal environmental impact is as important, if not more so, than the creation of a new green architecture. There are not sufficient resources in the world for the built environment to be reconstructed anew for each generation. Nor would it be right that it should be so, unless the layers of history that each generation applies to the built environment are to be ignored. So much of what is admired about buildings comes from associations, and an objective assessment of a building with no prior knowledge of its purpose or ownership is virtually impossible. Even a demolished building leaves traces, such as an entry point into the site, that need to be respected by anything new put in its place.

Re-use can take the form of recycling materials or recycling spaces. The recycling of both buildings and building components is part of the history of architecture. St Albans Abbey, for example, which was rebuilt between 1077 and 1115 on the site of a Benedictine Abbey used Roman bricks taken from the ruins of Verulanium at the bottom of the hill to reinforce the walls of flint, the only building stone available in Hertfordshire. The prefabrication methods of later medieval timber frame buildings whereby the pieces were cut and fitted together in the carpenters' yard, marked, disassembled and moved to site, meant that portions of medieval buildings could be moved if required, and even now often turn up in unexpected places. Sometimes whole structures have been moved to find a new purpose. For example, upon construction of the Victoria and Albert Museum in London, the Brompton Boilers building was no longer required, and in 1865 the iron building was offered to the authorities of North, South and East London with the intention that it should be re-erected as a local museum. East London accepted it, and the building opened as the Bethnal Green Museum in 1872. It is today the Museum of Childhood.

Often those whose access to resources is least have demonstrated the way in which structures designed for one purpose can be adapted to suit a different need. However, the alterations necessary can often more or less obliterate the original form of the structure or building. This produces a dilemma for those concerned with the conservation of buildings that has to be acknowledged. Should a building be preserved unchanged because it was once important, or should it, because it can still be made useful, be conserved in a changed state? A green approach to the problem might suggest that the question be decided on resources alone. If the

resources required to alter a building are less than for demolition and rebuilding then the former course is adopted. This, however, fails to acknowledge the historical importance of the structure, which might suggest that other values need to be considered. This dimension to the problem of changing existing buildings to make them meet present needs, especially in terms of upgrading buildings to improve thermal performance which may alter appearance, can be summarized as the paradox of Venice. If global warming produces even a small alteration in sea levels the future of Venice is again imperiled. . . . To preserve Venice, it may be necessary to conserve buildings in an altered state. It is important that these issues now form part of the debate of building conservation.

Some extraordinary schemes have been created through the re-use of large redundant buildings even without thought to thermal performance, such as the Gare d'Orsay in Paris, which is now the Museum of the Nineteenth Century. The difficulty of inserting a series of spaces to illustrate nineteenth-century paintings and sculpture in a vast volume that was intended to accommodate the then-new electric trains poses architectural problems which are, perhaps, not necessarily happily resolved in this instance. However, the benefits accruing from re-using a large piece of the urban fabric that has a presence in the city can override the internal considerations. The refurbishment of existing housing areas in cities and towns can also offer a considerable saving in resources over demolition and rebuilding, and avoid disruption of the community.

[. . .]

PRINCIPLE 4: RESPECT FOR USERS

A green architecture recognizes the importance of all the people involved with it.

This principle may appear to have little relevance to issues of pollution, global warming and the destruction of the ozone layer, but a green approach to architecture that includes respect for all the resources that contribute to making a building will not exclude human beings. All buildings are made by hand, but in some architectures this fact is acknowledged and appreciated, whereas others have attempted to deny the human dimension of the building process. Only in Japan have robots been developed to take over some of the human roles on building projects, but for a robot to work effectively the project-scale must be such that the same task, or a limited series of related tasks, can be repeated many times. At a very different scale, the small builder still has to rely on personal skill for a number of unrelated tasks, drawing on expert subcontractors where there is no alternative.

A greater respect for human needs and labor can be evidenced in two separate ways. For the professional builder, it is essential that the materials and processes that form the building are as little polluting and dangerous to the individual worker or user as they are to the planet. Architects have begun to realize the extent of the global or human poisons that may be found on building sites, and that it is no longer feasible to use insulating materials that contain CFCs, or to use methods of timber treatment that are carcinogenic. Alternative methods of detailing to protect timber physically become preferable to the chemical approach. Perhaps were designers to regard timber as living wood, as in the vernacular tradition, rather than as some squared and dimensionally stable material, the temptation to cover everything with chemicals might be lessened.

[. . .]

The other form of human involvement that needs to be considered is the positive involvement of the users in the design and construction process. If this energy remains uninvolved, a resource is being wasted which could inform the finished product and extend its usefulness. A range of buildings, some built by the owners, some involving large groups of people, have made use of this resource and the result has been a high level of satisfaction with the buildings created.

PRINCIPLE 5: RESPECT FOR SITE

A building will "touch-this-earth-lightly."

The Australian architect Glenn Murcutt quotes an Aboriginal saying, that "One must touch-this-earth-lightly."[3] This saying embodies an attitude to the interaction of a building and its site that is essential to a green approach, but it also implies wider concerns. A building that guzzles energy, creates pollution, and alienates its users does not "touch-this-earth-lightly."

The most direct interpretation of the phrase to "touch-this-earth-lightly" would be the idea that a building could be removed from its site and leave it in the condition it was before the building was placed there. This relationship to site is seen in the traditional dwellings of nomads, but their lightness of touch is not just a matter of moving their homes, it also concerns the materials with which they build, and the possessions they carry with them.[4] The black tent of the Bedouin is woven from the hair of their goats, sheep, and camels. When erected, the tent cloth adopts a low, aerodynamically efficient profile to avoid damage by high winds; it is kept in place by long ropes, also woven from hair, and supported by a very few wooden poles, because wood is a scarce resource in the desert.

The Netsilik Inuit people of northern Canada carry their tents in the summer when they need to follow the game that are their food, but in the winter the skins that form the tent cover are dipped in water and wrapped round frozen fish. The long bundles freeze solid, and are joined in pairs with caribou bones. A mixture of moss and slush is rubbed in and allowed to freeze smooth to turn the tent skins and frozen fish into a sledge to carry the Inuit and their possessions over the snow. The parts of a nomadic structure must serve several functions, because only a minimum of possessions can be carried from place to place. Over generations the items necessary for survival, comfort, and the continuation of culture have been determined.

While societies have abandoned the nomadic life for one in fixed dwellings and architecture has come into being, there is still a continuing demand for temporary structures for exhibitions, performances, and other cultural manifestations. These structures frequently take the form of tents; however, an interesting example using very different technology is the sculpture pavilion designed by the Dutch architects Benthem Crouwel for the Sonsbeek '86 festival.[5] This building was designed to protect fragile works of sculpture placed outside, and so the structure was intended to be almost invisible. It used only four materials: precast concrete for the footings, laminated glass for the walls and roof, steel for trusses and connections, and silicone mastic to stick the panes of glass together. Fins of glass glued to the glass walls gave additional rigidity, and provided a place of attachment for the light steel trusses that carried the flat glass roof. The floor was the earth, merely covered with wood shavings to prevent it from becoming muddy. At the end of the event the building was unbolted and removed, the foundations were lifted and the soil replaced, the shavings were raked up, and the site was completely unaltered by the events that had taken place on it. The building could be taken away to be used elsewhere for another exhibition, or recycled into another structure.

PRINCIPLE 6: HOLISM

All the green principles need to be embodied in a holistic approach to the built environment.

It is not easy to find buildings that embody all the principles of green architecture, for a green architecture is yet to be realized. . . . A green architecture involves more than the individual building on its plot; it must encompass a sustainable form of urban environment. The city is far more than a collection of buildings; rather it can be seen as a set of interacting systems – systems for living, working, and playing – crystallized into built forms. It is by looking at systems that we can find the face of the city of tomorrow.

NOTES

1 K. Butti and J. Perlin, *A Golden Thread*, Cheshire Books, Palo Alto, CA, 1980.
2 Ibid.
3 G.L. Murcutt in Foreword to P. Drew, *Leaves of Iron: Glenn Murcutt: Pioneer of an Australian Architectural Form*, The Law Book Company Ltd., North Ryde, New South Wales, 1985.
4 T. Faegre, *Tents: Architecture of the Nomads*, Anchor Press/Doubleday, Garden City, NY, 1979.
5 P. Buchanan, "Barely There," *Architectural Review*, No. 1087, 1987.

"Sustainability and Building Codes"

from *Environmental Building News,*
10(9) 1, 8–15 (2001)

David Eisenberg and Peter Yost

Editors' Introduction

Visionary ecological design principles are often confronted by the cold, hard reality of building codes and zoning regulations, which forbid many innovative design practices and materials. Most of these regulatory mechanisms didn't exist a century ago, but have become ever more complex and demanding in recent decades. They help ensure human health, safety, and welfare, to be sure, but also mandate particular modes of building that are now seen as unsustainable. For example, graywater systems (collecting sink or shower water for reuse in toilets or irrigation) and alternative building materials such as straw bale or rammed earth have been prohibited by codes in many locations until recently. Codes also often set minimum room sizes and require unnecessarily expensive construction materials and practices. Meanwhile, zoning regulations frequently require large amounts of parking, large lot sizes, substantial building setbacks from lot lines, and low building heights. All of these requirements constrain what ecological designers can do.

The following piece from the journal *Environmental Building News* (www.buildinggreen.com) addresses building codes specifically, tracing their development and some ways they might be changed to better promote sustainability. David Eisenberg is director of the Development Center for Appropriate Technology (DCAT) in Tucson, Arizona, and a member of the International Conference of Building Officials. His work has ranged from the steel and glass cover for Biosphere 2 to adobe, rammed-earth, and straw-bale structures. Co-author of *The Straw Bale House* (White River Junction, VT: Chelsea Green, 1994; with Athena Swentzell Steen, Bill Steen, and David Bainbridge), he helped write the first load-bearing straw-bale construction code for the City of Tucson and the County of Pima, Arizona, and led DCAT in a collaborative effort called "Building Sustainability into the Codes." Peter Yost is a former editor of *Environmental Building News* and a research associate with the Building Science Corporation, a Boston-based architecture and building science consulting firm. Further information on this subject is available from the Developmental Center for Appropriate Technology in Tuscon, Arizona, at www.dcat.net.

Shallow frost-protected foundation, straw-bale walls, composting toilet, graywater system, rainwater harvesting. . . . An impressive array of green building features! From the foundation to the roof, these are exemplary systems and materials. But there is another commonality to these features: each represents a potential – if not likely – regulatory challenge. It can be frustrating to have the knowledge

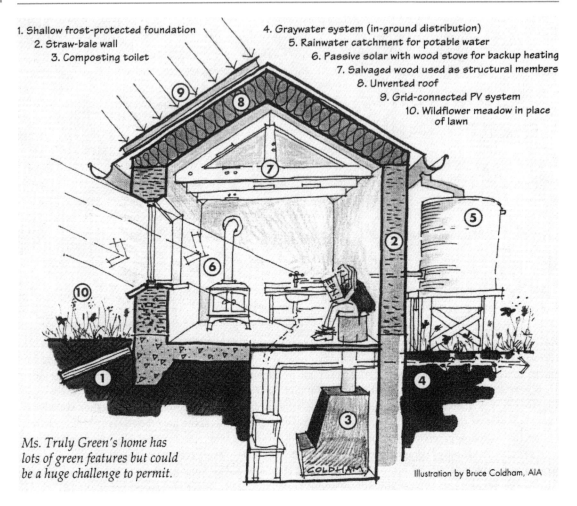

1. Shallow frost-protected foundation
2. Straw-bale wall
3. Composting toilet
4. Graywater system (in-ground distribution)
5. Rainwater catchment for potable water
6. Passive solar with wood stove for backup heating
7. Salvaged wood used as structural members
8. Unvented roof
9. Grid-connected PV system
10. Wildflower meadow in place of lawn

Ms. Truly Green's home has lots of green features but could be a huge challenge to permit.

Illustration by Bruce Coldham, AIA

Figure 1. Code obstacles to green design.

and skills required for building green, yet lack the approvals to do it.

This article takes an in-depth look at the inherent but largely unrecognized relationship between sustainability and building codes, and efforts under way to change this relationship. It also presents a process for professionals to use in gaining approvals for alternative designs, systems, and materials within the existing regulatory framework. A sampling of code success stories demonstrates what is possible when this process is employed.

Though it is beyond the scope of this article, the issue of regulatory hurdles with green building is not restricted to buildings and building codes; a new approach is needed as well for the larger issues of land development, zoning, and planning.

A BRIEF HISTORY OF BUILDING CODES

Building codes have long been used by societies to protect individual and general welfare, and to hold practitioners accountable for their work. As long ago as 1750 BC, Hammurabi, the Babylonian king of Mesopotamia, created his famous Code of Laws covering a wide range of public and private matters. Number 229 of this Code states: "If a builder build a house for someone, and does not construct it properly, and the house which he built fall in and kill its owner, then that builder shall be put to death." This type of "performance" code must certainly have had an impact on quality of construction, but it very likely stifled innovation!

There were many intermediate steps on the way to our present codes. In 1189 AD, the city of London adopted regulations for the construction of common walls, rights to light access, drainage, and safe egress in case of fire. Historically, fire has been the most common concern driving interest in building regulations. Early in the Colonial period of the United States, concern about fire resulted in a ban on wood chimneys and thatch roofs. In 1860 the City of New York appointed a Superintendent of Building and provided staff for code enforcement. In 1867, the Tenement House Act was enacted to regulate conditions in existing buildings, covering such things as fire escapes, ventilation, water supply, toilets, and stair railings. In 1905, the National Board of Fire Underwriters, an insurance industry group, wrote the first National Building Code.

This code led to the formation of organizations for building code officials and the next stage of code development in the United States. By 1940, three model code organizations were established: the Building Officials and Code Administrators International, Inc. (BOCA) in the northeastern US, which produced the National Building Code; the International Conference of Building Officials (ICBO) covering the western half of the United States, which produced the Uniform Building Code; and the Southern Building Code Congress International (SBCCI) in the southeastern United States, which published the Standard Building Code. Reflecting regional differences and different code philosophies, the three model codes also embodied variations that have made code compliance difficult for designers, builders, and manufacturers working across different code-enforcement areas.

Efforts to harmonize the three codes, initially through the Council of American Building Officials (CABO) and more recently by its successor, the International Code Council (ICC), have now resulted in the creation of a single national building code – or family of codes. The ICC codes (including the International Building Code, International Residential Code, and "International" versions of the Mechanical, Plumbing, Fire, and Energy Conservation codes) are replacing the BOCA, SBCCI, ICBO, and CABO codes, which are no longer being maintained. Instead, these groups now support and maintain the ICC codes, the first full edition of which was published in 2000. (Recently, the NFPA

dealt a blow to this consolidation effort when it split from the ICC process and began developing its own building code to compete with the International family of codes.)

An important new development in the ICC process is creation of the International Performance Code (IPC). This code differs from the other International codes in that it is based on stating what must be accomplished, rather than describing in detail what must be done and how to do it. While the more typical *prescriptive* approach is straightforward and relatively easy to implement for both builder and code official (because everyone knows what must be done), it can also be confining and thus limit innovation.

Though new to the US, the experience of other countries using performance codes has shown that they are viable. The greater flexibility provided by performance requirements is both liberating and problematic. The added freedom comes at a price because the performance approach requires that the proposed designs, materials, or methods be supported by calculation, test results, or other demonstrations of adequate performance. That often means more engineering services, testing, and time – both for designer and plan reviewer. It adds a burden for the building department because building officials must be able to analyze the project rather than just making sure it conforms to common practices with which they are familiar.

BUILDING CODES IN ACTION

One might assume that the creation of a single family of codes would bring about complete consolidation of building codes across the United States, but for several reasons this is not the case. First, unlike in many countries where code adoption takes place at the national level, in the United States it occurs at the local, county, or state level. Codes derive their legal authority from their enactment as laws, ordinances, or statutes. While it appears likely that most US jurisdictions will eventually employ the ICC system, in most cases each jurisdiction makes its own determination of which codes and which versions of those codes it will adopt. Some jurisdictions are still without any building codes.

Complicating the matter further, nearly all jurisdictions reserve – and often exercise – the right to add to or amend the codes they adopt. Local amendments may be in response to conditions such as high winds, wildfires, or earthquakes, and additions often include appendix chapters for traditional or regional building approaches – for example, adobe and rammed-earth in the southwestern United States.

At the other end of the spectrum, state or federal government can, as public policy, pass legislation or develop programs that either directly or indirectly supersede local codes. Two examples are the low-flow toilet requirements included in the 1992 Energy Policy Act[1] and the recent code requirement by the city of Frisco, Texas that all new homes be EPA ENERGY STAR-compliant.[2]

Just as important as the process by which codes are adopted is the process by which building codes are developed, changed, and enforced. Few people are aware that the building code development and code change processes are open to the public. Anyone – a business, interest group, or individual – can propose changes to the codes. On an annual basis, all filed proposals go through the same process – committee review, scheduled hearing, and voting. This process results in many changes to codes every year. Typically, supplements are published annually and then consolidated into a new edition of the code every three years.

At the other end of the process are local building officials who have the authority, granted by provisions in the codes, to approve alternative designs, materials, and methods of construction as long as they are deemed adequate to meet the intent of the building code. All codes have such provisions for dealing with building practices, materials, and systems not specifically addressed in the code. Understanding how to use this process can be of enormous benefit when proposing alternatives to standard practice.

THE CASE FOR INTEGRATING SUSTAINABILITY INTO THE CODES

A key to shifting the building regulatory system towards greater acceptance of more sustainable, alternative approaches is to create a context in which those alternatives can be seen both as positive and as representing a reduction of risk, rather than an increase in risk. That requires developing awareness of the inherent risk in the status quo: what is likely to happen or is already happening if we maintain our current practices. To see the risk requires shifting from the details of the codes to the larger context and intent of the codes – understanding how current practice jeopardizes the public welfare that the regulatory system was established to protect.

Historically, building codes were developed as a reaction to disasters and building failures. They derived their authority from a societal expectation that the public must be protected from these threats. This led to a focus on the protection of people in and around buildings and secondarily on protection of property. Over time, this focus has become ever more detailed and has expanded into nearly every aspect of buildings and their components and systems. It is no surprise that this focus, combined with our slow awakening to the scope and magnitude of the environmental impacts of the building industry, has resulted in a lack of concern for impacts that occur *away* from the actual building site, impacts that are cumulative or difficult to measure (such as climate impacts or the health effects of indoor air quality or toxicity of materials), or that extend into the future.

The idea of addressing such aggregated impacts through codes, though relatively new, has precedents in such areas as sewage systems, building energy codes, and water-efficiency requirements. Building energy codes provide a valuable, though still somewhat controversial, precedent for incorporating into building codes the larger, more distant, and cumulative consequences of buildings. It has been argued that energy-efficiency is not a safety issue and therefore has no place in the building codes. "I thought that [insulation requirements in building codes] was the dumbest idea I'd ever heard and that it had no place in the codes," admitted Bob Fowler in an interview in *Building Standards*. Good arguments were made for minimum insulation requirements for buildings exposed to extreme temperatures as part of the concern for health and safety of the occupants or users of buildings, and thus they were developed. But it took a combination of economic, environmental, health, and even national security issues to finally propel building energy codes into existence and widespread adoption. "Looking back," reflected Fowler in the same interview, "I see that the energy-efficiency requirements set a very important precedent for our learning to take responsibility for the full range of the consequences of our buildings. We now need to continue that learning process and open our eyes and our minds to the work of creating sustainable buildings."

The larger, ecologically based risks to public welfare must eventually be seen as risks that demand responsibility for protecting public welfare as much as structural integrity, fire safety, or means of egress. The current regulatory system requires a high degree of safety and certainty in each building project, while ignoring the unintended role it plays in encouraging the depletion of natural resources and the demise of the natural systems upon which everyone's health, safety, and survival ultimately depend.

It is not difficult to find evidence to support concerns about the environmental impacts of the built environment:[3]

- Over 40 per cent of the material resources entering the global economy today are related to the building industry.
- Modern buildings use tremendous quantities of energy – in the United States (with less than 5 per cent of the world's population) buildings alone account for a staggering 10 per cent of *global* energy use.

Such statistics are all the more remarkable when one realizes that only about two billion of the world's more than six billion people live and work in resource-consumptive buildings – the sort of buildings described by modern building codes. The rest of the world's people today live in earthen buildings (adobe, rammed- or puddled-earth, cob, wattle-and-daub) or other types of indigenous buildings, shelters made of scavenged materials, or no buildings at all. Yet all over the world modern building methods, with their greater impacts and resource consumption, are replacing traditional – and often far more sustainable – ways of building. It is important not to romanticize indigenous buildings or dismiss the very real problems that are often associated with them

(poor earthquake resistance, lack of insulation, etc.), but to recognize the value and viability of simple, low-tech materials and building methods when used wisely. At the same time, modern materials and building systems must be viewed with the same critical eye, acknowledging their real costs and impacts, not just their benefits. With projections of the world's population reaching at least eight or nine billion this century and with the needed development and construction that must accompany such growth, these issues cannot be ignored much longer....

Relationships with leaders in the building codes community are important, but creating similar relationships locally and regionally is required in order to achieve the needed changes. That can only happen through the engagement of the environmental design and building community in a proactive, constructive partnership with their building code officials, based on a very real, mutual interest in creating safe buildings. Then the definition of public health, safety, and welfare related to buildings can be expanded to include this larger set of responsibilities....

The environmental design and construction community must become actively engaged in writing code change proposals and encouraging funding, research, and testing to support those changes. Additionally, standards-development activities, such as those in ASTM and ASHRAE, often result in requirements less than satisfactory in terms of the environment. The green building community needs to share their direct experience in contending with the realities of those standards by participating more fully in the standards-development process.

It is also time for the environmental design and construction community to seek representation on relevant building code development committees. The ICC code development process is now opening up representation on their committees to the public and industry. Organizations such as The American Institute of Architects Committee on the Environment (AIA-COTE), US Green Building Council (USGBC), Sustainable Building Industry Council (SBIC), Energy and Environmental Building Association (EEBA), and New Buildings Institute need to come together and focus on how to gain such representation. Other interest groups are well organized and funded to represent their interests;

the green building community needs to take responsibility for bringing about changes, rather than simply lamenting the status quo.

Finally, local green building programs provide an ideal forum for education and exchange about alternative designs, materials, and methods and the building codes. Local code officials could be brought into these programs to share their existing skills and experience as well as for their education and enlightenment. Everyone would benefit from such an exchange.

NOTES

1 See *Environmental Building News*, 2(1).
2 See *Environmental Building News*, 10(6).
3 See *Environmental Building News* feature, 10(5).

PART THREE

Tools for sustainability planning

INTRODUCTION TO PART THREE

Although the field of urban planning contains many existing mechanisms and tools, many of these may need to be rethought and new tools developed to help bring about more sustainable cities and towns. An improved ability to analyze long-term trends and monitor progress toward healthier urban environments is essential, for one thing, as is development of processes for constructive public participation and achievement of a more constructive urban politics in general. This section highlights several tools that have been influential to date within the movement for sustainable urban development, and considers what a more satisfactory political context for sustainability planning might look like.

We start with the topic of sustainability indicators, on which much work is being done around the world. Virginia Maclaren's article from the *Journal of the American Planning Association* defines some key characteristics of these measures and discusses several leading examples. Next, we learn about the concept of ecological footprint analysis from two of its originators, William Rees and Mathis Wackernagel. This method has been used to dramatize the resource impacts of cities and metropolitan regions. In our third selection, from Allan Jacobs' book *Looking at Cities*, we highlight careful, firsthand observation as a much-neglected tool through which planners and citizens can analyze urban environments and develop knowledge of their historical development and their future possibilities. Finally we consider the ultimate mechanism for sustainability planning – a constructive politics that can truly respond to long-term human and ecological needs. Though such a situation may be far off, Michael Lerner argues that many individuals are already working toward development of such a "politics of meaning," and that this must be based on new understandings of interdependency and renewed spiritual commitment.

The large array of existing urban planning tools – general plans, specific plans, regional plans, development permitting processes, design charettes, environmental impact reports, geographic information systems (GIS), and so forth – will of course also be essential to sustainable urban development. Readers can gain more background on these through many standard urban planning texts, such as the International City/County Management Association's "bible" entitled *The Practice of Local Government Planning* (Washington, DC: ICMA, 2000), and materials published by the American Planning Association (www.planning.org), the Canadian Institute of Planners (www.cip-icu.ca), the Royal Town Planning Institute (www.rtpi.org.uk), and other professional associations. However, bringing about more sustainable communities will require rethinking some of these traditional planning methods – at least to make them more effective at meeting goals such as the three Es of sustainability.

"Urban Sustainability Reporting"

from the *Journal of the American Planning Association* (1996)

Virginia W. Maclaren

Editors' Introduction

One of the main questions facing those interested in bringing about more sustainable communities is: how do we recognize progress toward sustainability? Some method of measuring the direction of current trends and success or failure of particular initiatives is crucial. This is not a new topic within urban planning – performance measurement and evaluation techniques have been around for years – but the subject takes on a new urgency when the aim is to radically change current ways of developing cities and to justify substantial new initiatives. Indicators can also be extremely useful in educating the public about the direction of current trends, and in developing political support for change.

For such reasons sustainability indicators have become one of the central tools for sustainable urban development. Examples range from Sustainable Seattle's community-based indicator set, developed by a group of citizens who convened public meetings with local leaders in the mid-1990s, to the United Nations' Human Development Index, developed by a large international agency relying on national-level data from 125 countries. (For further information on both of these, see www.sustainableseattle.org and www.undp.org.)

The following review of sustainability indicator approaches by Canadian planner Virginia Maclaren is excerpted from the *Journal of the American Planning Association* (62(2), pp. 184–202). The author teaches at the University of Toronto and has also written on topics of waste management and regional economics. Here she emphasizes the distinctive characteristics and types of sustainability indicators. For further examples of local, regional, or national indicator programs, interested readers might refer to the website of the International Sustainability Indicators Network (ISIN), www.sustainabilityindicators.org. Other helpful resources include Maureen Hart's *Guide to Sustainable Community Indicators* (North Andover, MA: Sustainable Measures, 2001; available from www.sustainablemeasures.com), the World Resources Institute's publication *Environmental Indicators* (Washington, DC: World Resources Institute, 1995), Simon Bell and Stephen Morse's book *Sustainability Indicators: Measuring the Immeasurable?* (London: Earthscan, 1999), and *Towards Sustainable Development: Environmental Indicators* (Paris: Organization for Economic Co-operation and Development, 1998).

The concept of sustainability is starting to have a significant influence on planning and policy at the local level. Previous research has identified numerous examples of urban sustainability initiatives in North America. A certain number of communities are starting to adopt sustainability as a goal in comprehensive plans and other planning activities (Maclaren 1993, Oullet 1993, Beatley 1995). Now, the important next step for sustainability initiatives at the local level is to determine whether or not these actions are leading a community to become more sustainable. A significant barrier to accomplishing this task is the absence of a clearly articulated method of reporting on urban sustainability.

Urban sustainability reports include a range of information about environmental, economic, and social conditions and policies in the local community and use that information to make judgments about whether the community is making progress towards sustainability. Evidence of positive progress is important for justifying past expenditures on sustainability initiatives and building support for new initiatives. Evidence of a lack of sustainability can provide ammunition for community groups in local government, other levels of government, or the private sector. Individuals in the community also can use sustainability reports to educate themselves about sustainability trends and evaluate how their own actions may improve sustainability.

The purpose of this paper is to present a structured process for urban sustainability reporting that improves upon the *ad hoc* reporting processes currently in use, and to explore some of the characteristics of urban sustainability indicators. In researching this paper, I examined some of the first efforts at urban sustainability reporting in North America and Europe and drew on local experiences with related types of reporting, namely state of the environment reporting, healthy city reporting and quality of life reporting. State of the environment (SOE) reports describe and analyze environmental conditions and trends of significance. Social or economic conditions are discussed only insofar as they relate to the biophysical environment (Campbell and Maclaren 1995). Thus SOE reporting is not broad enough to be called sustainability reporting. In contrast, healthy city reporting has just as broad a focus as sustainability reporting, but with a much stronger emphasis on human health. (See, for example, Healthy City Toronto 1993.) Quality of life reporting has evolved to the point where it, too, has become very similar to sustainability reporting in that it examines economic, environmental, and social conditions and the linkages among them (e.g., Murdie *et al.* 1992); but quality of life reporting does not have the same concern for issues of intergenerational equity.

The examples of urban sustainability reports that are referred to in this paper come from three different levels of government: (1) the city of Seattle, Washington; (2) the Regional Municipality of Hamilton-Wentworth, Ontario; and (3) the province of British Columbia. Each of these cases is described briefly below.

Sustainable Seattle is the name of a multistakeholder group that was established in 1990 as a volunteer network and civic forum for the promotion of community sustainability. It is administered by the local YMCA and governed by an independent board of trustees. In 1993, the group released an urban sustainability report for Seattle containing 20 sustainability indicators and an evaluation of Seattle's progress towards sustainability (Sustainable Seattle 1993). An additional 20 indicators were released two years later. The target audience for the report was primarily individual members of the community and the media, with businesses and local government being a secondary target.

The Sustainable Community Indicators project in the Regional Municipality of Hamilton-Wentworth, Ontario, is a continuation of the region's Sustainable Community Initiative, which began in 1990. At that time, the Regional Council appointed a citizen's Task Force on Sustainable Development with a mandate to examine the concept of sustainable development as a basis for reviewing all regional policies. In 1992, after consultation with over 400 individuals and 50 community groups, the Task Force released a document entitled "Vision 2020," describing the type of community that Hamilton-Wentworth could be in the year 2020 if it followed the principles of sustainable development (Regional Municipality of Hamilton-Wentworth 1992). As a follow-up to this document, the Council launched the Sustainable Community Indicators project in 1994, with the goal of developing sustainability indicators for measuring the region's progress towards Vision 2020. The output of the project will be an annual report card that identifies the status of the indicators as well as the way in which they can be influenced by individuals, organizations, business, local government, and the community as a whole.

The British Columbia Round Table's State of Sustainability Report examines urban sustainability at the provincial level. The Round Table is a multi-stakeholder group, funded by the provincial government, and was responsible for developing the province's first sustainability strategy. For its urban sustainability report, the Round Table chose a sample of five cities, accounting for over 60 per cent of the province's population, to represent the broad regions of the province as well as a variety of economic, environmental, and social conditions. The report, containing over 90 urban sustainability indicators, was released in 1994 (British Columbia Round Table 1994). Like the Hamilton-Wentworth initiative, the British Columbia report is meant to be a guide for both modifying personal behavior and informing planning and policy decisions. . . .

DEFINING URBAN SUSTAINABILITY

What is the meaning of the term "urban sustainability"? It may help to first compare it to "sustainable urban development." The meanings of these two terms are very close and are often used interchangeably in the literature (cf. Richardson 1994). One way of distinguishing them, however, is to think of sustainability as describing a desirable *state* or set of conditions that persists over time. In contrast, the word "development" in the term "sustainable urban development" implies a *process* by which sustainability can be attained.

Some of the key characteristics of urban sustainability that are often mentioned in the literature and in policy documents are: intergenerational equity, intragenerational equity (including social equity, geographical equity,[1] and equity in governance), protection of the natural environment (and living within its carrying capacity), minimal use of nonrenewable resources, economic vitality and diversity, community self-reliance, individual well-being, and satisfaction of basic human needs.[2]

There is considerable debate within the academic community, planning agencies, and other organizations over the relative importance of each of these urban sustainability characteristics, and there is even disagreement on whether all of them should be included when developing sustainability goals. Almost everyone who has tried to define urban sustainability agrees, however, that the concept points to the necessity of introducing environmental considerations to the policy debate over the future of our cities. Some maintain that environmental

considerations should now be paramount in this debate, while others call for a more holistic approach that balances environmental, economic, and social concerns.

For the purposes of urban sustainability reporting, I contend that there is no single "best" definition of urban sustainability, since different communities are likely to develop slightly, or even significantly, different conceptualizations of urban sustainability, depending on their current economic, environmental, and social circumstances and on community value judgments. As a consequence, a set of indicators designed to measure progress towards achievement of one community's sustainability goals may not necessarily be appropriate for measuring progress in another community. Nevertheless, there are certain fundamental properties of sustainability indicators that all communities will wish to consider. These are described in the next section.

WHAT IS AN URBAN SUSTAINABILITY INDICATOR?

One definition of urban sustainability indicators is that they are "bellwether tests of sustainability and reflect something basic and fundamental to the long term economic, social or environmental health of a community over generations" (Sustainable Seattle 1993: 4). This definition provides a good starting point, but it requires considerable elaboration. Looking first at the "indicator" component of "urban sustainability indicators," it is important to remember that most indicators are simplifications of complex phenomena. The term "indicator" should therefore be taken literally in the sense that it provides only an indication of conditions or problems (Whorton and Morgan 1975, Clarke and Wilson 1994). Since a single indicator will seldom be able to give the full picture, it is often useful to employ a wide range of indicators to characterize the different dimensions or aspects of a situation. Unfortunately, this requirement can conflict with the need to identify a fairly limited set of indicators for purposes of decision-making, and to minimize double-counting.

Urban sustainability indicators can be distinguished from simple environmental, economic, and social indicators by the fact that they are:

- integrating
- forward-looking
- distributional
- developed with input from multiple stakeholders in the community.

All sustainability indicators should possess the last characteristic. It may not be possible to develop individual sustainability indicators that possess all of the first three characteristics, but they should possess at least one, and within a given set of sustainability indicators, all of these characteristics should be represented.

Integrating indicators

Sustainability indicators are integrating in the sense that they attempt to portray linkages among the economic, environmental, and social dimensions of sustainability. One example of an integrating indicator might be the amount of "brownfield" land found in an urban area. This could be considered both as an indicator of industrial activity loss and as an indicator of environmental constraints on redevelopment (if the lands are contaminated). Still another integrating indicator would be the unemployment rate, since it is a measure of both economic stress and social stress.

One of the integrating indicators used by Sustainable Seattle is the number of salmon returning to spawn in a representative sample of local salmon runs. This indicator is relevant for both an environmental condition (water quality) and an economic vitality condition (survival of one of the Seattle area's most important industries).

Composite indicators, which combine two or more individual indicators, can also be useful as integrative indicators. For example, the cost of recycling per ton of waste recycled is a simple composite indicator that integrates economic and environmental considerations. Unfortunately, the construction of more complex composite indicators faces a number of methodological problems, including such issues as deciding how to weight the individual indicators, how to standardize different measurement units, and whether to choose a multiplicative or additive aggregation technique (Ott 1978, Innes 1990). Despite these problems, some composite indicators, such as the Human Development Index,[3] have gained considerable popularity because they reduce the information contained in several individual indicators down to a single number.

Forward-looking indicators

A second important characteristic of sustainability indicators is that they must be forward-looking if they are to be used in measuring progress towards achieving intergenerational equity. There are several different ways in which an indicator might be considered forward-looking. The simplest type of forward-looking indicator is a "trend indicator." A trend indicator describes historical trends and provides indirect information about future sustainability. For example, it is often obvious from examining historical trends that a development path followed in the past cannot possibly be sustainable into the future. However, because trend indicators provide only indirect information about the future, they are more useful for reactive than for proactive policy-making (Ruitenbeek 1991).

The forward-looking capabilities of trend indicators can be enhanced if they are linked to reference points that define intermediate or final steps in the move towards meeting sustainability goals. The two main types of reference points are targets and thresholds. Whereas targets are levels that must be met in the future if sustainability is to be achieved, thresholds are levels that should not be exceeded. Thresholds are scientifically determined and may possess regulatory status. Examples include air and water quality standards. Targets can be set in a fairly arbitrary manner either by using easily recognized numbers (e.g., reduce solid waste by 50 per cent by the year 2000), by comparison to higher order jurisdictions (e.g., national or state means), or by norms (e.g., the poverty level). A threshold, such as an air quality standard, also can be part of a target (e.g., zero exceedances of the standard by the year 2020).

The Oregon "benchmarks" are a well-known application of the use of targets for reviewing government accountability. In 1991, the Oregon Progress Board released its first benchmarks report, in which it identified 272 indicators of environmental, social, and economic well-being in that state (Oregon Progress Board 1991). The Board also specified a series of targets for each indicator, to be met at regular intervals up to the year 2010. They referred to these targets as benchmarks. The indicators in the report are primarily output indicators (e.g., number of households with drinking water that does not meet government standards) rather than input indicators (e.g., expenditures on water treatment facilities), and are being used to help set a broad range of program and budget priorities.

Both targets and thresholds are present in the Netherlands' national environmental policy indices. Each index has one or more policy targets set for specified future dates (e.g., the years 2000, 2010), and in some cases the index includes a longer-term "sustainability level" that is scientifically determined. For example, the Eutrophication Index, which measures releases of phosphates and nitrogen compounds to the environment, will reach a sustainable level when the excessive supply of phosphates and nutrients has been reduced enough that a balance has been achieved between supply and the removal from the environment of these two major contributors to eutrophication (Adriaanse 1993).

Another type of forward-looking indicator is the "predictive indicator." Predictive sustainability indicators rely on mathematical models for the future state and development of variables describing the environment, the economy, and society, or the linkages among them. Population levels and population growth are commonly used predictive indicators found in planning reports. Bratt (1991) notes that since all predictions are inherently disputable, the best that predictive indicators can do is to provide plausible information about future conditions. Only trend indicators provide scientifically reliable information, assuming that the data collection methods were reliable.

The uncertainty inherent in predictive indicators points to the need for a third type of forward-looking sustainability indicator known as the "conditional indicator." Conditional indicators depend on a form of scenario development; they answer the question: "If a given indicator achieves or is set at a certain level, what will the level of an associated indicator be in the future?" This type of indicator attempts to overcome the difficulty that predictive indicators have in forecasting, by developing a range of forecasts or predictions. Table 1, taken from the British Columbia Round Table's State of Sustainability Report (1994), provides an example of a conditional indicator of urban form. The "if" indicator is future residential density. The "then" indicator is the total amount of land that will be needed to accommodate the expected urban population of British Columbia in 2021 at each of these density levels. Two different measures of the land-area indicator are presented: the amount of land in hectares and the equivalent amount of land currently occupied by the City of Vancouver. The former measure may be most useful for planners, and the later measure is probably more meaningful to the general public.

Housing (units per hectare)	Area needed for housing (hectares)	Area needed for other urban functions	Total area needed (hectares)	Density* City of Vancouver equivalents
1.4	479,000	240,000	719,000	64
2.3	290,000	145,000	435,000	38
6.5	103,000	52,000	155,000	14
9.5	70,000	35,000	105,000	9
18	37,000	19,000	56,000	5

Table 1 Land area needed for cities to serve additional British Columbia residents in the year 2021 at various residential densities

Source: BC Round Table on Environment and Economy (1994).

Note: *From lowest to highest, these are the current densities, respectively, for the City of Kelowna, the City of Cranbrook, Greater Victoria, Greater Vancouver Regional District, and the City of Vancouver.

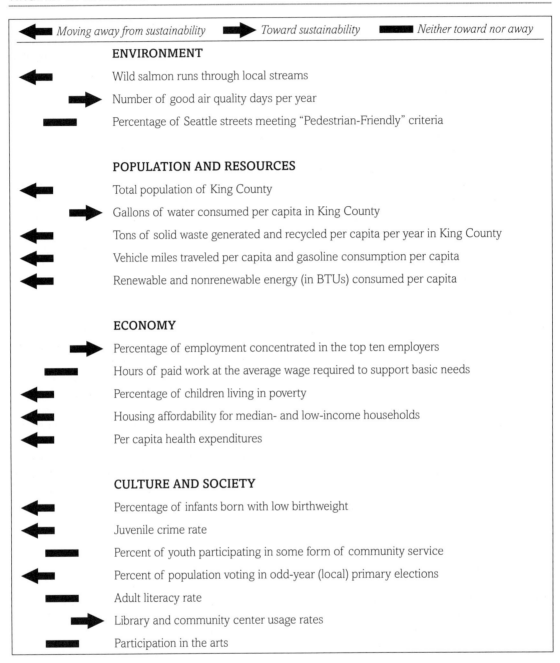

Figure 1. Sustainable Seattle indicators.

Source: Sustainable Seattle (1993).

Distributional indicators

Sustainability indicators must be able to measure not only intergenerational equity but also intra-generational equity. They should be able to take into account the distribution of conditions (social, economic, environmental) within a population or across geographic regions. Typically, spatially aggregated indicators fail to account for distributive effects. An example is GNP, which may increase even though economic conditions for many groups or different regions in the country are declining (Liverman *et al.* 1988). Disaggregating certain indicators for a community by such factors as age, gender, and location can help to overcome this problem.

Sustainability indicators should also be able to distinguish between local and nonlocal sources of environmental degradation, and between local and nonlocal environmental effects. A downstream community may generate very little pollution and display all the characteristics of a sustainable community – except for the fact that it suffers from significant upstream water pollution or upwind air pollution. The development of indicators that can identify pollution sources outside the local community's control will facilitate the formulation of appropriate policy responses to geographical inequities. Similarly, sustainability indicators should also measure the extent to which a local community contributes to environmental degradation in other communities, regions, or the world at large.

Multi-stakeholder input

A final characteristic that distinguishes sustainability indicators from other types of indicators is the manner in which they are developed. The history of the social indicator movement suggests that the most influential, valid, and reliable indicators have been those that were developed with input from a broad range of participants in the policy process (Innes 1990). This lesson is especially applicable to the development of sustainability indicators, since sustainability is such a value-laden and context-sensitive concept. It therefore makes sense to seek input on sustainability concerns and priorities from a broad range of stakeholders. This can be accomplished by assigning significant responsibil-

ity for selecting sustainability indicators to a broadly-based, multi-stakeholder group or by consulting in some other way with multiple stakeholders from the earliest stages of indicator development.

NOTES

1 Geographical equity is a term coined by Haughton and Hunter (1994) to emphasize the undesirability of achieving economic growth of a higher quality of life in one community at the expense of environmental degradation in another. They assert that this form of development is inequitable unless some form of reparation or compensation takes place between the communities.

2 See, e.g., Alberta Round Table 1993, Jacobs 1991, Hardoy *et al.* 1992, Richardson 1992, British Columbia Round Table 1994, Haughton and Hunter 1994, Beatley 1995.

3 The Human Development Index was developed by the United Nations Development Program (UNDP) for comparing human welfare levels in different countries. The index is an aggregation of four indicators: life expectancy at birth, adult literacy, average years of schooling, and GDP per capita. The UNDP publishes the Index for all members of the United Nations in its annual "Human Development Report."

REFERENCES

Adriaanse, A. 1993. *Environmental Policy Performance Indicators: A Study of Indicators for Environmental Policy in the Netherlands.* The Hague: Sdu Uitgeverji Koninginnegracht.

Alberta Round Table on Environment and Economy. 1993. *Steps to Realizing Sustainable Development.* Edmonton: ARTEE.

Beatley, T. 1995. Planning and Sustainability: The Elements of a New (Improved?) Paradigm. *Journal of Planning Literature* 9, 4: 383–395.

Bratt, L. 1991. The Predictive Meaning of Sustainability Indicators. In *In Search of Indicators of Sustainable Development*, edited by O. Kuik and H. Vergruggen. Boston: Kluwer Academic Publishers, pp. 57–70.

British Columbia Round Table on the Environment and the Economy. 1994. *State of Sustainability: Urban Sustainability and Containment.* Victoria: British Columbia Round Table on the Environment and the Economy.

Campbell, M. and V.W. Maclaren. 1995. *Municipal State of the Environment Reporting in Canada: Current Status and Future Needs.* Ottawa: Occasional Paper Series No. 6, State of the Environment Reporting, Environment Canada.

Clarke, G.P. and A.G. Wilson. 1994. Performance Indicators in Urban Planning: The Historical Context. In *Modelling the City: Performance, Policy and Planning*, edited by C.S. Bertuglia, G.P. Clarke and A.G. Wilson. London: Routledge, pp. 4–19.

Hardoy, J.E., D. Mitlin, and D. Satterthwaite. 1992. *Environmental Problems in Third World Cities.* London: Earthscan.

Haughton, G. and C. Hunter. 1994. *Sustainable Development and Geographical Equity.* Paper presented at the Annual Conference of the Association of American Geographers, Chicago.

Healthy City Toronto. 1993. *A Strategy for Developing Healthy City Indicators.* Toronto: Healthy City Toronto.

Innes, J.E. 1990. *Knowledge and Public Policy: The Search for Meaningful Indicators*, 2nd edn. New Brunswick, NJ: Transaction Publishers.

Jacobs, M. 1991. *The Green Economy.* London: Pluto Press.

Liverman, D., M.E. Hanson, B.J. Brown, and R.W. Meredith. 1988. Global Sustainability: Towards Measurement. *Environmental Management* 12, 2: 133–143.

Maclaren, V.W. 1993. *Sustainable Urban Development in Canada: From Concept to Practice.* Volumes I–III. Toronto: ICURR Press.

Murdie, R.A., D. Rhyne, and J. Bares. 1992. *Modelling Quality of Life Indicators in Canada: A Feasibility Analysis.* Ottawa: Canada Mortgage and Housing Corporation.

Ontario Round Table on Environment and Economy. 1995. *Sustainable Communities Resource Package.* Toronto: Ontario Round Table on Environment and Economy.

Oregon Progress Board. 1991. *Oregon Benchmarks: Setting Measurable Standards for Progress.* Report to the 1991 Legislature. Salem: Oregon Progress Board.

Ott, W. 1978. *Environmental Indices.* Ann Arbor, MI: Ann Arbor Science Publishing Inc.

Oullet, P. 1993. *Environmental Policy Review of 15 Canadian Municipalities.* Toronto: ICURR Press.

Regional Municipality of Hamilton-Wentworth. Regional Chairman's Task Force on Sustainable Development. 1992. *Vision 2020: The Sustainable Region.* Hamilton: Regional Municipality of Hamilton-Wentworth.

Richardson, N. 1992. Canada. In *Sustainable Cities: Urbanization and the Environment in International Perspective*, edited by R. Stren, J.B.R. Whitney, and R. White. Boulder, CO: Westview Press, pp. 145–168.

Richardson, N. 1994. Making Our Communities Sustainable: The Central Issue is Will. In *Sustainable Communities Resource Package,* Ontario Round Table on Environment and Economy. Toronto: Ontario Round Table on Environment and Economy, pp. 15–44.

Roseland, M. 1992. *Toward Sustainable Communities.* Ottawa: National Round Table on the Environment and the Economy.

Ruitenbeek, J.H. 1991. *Indicators of Ecologically Sustainable Development: Towards New Fundamentals.* Ottawa: Canadian Environmental Advisory Council.

Sustainable Seattle. 1993. *Sustainable Seattle Indicators of Sustainable Community: A Report to Citizens on Long Term Trends in Their Community.* Seattle, WA: Sustainable Seattle.

Tomalty, R. and D. Pell. 1994. *Sustainable Development and Canadian Cities: Current Initiatives.* Ottawa: Canada Mortgage and Housing Corporation and the Royal Society of Canada.

Whorton, J.W., Jr. and D.R. Morgan. 1975. *Measuring Community Performance: A Handbook of Indicators.* Norman, OK: University of Oklahoma, Bureau of Government Research.

"What *Is* an Ecological Footprint?"

from *Our Ecological Footprint* (1996)

Mathis Wackernagel and William Rees

Editors' Introduction

One of the most intriguing tools with which to measure the sustainability of particular places or lifestyles is ecological footprint analysis, developed by University of British Columbia professor William Rees and his colleagues and graduate students. By converting resource needs and pollution into the equivalent land area that would be required to produce or offset these, the footprint model provides a dramatic indication of the impacts of modern life. Footprints can be calculated for individuals, entire cities, regions, and nations. Readers can create a simple, personal footprint online at the website of Redefining Progress, a San Francisco-based NGO that has worked extensively to develop new measurements of social and ecological health (www.rprogress.org).

Ecological footprint analysis may be most useful as an educational tool, to give a general indication of the sustainability of particular places or modes of living. Although researchers are attempting to calculate extremely detailed footprints for some cities and nations, these efforts run into the same problems that most ecological economics encounter, such as how to place a value on social and environmental costs in terms of either dollars or land area. Assumptions must be made about how to translate resource use and pollution into these variables, and these are open to question. Still, the ecological footprint method has an intuitive appeal, similar to more general notions of "carrying capacity," that has made it attractive to many people.

This selection is drawn from Wackernagel and Rees' book *Our Ecological Footprint: Reducing Human Impact on the Earth* (Gabriola Island, BC: New Society Publishers, 1996). Swiss engineer and planner Mathis Wackernagel is director of the Redefining Progress organization in San Francisco and also directs the Centre for Sustainability Studies at Anáhuac University of Xalapa, Mexico. Canadian urban planning professor William Rees researches ecological economics and biodiversity issues in addition to his work in developing the ecological footprint model. For additional information on ecological or social accounting, see *Sharing Nature's Interest: Ecological Footprints as an Indicator for Sustainability* (London: Earthscan, 2000) by Nicky Chambers, Craig Simmons, and Mathis Wackernagel, the *Living Planet Report* (Gland, Switzerland: World Wildlife Federation, 2002; available at http://www.panda.org/livingplanet), materials from the Best Foot Forward organization in Oxford, UK (www.bestfootforward.com), and Madeline Warring's *If Women Counted: A New Feminist Economics* (San Francisco: Harper & Row, 1988).

Ecological footprint analysis is an accounting tool that enables us to estimate the resource consumption and waste assimilation requirements of a defined human population or economy in terms of a corresponding productive land area. Typical questions we can ask with this tool include: how dependent is our study population on resource imports from "elsewhere" and on the waste assimilation capacity of the global commons?, and will nature's productivity be adequate to satisfy the rising material expectations of a growing human population into the next century? William Rees has been teaching the basic concept to planning students for 20 years and it has been developed further since 1990 by Mathis Wackernagel and

other students working with Bill on UBC's Healthy and Sustainable Communities Task Force.

To introduce the thinking behind ecological footprint analysis, let's explore how our society perceives that pinnacle of human achievement, "the city." Ask for a definition, and most people will talk about a concentrated population or an area dominated by buildings, streets and other human-made artifacts (this is the architect's "built environment"); some will refer to the city as a political entity with a defined boundary containing the area over which the municipal government has jurisdiction; still others may see the city mainly as a concentration of cultural, social and educational facilities that would simply not be possible in a smaller

Figure 1. The Ecological Footprint is a measure of the "load" imposed by a given population on nature. It represents the land area necessary to sustain current levels of resource consumption and waste discharge by that population.

Source: Illustration by Phil Testemale.

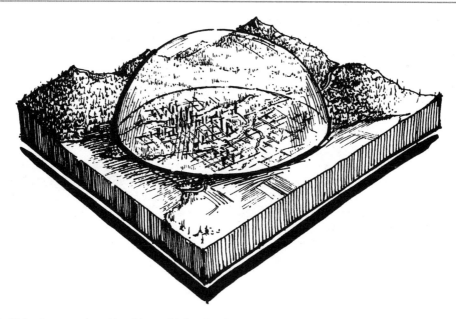

Figure 2. Living in a terrarium. How big would the glass hemisphere need to be so that the city under it could sustain itself exclusively on the ecosystems contained?

Source: Illustration by Phil Testemale.

settlement; and, finally, the economically-minded see the city as a node of intense exchange among individuals and firms and as the engine of production and economic growth.

No question, cities are among the most spectacular achievements of human civilization. In every country cities serve as the social, cultural, communications and commercial centers of national life. But something fundamental is missing from the popular perception of the city, something that has so long been taken for granted it has simply slipped from consciousness.

We can get at this missing element by performing a mental experiment based on two simple questions designed to force our thinking beyond conventional limits. First, imagine what would happen to any modern city or urban region – Vancouver, Philadelphia or London – as defined by its political boundaries, the area of built-up land, or the concentration of socioeconomic activities, if it were enclosed in a glass or plastic hemisphere that let in light but prevented material things of any kind from entering or leaving – like the "Biosphere II" project in Arizona (Figure 2). The health and integrity of the entire human system so contained would depend entirely on whatever was initially trapped within the hemisphere.

It is obvious to most people that such a city would cease to function and its inhabitants would perish within a few days. The population and the economy contained by the capsule would have been cut off from vital resources and essential waste sinks, leaving it both to starve and to suffocate at the same time! In other words, the ecosystems contained within our imaginary human terrarium would have insufficient "carrying capacity" to support the ecological load imposed by the contained human population. This mental model of a glass hemisphere reminds us rather abruptly of humankind's continuing ecological vulnerability.

The second question pushes us to contemplate this hidden reality in more concrete terms. Let's assume that our experimental city is surrounded by a diverse landscape in which cropland and pasture, forests and watersheds – all the different ecologically productive land-types – are represented in proportion to their actual abundance on the Earth, and that adequate fossil energy is available to support current levels of consumption using prevailing technology. Let's also assume our imaginary glass enclosure is elastically expandable. The question now becomes: how large would the hemisphere have to become before the city at its center could sustain itself indefinitely and exclusively on the land

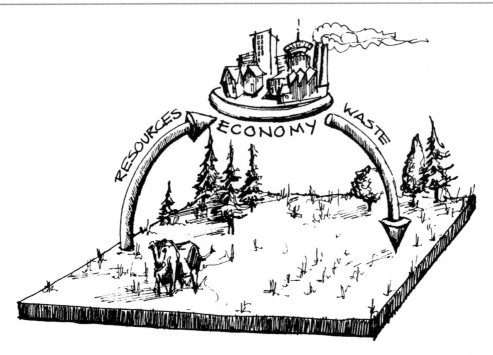

Figure 3. What is an Ecological Footprint? Think of an economy as having an "industrial metabolism." In this respect it is similar to a cow in its pasture. The economy needs to "eat" resources, and eventually, all this intake becomes waste and has to leave the organism – the economy – again. So the question becomes: How big a pasture is necessary to support that economy – to produce all its feed and absorb all its waste? Alternatively, how much land would be necessary to support a defined economy sustainably at its current material standard of living?

Source: Illustration by Phil Testemale.

and water ecosystems and the energy resources contained within the capsule? In other words, what is the total area of terrestrial ecosystem types needed continuously to support all the social and economic activities carried out by the people of our city as they go about their daily activities? Keep in mind that land with its ecosystems is needed to produce resources, to assimilate wastes, and to perform various invisible life-support functions. Keep in mind too, that for simplicity's sake, the question as posed does not include the ecologically productive land area needed to support other species independent of any service they may provide to humans.

For any set of specified circumstances – the present example assumes current population, prevailing material standards, existing technologies, etc. – it should be possible to produce a reasonable estimate of the land/water area required by the city concerned to sustain itself. By definition, the total ecosystem area that is essential to the continued existence of the city is its *de facto*

Ecological Footprint on the Earth. It should be obvious that the Ecological Footprint of a city will be proportional to both population and *per capita* material consumption. Our estimates show for modern industrial cities the area involved is orders of magnitude larger than the area physically occupied by the city. Clearly, too, the Ecological Footprint includes all land required by the defined population wherever on Earth that land is located. Modern cities and whole countries survive on ecological goods and services appropriated from natural flows or acquired through commercial trade from all over the world. The Ecological Footprint therefore also represents the corresponding population's total "appropriated carrying capacity."

By revealing how much land is required to support any specified lifestyle indefinitely, the Ecological Footprint concept demonstrates the continuing material dependence of human beings on nature. For example, Table 1 shows the Ecological Footprint of an average Canadian, i.e., the amount

Cell entries = ecologically productive land in ha/capita	A Energy	B Degradation	C Garden	D Crop	E Pasture	F Forest	Total
1 *Food*	0.33		0.02	0.60	0.33	0.02	1.30
11 fruit, vegetables, grain	0.14		0.02	0.18		0.01?	
12 animal products	0.19			0.42	0.33	0.01?	
2 *Housing*	0.41	0.08	0.002?			0.40	0.89
21 construction/ maintenance	0.06					0.35 0.05	
22 operation	0.35						
3 *Transportation*	0.79	0.10					0.89
31 motorized private	0.60						
32 motorized public	0.07						
33 transport of goods	0.12						
4 *Consumer Goods*	0.52	0.01		0.06	0.13	0.17	0.89
40 packaging	0.10					0.04	
41 clothing	0.11			0.02	0.13		
42 furniture and appliances	0.06					0.03?	
43 books/magazines	0.06					0.10	
44 tobacco and alcohol	0.06			0.04			
45 personal care	0.03						
46 recreational equipment	0.10						
47 other goods	0.00						
5 *Services*	0.29	0.01					0.30
51 government/military	0.06						
52 education	0.08						
53 healthcare	0.08						
54 social services	0.00						
55 tourism	0.01						
56 entertainment	0.01						
57 bank/insurance	0.00						
58 other services	0.05						
Total	2.34	0.20	0.20	0.66	0.46	0.59	4.27

Table 1 The consumption–land-use matrix for the average Canadian (1991 data)

Notes:
(0.00 = less than 0.005 ha or 50 m²; blank = probably insignificant; ? = lacking data)
A *Energy* = fossil energy consumed expressed in the land area necessary to sequester the corresponding CO_2
B *Degratation* = degraded land or built-up environment
C *Garden* = gardens for vegetable and fruit production
D *Crop* = cropland
E *Pasture* = pastures for dairy, meat and wool production
F *Forest* = prime forest area. An average roundwood harvest of 163 m³/ha every 70 years is assumed

of land required from nature to support a typical individual's present consumption. This adds up to almost 4.3 hectares, or a 207 meter square. This is roughly comparable to the area of three city blocks. The column on the left shows various consumption categories and the headings across the top show corresponding land-use categories. "Energy" land as used in the table means the area of carbon sink land required to absorb the carbon dioxide released by per capita fossil fuel consumption (coal, oil and natural gas) assuming atmospheric stability as a goal. Alternatively, this entry could be calculated according to the area of cropland necessary to produce a contemporary biological fuel such as ethanol to substitute for fossil fuel. This alternative produces even higher energy land requirements. "Degraded Land" means land that is no longer available for nature's production because it has been paved over or used for buildings. Examples of the resources in "Services" are the fuel needed to heat hospitals, or the paper and electricity used to produce a bank statement.

To use Table 1 to find out how much agricultural land is required to produce food for the average Canadian, for example, you would read across the "Food" row to the "Crop" and "Pasture" columns. The table shows that, on average, 0.95 hectares of garden, cropland and pasture is needed for the typical Canadian. Note that none of the entries in the table is a fixed, necessary or recommended land area. They are simply our estimates of the 1990s ecological demands of typical Canadians. The Ecological Footprints of individual and whole economies will vary depending on income, prices, personal and prevailing social values as they affect consumer behavior, and technologic sophistication – e.g., the energy and material content of goods and services.

SO WHAT? – THE GLOBAL CONTEXT

Our economy caters to growing demands that compete for dwindling supplies of life's basics.

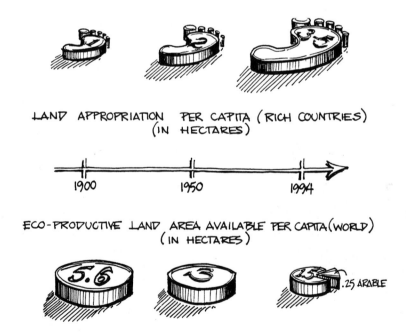

Figure 4. Our Ecological Footprints keep growing. Since the beginning of this century, the available ecologically productive land has decreased from over 5 hectares to less than 1.5 hectares per person in 1994. At the same time, the average North American's footprint has grown to over four hectares. These opposing trends are in fundamental conflict: the ecological demands of average citizens in rich countries exceed per capita supply by a factor of three. This means that the Earth could not support even today's population of 5.8 billion sustainably at North American material standards.

Source: Illustration by Phil Testemale.

The Ecological Footprint of any population can be used to measure its current consumption and projected requirements against available ecological supply and point out likely shortfalls. In this way, it can assist society in assessing the choices we need to make about our demands on nature. To put this into perspective, the ecologically productive land "available" to each person on Earth has decreased steadily over the last century (Figure 4). Today, there are only 1.5 hectares of such land for each person, including wilderness areas that probably shouldn't be used for any other purpose. In contrast, the land area "appropriated" by residents of richer countries has steadily increased. The present Ecological Footprint of a typical North American (4–5 ha) represents three times his/her fair share of the Earth's bounty. Indeed, if everyone on Earth lived like the average Canadian or American, we would need at least three such planets to live sustainably. Of course, if the world population continues to grow as anticipated, there will be 10 billion people by 2040, for each of whom there will be less than 0.9 hectares of ecologically productive land, assuming there is no further soil degradation.

Figure 5. *Wanted*: two (phantom) planets. If everybody lived like today's North Americans, it would take at least two additional planet Earths to produce the resources, absorb the wastes, and otherwise maintain life-support. Unfortunately, good planets are hard to find. . . .

Source: Illustration by Phil Testemale.

Such numbers become particularly telling when used to compare selected geographic regions with the land they actually "consume." For example, we estimate the Ecological Footprint for the Lower Fraser Valley, east of Vancouver to Hope, BC. This valley bottom has 1.8 million inhabitants for a population density of 4.5 people per hectare. In short, the area is far smaller than needed to supply the ecological resources used by its population. If the average person in this basin needs the output of 4.3 hectares (Table 2), then the Lower Fraser Valley depends on an area 19 times larger than that contained within its boundaries for food, forestry products, carbon dioxide assimilation and energy. Similarly, Holland has a population of 15 million people, or 4.4 people per hectare, and although Dutch people consume less than North Americans on average, they still require about 15 times the available land within their own country for food, forest products and energy use (Figure 6). In other words, the ecosystems that actually support typical industrial regions lie invisibly far beyond their political or geographic boundaries.

A world upon which everyone imposed an over-sized Ecological Footprint would not be sustainable – the Ecological Footprint of humanity as a whole must be smaller than the ecologically productive portion of the planet's surface. This means that if every region or country were to emulate the economic example of the Lower Fraser Basin or The Netherlands, using existing technology, we would all be at risk from global ecological collapse.

The notion that the current lifestyle of industrialized countries cannot be extended safely to everyone on Earth will be disturbing to some. However, simply ignoring this possibility by blindly perpetuating conventional approaches to economic development invites both eco-catastrophe and subsequent geopolitical chaos. To recognize that not everybody can live like people do in industrialized countries today is not to argue that the poor should remain poor. It is to say that there must be adjustments all round and that, if our ecological analyses are correct, continuing on the current development path will actually hit the less fortunate hardest. Blind belief in the expansionists' cornucopian dream does not make it come true – rather it side-tracks us from learning to live within the means of nature and ultimately becomes ecologically and socially destructive.

To keep things simple, we consider only four important categories of domestic consumption: built-up land, food, forest products, and fossil energy. This avoids any significant double counting, yet is sufficient to illustrate the strength of ecological footprint analysis.

Basic data (The Netherlands)

- 1991 population: 15,050,000; land area: 33,920 square kilometres.
- Built up land: 538,000 hectares.
- Commercial energy consumption in 1991: 3,197 PJ – 36 PJ from non-fossil fuel sources (mainly nuclear energy). Therefore, for this calculation ((3197 – 36) [PJ] / 15 million Dutch) = 210 GJ/cap/yr is used to represent the fossil fuel consumption.

Calculations

- **forest:** assuming a consumption of 1.1 m^3/cap/yr and a forest productivity of 2.3 m^3/ha/yr, this consumption corresponds to (1.1 [m^3/cap/yr] / 2.3 [m^3/ha/yr]) = 0.47 [ha/cap] of forest land.
- **fossil fuel:** 210 [GJ/cap/yr] corresponds to (210 [GJ/cap/yr] / 100 [GJ/ha/yr] = 2.10 [ha/cap].

Results

food:	cropland	0.45 [ha/cap]
rangeland:		0.26 [ha/cap]
forest:	1.1 [m^3/cap/yr] corresponds to	0.47 [ha/cap]
fossil fuel:	210 [GJ/cap/yr] corresponds to	2.10 [ha/cap]
degraded land	(settlements and roads):	
	(538,000 [ha] / 15,000,000 [Dutch people])	0.04 [ha/cap]
Total individual footprint:		**3.32 [ha/cap]**

Table 2 Assessing the footprint of The Netherlands

Notes:
The Netherlands' aggregate Ecological Footprint is:
(15,000,000 [Dutch people] × 3.32 [ha/cap] × 0.01 [ha/km^2] = 498,000 square kilometres.
This is almost 15 times larger than the Dutch territory of 33,920 square kilometres.

Figure 6. The Ecological Footprint of The Netherlands. For urbanization, food, forest products and fossil fuel use, the Dutch use the ecological functions of a land area over 15 times larger than their country.

Source: Illustration by Phil Testemale.

"Seeing Change"

from *Looking at Cities* (1985)

Allan B. Jacobs

Editors' Introduction

One of the most basic tools of planning is an ability to look at the landscape around us – whether urban, suburban, or rural – and see what is going on. This is particularly important for sustainability planning, when it is important to be able to recognize how places have evolved in the past and how they might become more sustainable in the future. Clues always exist about the history of a place, the social and economic dynamics within it, the ways that people are actually using it, and the ways that land use, buildings, and natural environments have changed over time. An experienced urban planner can also quickly determine important pieces of information such as residential density, the dimensions of streets, lots, and buildings, and the presence of culverted streams or remnant pieces of wildlife habitat. A sensitivity to the subjective experience of place is also extremely useful – in other words, being able to analyze what cities, neighborhoods, or public spaces feel like to different groups of people.

Unfortunately, "looking at cities" is a skill that has often been ignored in planning education and practice. Instead planners have frequently spent their time in front of computers, relying on abstract concepts, numerical data, satellite imagery, and information gathered from secondary sources. These types of information are important, but too often the inability of architects, landscape architects, and urban planners to understand the on-the-ground context of development has helped create sterile places that do not meet the needs of human communities or the natural environment. Luckily, since the early 1990s many planners and citizens have recognized that a new attention to the subjective experience of place is important, as well as detailed, firsthand analysis of urban environments as a basis for planning and design. One proponent of this approach has been Allan Jacobs, the former planning director of San Francisco, who has taught at the University of California at Berkeley. In this selection from his book *Looking at Cities* (Cambridge, MA: Harvard University Press, 1985), Jacobs describes an approach he has taken with some of his classes of simply walking through cities observing the urban environment firsthand. Jacobs' other work has focused on how to create vibrant, walkable streets. His books include *Great Streets* (Cambridge, MA: MIT Press, 1993) and *The Boulevard Book: History, Evolution, Design of Multiway Boulevards* (Cambridge, MA: MIT Press, 2002; with Elizabeth Macdonald and Yodan Rofe).

Other urban designers have also written on this theme. Some are affiliated with the philosophical viewpoint known as phenomenology, which emphasizes direct experience as a basis for knowledge. Examples include David Seamons' volume *Dwelling, Seeing, and Designing: Toward a Phenomenological Ecology* (Albany: State University of New York Press, 1993), Edward Relph's book *The Modern Urban Landscape* (London: Croom Helm, 1987), Tony Hiss' *The Experience of Place* (New York: Vintage Books, 1990), and Jack Nazar's *The Evaluative Image of the City* (Thousand Oaks, CA: Sage Publications, 1998). MIT urban theorist Kevin Lynch laid much of the groundwork for this point of view in works such as *The Image of the City* (Cambridge, MA: MIT Press, 1960) and *Good City Form* (Cambridge, MA: MIT Press, 1981).

As parents always say to their children, "It's in front of you. Use your eyes." We take messages – or we fail to take them – from urban environments by looking, and we act upon those messages to maintain or change or create places in ways that seem appropriate responses to urban problems and opportunities.

This book calls for getting involved with what we see: learning from what we observe in the urban environment; employing observation more consciously and regularly as an analytic and decision-making tool; and using what we learn to help people live in concert with one another and with the land. If conscious, systematic observation, as opposed to haphazard visual experiencing, does nothing more than help avoid unfortunate decisions and actions that affect people's lives, it will have served well. But it can do much more than that.

I want to summarize briefly some of the more significant findings about observation and to offer additional ideas that can help any interested city dweller go out and do it.

There is nothing quite like walking as a way to observe and get to know a city. Much more than any other mode of transportation, walking allows the observer to control the pace of observation, and there are fewer distractions than there are when driving or riding a bike. It is possible to get to otherwise inaccessible places. Most important, walking allows the observer to be in the environment more fully, and the deliberate pace permits one to integrate what is seen with the knowledge and experiences stored in one's mind. I also think it facilitates recall.

There are problems, to be sure, not the least of which is that the observer feels like an intruder in an unfamiliar environment and therefore is uncomfortable. Because of that feeling, the observer may see things differently, may look too rapidly, may come to conclusions that reflect the discomfort. Women, who can be targets of overt observation, verbal confrontation, and sometimes even physical abuse, are more likely to be uncomfortable as walking observers, a problem that has yet to be overcome. A short, simple explanation of what one is doing can be an adequate response to the "Who are you? What are you doing here?" questions, even when asked with hostility. Once people know what the observer is doing, they are often pleased to talk about or show their neighborhoods.

For some purposes and at certain scales, walking may not be appropriate. A helicopter trip is a good way to find out quickly where major new development is happening or is likely to happen in the future. One flight over the Phoenix area clearly shows that nothing will stop the development of all the presently cultivated land, and maybe more, if someone wants to do so. Donald Appleyard and Kevin Lynch reconnoitered all of the San Diego metropolitan area in one afternoon by helicopter, and the messages they read became an important part of their subsequent work there.

Low-density suburban areas, which were designed for driving, not walking, invite observation by car. The windshield survey has its uses, particularly for getting a general impression of the nature of development and the income status of residents. But I would always suggest getting out of the car at some point and taking a walk, even if only for ten minutes. One begins to experience the area differently.

Buses and other public transportation, bicycle, boat – all can be appropriate for specific purposes. When walking is not possible or appropriate, one should look for clues that are consistent with the speed of looking and the distance from what is being looked at. In a car or helicopter, for example, one should not try to see or interpret detail. Look for the physical qualities of large areas, not the dynamics within them.

Do not try to take photos and observe at the same time. Taking a picture interrupts observing and thinking about and questioning what you see. The photographer is concerned with focus, lens openings, composition, light, and shadow, with how the picture will look. Come back to take pictures later.

Sketching, however, helps looking, makes one more observant. One can think about what one is seeing while sketching, how the elements are arranged and fit together. Sketching facilitates measuring, which, as I have noted, is crucial in making comparisons and understanding the meaning of small and large, good and bad, a lot and a little. Drawing skills are not important, because the sketch will not be shown to anyone.

If possible, walk an area at a time when it is busy. Seeing more people means seeing more clues, and also seeing how people use their city, what is important to them and what is less so. At the same time, the observer should be aware that this is an active time and should try to imagine what it is like at other times. An area may have a very different character in summer and winter, sun and rain, day and night. Understanding these differences is the next best thing to repeated and prolonged observation.

There is no best path, no best place to start or stop a walk. If there is a best way, it is to follow a number of different, overlapping paths, including those that go in back of buildings, along alleys and service lanes. Backs sometimes tell more about maintenance, condition, and space than fronts. People are always surprised at the spaciousness of rear yards in many densely populated eastern cities and in San Francisco.

In this business of looking and interpreting, two people seem to be better than one. Two people can question each other, develop and challenge more hypotheses, bring more knowledge to a situation. Two people may also be an answer to the safety problem women face when alone in some areas. If there is a drawback to having two or more observers, it is that it takes more time to express verbally what one is seeing or thinking, and in that time the observer is less aware of the environment.

When I am observing, I talk with anyone who speaks to me in a friendly way or who, after eye contact and a nod or smile, seems willing to talk. Firemen at leisure, storekeepers, real estate brokers, and librarians know a lot about their areas; so may a person walking along the street. The observer uses everything he can get his hands on to understand and plan for a community; the residents are a very good source of information.

Remember that observation is not a test. No one is forcing the observer to come up with conclusions, except perhaps himself. Don't try to cover too much ground at one time, because one sees less when tired.

A single clue cannot answer questions about an area's historic development and evolution, present state, and problems that exist or may unfold. Taken together, clues are more meaningful, but even then their meanings are more "iffy" than precise. That iffiness is not necessarily a problem; it is a reality that is also true of other research and diagnostic techniques. The lack of certainty may lead to a number of alternative hypotheses about an area, which can be tested if they are important enough. Rather than being a problem, the unsureness of observation makes an area more real, alive, breathing.

Observers see things differently, even to the extent of seeing different clues. But it should come as no surprise that any number of different clues may lead one to similar conclusions about an area. Seeing a rash of bicycles, basketball hoops, caution signs for drivers, all within a tract of ten-year-old three-bedroom homes may lead an observer to conclude that there are many school-aged children, that the families have a particular lifestyle, and even that certain problems will accompany this population group. But an observer who sees none of those clues but sees a neighborhood school with many students might well come to the same conclusions. . . . One need have no fear of not seeing "the right things" – there may be no right things. There is plenty to see, plenty from which to take messages and form hypotheses.

The knowledge one brings to observation can help narrow down the many possible interpretations of what is seen. What knowledge is most helpful? The *social and economic history* of a culture and urban area is crucial; that knowledge is the context for what is observed. When did important social and economic movements take place? What was life like for people here in different periods? What was the timing of reform movements? How have welfare concepts and programs, government roles, technology, and political movements and philosophies changed over time? This knowledge is just as important locally as on a regional or national scale.

Urban planners and others involved in city conservation and development should know *how cities have grown and developed physically*. They should be able to relate that knowledge to the social and economic history of the culture. One ought to know, for example, how streetcars and railroads and highways have structured city development.

Some knowledge of *architectural styles* and their history is important. Experience suggests that to be useful, style periods need not be precise and that they can be longer the further removed they are from the present day: pre-Civil War, late 1800s, turn of the century, the 1920s, the Depression, pre-World War II, 1950s, 1960s, post-1960s. Most people know more than they think they do about when different styles of buildings were constructed. Without studying the subject, however, they are not likely to know enough to consistently understand what these styles tell us about urban areas.

In the same way, knowledge of *artifact history* is profoundly useful. By this I mean the time periods when different types of curbs, street lights, paving materials, signs, curtains, blinds, and building materials were used. This knowledge is more difficult to acquire; there are so many details, and the evolution of any one kind is rarely documented or easily discovered. Reading old photographic journals and technical manuals and looking for dates on the artifacts themselves helps. Perhaps more than for any other category of clues, this information is best learned from experienced professionals. Once a person becomes aware of and starts thinking about the history of a detail, say of the different kinds of curbs that have been used in a city, it becomes an enjoyable pastime to find out more about it.

It is critical to know something of *construction and maintenance.* The condition and maintenance of buildings are important clues to problems and changes that are taking place. Nonspecialist observers often do not understand building construction and what it takes to keep a building in good condition. This can be learned, if not from taking academic courses then from reading books and manuals on construction and renovation, from being a truly attentive sidewalk superintendent at construction sites, or from actually building and maintaining a house.

Almost all of this knowledge can be learned. Effective observation and diagnosis require no special gift, but they are facilitated by all the knowledge one has accumulated and by constant conscious questioning of what is observed.

The next question is, what can you do with what you find out by looking? There are situations where observation may be the only tool available to suggest what to do. A group or an official or a potential client may need to know quickly how to go about planning for a specific area. There may be only enough time for one site visit before making some initial decisions. I was once told by the officials of a large city that they were strongly considering a major development project, which they would be announcing soon. What did I think about their intentions? In the two to three hours I spent in the area I came up with what seemed *obvious* questions about dislocation of people and businesses, traffic circulation, the market for what they were considering, and more. Some of those

questions had not been obvious to the officials; at least they had not thought about them. They decided to find out a great deal more about the area before proceeding.

Usually observation is used, less dramatically, with other research tools in a continuous, back and forth manner. Today a team of observers recognizes that downtown seems to be pushing into a neighboring residential area. This observation generates economic and demographic research, with implications for public policies and programs. Tomorrow some traffic data calls for a field trip to see what the actual conditions are like. Often the local residents' concerns about a particular issue generate coordinated research, including observation. In any case, field observation is used along with other research methods.

Early knowledge of a problem permits early action, if that is appropriate, or early preparation for action. Observation may reduce the number of surprises to be faced. If one knows by looking that a large unused railroad yard near a busy downtown area is a likely site for development, then one can prepare for it. After looking at an area south of San Francisco's downtown, an area of marginal economic uses and boardinghouses for poor, transient men, I was able to advise a potential purchaser of land for a small new office building that the location was presently inappropriate. But having also seen signs that the downtown area was rapidly expanding in that direction, I was also able to advise the buyer that depending on the price and the length of time he could hold an empty lot, the site would soon have potential value for what he had in mind.

Observation also enables planners to take early direct action in response to problems and opportunities. In 1968, toward the end of the Johnson administration in Washington, San Francisco had an unexpected opportunity to receive federal funds for small, neighborhood parks. Within thirty days the city would have to propose specific sites where the money could be appropriately used. Intimate knowledge of the city, gained in large measure through looking, enabled a handful of staff to come up with over a hundred sites almost overnight, from which the final thirty-three locations were then chosen....

I think also that any meaningful plan for a city, including small building projects, should start with an understanding of the nature of the place and should call for respecting and improving the existing physical character of the community. It should respond to important social and economic issues within that framework. . . .

In the end, the whole process of looking, questioning, trying to gain understanding makes a person a more intimate, respectful part of any environment and therefore more likely to be caring of it. That is the basis for good planning and beneficial action.

"A Progressive Politics of Meaning"

from *The Politics of Meaning: Restoring Hope and Possibility in an Age of Cynicism* (1993)

Michael Lerner

Editors' Introduction

Beyond specific tools such as sustainability indicators and ecological footprint analysis, and even beyond the more general ability to understand the past and potential future of urban environments through firsthand observation, lies the question of how to work effectively for political and social change. How might planners and activists best position themselves to bring about more sustainable communities? What strategies can make the most difference in the long run? How can they view their own work as satisfying and meaningful in the face of frequently overwhelming obstacles?

Such questions have no easy answers, but planning theorists and social reformers have developed a number of approaches toward addressing them in recent years.

One strategy has been to reform planning processes by making them more open and participatory, in contrast to past decades when politicians or local government staff often made decisions with little public involvement. This emphasis on participation has overlapped with a growing theoretical view of planning as a process of communication and social learning. So-called "communicative" planning theorists see the role of planners as being to network with others, communicate information, structure participatory processes, mediate conflicts, and help many diverse stakeholder groups gain understanding of solutions in a way that can benefit all. Leading writings in this field include John Forester's *Planning in the Face of Power* (Berkeley: University of California Press, 1989), Patsy Healey's *Collaborative Planning* (Vancouver: University of British Columbia Press, 1997), and Judith Innes' "Planning Theory's Emerging Paradigm: Communicative Action and Interactive Practice" (in the *Journal of Planning Education and Research*, 14(3), pp. 128–35, 1995). These writers in turn build on philosophers such as Jürgen Habermas, who pioneered concepts such as "communicative action" in works such as *Communication and the Evolution of Society* (Boston: Beacon Press, 1979), and Anthony Giddens, who emphasized the dynamic interaction of people with structures of rules, resources, and meaning within society in books, such as, *Beyond Left and Right: The Future of Radical Politics* (Palo Alto: Stanford University Press, 1994) and *The Third Way: The Renewal of Social Democracy* (Cambridge, UK: Polity Press, 1998). Political scientist John Dryzek has also developed the concept of "discursive democracy," in which more open and participatory political institutions would help encourage more reflective public decision-making, in works such as *Discursive Democracy* (New York: Cambridge University Press, 1990).

Another approach to the question of how planners and others can bring about change has been to stress the role of advocacy. A large literature has stressed the importance of urban social movements in bringing about change and in forming a balance to the power of business groups and social elites. Some have argued that planners themselves should help organize these movements. In a well-known article "Advocacy and Pluralism in Planning" (*Journal of the American Institute of Planners*, 1965), Paul Davidoff called for "advocacy planning" in which professionals provided their expertise to help empower disenfranchised communities. A related call for "equity planning" by Norman Krumholz and others asked for planners to actively address power and resource imbalances within society, and asserted that this was an ethical responsibility for the profession. John Friedmann discusses several of these approaches in his book *Planning in the Public Domain* (Princeton, NJ: Princeton University Press, 1987), contrasting in particular "social learning" and "social mobilization" perspectives.

In this selection from his book *The Politics of Meaning* (Reading, MA: Addison-Wesley, 1993), Michael Lerner takes an even broader view of the process of political change, agreeing that politics and institutions need to be changed, but arguing that at the same time the process must be one of personal and spiritual transformation. Similar to Giddens, he stresses reforming social institutions so as to be more supportive of individuals and community, and to emphasize compassion and caring. In his view this reform can come about both through developing a collective political agenda and movement, and through dedicated work by individuals based on their growing appreciation of the fundamental unity of all beings. Not surprisingly, Lerner's own roots are in spiritual traditions. A rabbi and psychologist, he is the founder and editor of *Tikkun* Magazine and a leader of the Jewish Renewal movement. Following the publication of *The Politics of Meaning* Lerner was consulted by candidate and then President Bill Clinton and his wife Hillary, and briefly became known in the media as "the guru of the White House." Although the Clintons failed to follow through on their interest in developing a politics of caring, and their successor George W. Bush relied on a rhetoric of "compassionate conservatism" which was anything but compassionate in practice, it could be argued that the simple fact that these politicians acknowledged the importance of such values represents a step toward Lerner's point of view. A great deal of work needs to be done to change conditions so that a truly compassionate politics can emerge – and so that the electorate will demand this – but in the long run such a politics is perhaps the most powerful tool we could have for bringing about sustainable development.

A BRIEF DEFINITION

A progressive politics of meaning is a political effort to accomplish the following five goals:

1. To create a society that encourages and supports love and intimacy, friendship and community, ethical sensitivity and spiritual awareness among people

Our economic, political and social arrangements make this kind of sensitivity and awareness more difficult to obtain and sustain. A politics of meaning does not seek to create a particular meaning system, but it does seek to create social and economic arrangements that will be friendly to meaning-oriented communities rather than harmful to their central concerns. In part, it means challenging the instrumental, utilitarian, mechanistic reductionism of thought and the disenchantment of our social experience. In part, it means creating institutions and economic practices that awaken within us our own ethical and spiritual sensitivity and our desire to treat one another with gentleness and compassion.

By spiritual awareness or sensitivity, I mean this: an awareness of the fundamental unity of all being and of our connectedness to one another and to the universe. When our unity and interconnectedness is fully appreciated, the arrogance and egotism that predominates in politics will dramatically decrease. Virtually every religious and spiritual system aims at this awareness. The politics of meaning does not seek to endorse any particular way of achieving it, but seeks to replace political and economic institutions that undermine this kind of spiritual awareness and that encourage human arrogance and ecological insensitivity.

2. To change the bottom line

In most Western societies, productivity or efficiency is measured by the degree to which any individual or institution or legislation or social practice increases wealth or power. To pay attention to the bottom line is thus defined as paying attention to the degree to which the person or the project in question succeeds in maximizing wealth and power. Other goals are ancillary – acceptable only if they help accomplish (or, at least, do not thwart) the material goal.

A progressive politics of meaning posits a new bottom line. An institution or social practice is to be considered efficient or productive to the extent that it fosters ethically, spiritually, ecologically, and psychologically sensitive and caring human beings who can maintain long-term, loving personal and social relationships. While this new definition of productivity does not reject the importance of material well-being, it subsumes that concern within an expanded view of "the good life": one that insists on the primacy of spiritual harmony, loving relationships, mutual recognition, and work that contributes to the common good.

3. To create the social, spiritual, and psychological conditions that will encourage us to recognize the uniqueness, sanctity, and infinite preciousness of every human being, and to treat them with caring, gentleness, and compassion

4. To create a society that gives us adequate time and encouragement to develop our inner lives

We seek a society that no longer counterposes the time needed for inner work to develop our spiritual, aesthetic, and psychological sensibilities with the time needed to make ourselves economically secure and successful.

5. To create a society that encourages us to relate to the world and to one another in awe and joy

Instead of rewarding our ability to dominate, manipulate, or control, we seek to build families, communities, and economic and political institutions that encourage our capacity to experience wonder and radical amazement at the grandeur of the universe, and to experience pleasure and celebration of one another as embodiments of the spiritual energy of the universe (or, in religious terms, as creatures made in the image of God).

Large-scale changes of this kind cannot be accomplished quickly. Nevertheless, there are many things we can do in the short run to move our society toward this goal. I discuss these possibilities in greater detail in the last section of this book, but will outline a few of them here.

For example, a program for a progressive politics of meaning in the next few decades would seek:

- Public-school curricula that integrate the teaching of empathy and caring for others, and reward schools according to their success in creating empathic human beings (a goal quite different from that of traditional liberal demands, which have focused on student–teacher ratios or teacher salaries, but have rarely addressed the values being taught).

- Restructured health-care systems, so that medical care more adequately reflects an understanding of how the frustration of people's needs – for meaningful and nonalienating work, mutual recognition, love and caring, and a spiritually and ethically grounded society – may underlie receptivity to disease. Certainly, a progressive politics of meaning would endorse a single-payer universal health-care plan, on the grounds that we are never going to build a society that rejects selfishness while denying equal access to health care. But what distinguishes a politics-of-meaning approach is that it also links physical health to our ethical and spiritual well-being. So, for example, it also maintains that meaningful work – affording workers the opportunity to use their intelligence, creativity, and cooperative ability, and providing space for their spiritual lives – will decrease vulnerability to disease.

- A pro-family agenda that gives families the social supports they need. In addition to the traditional liberal elements (economic viability, flextime, child care, and so on), a politics-of-meaning approach includes as equally central the psychological and spiritual needs of families. We want to create a society that is safe for love and intimacy – as opposed to contemporary societies that identify sophistication with cynicism, critical intelligence with moral detachment, and maturity with "healthy suspicion of others." Instead, we seek a society that will encourage people to be more caring, sensitive, and empathic to others. So, for example, we want workplaces that encourage cooperation and give all of us an opportunity to use our intelligence and creativity. We want an economy that encourages us to take into account the needs and interests of others. We see these things not primarily as "rights" of the lone individual, but as requisite for shaping societies that nourish loving relationships.

- Annual ethical-impact reports from government and private sector institutions to assess their effect on the ethical, spiritual, and psychological well-being of our society and on the people who work in and with these institutions.
- Reflection within every profession on activities and attitudes that would be possible if the goals were to serve the common good; to heighten ethical, spiritual, and ecological sensitivity; and to reward loving and caring behavior. Such reflection, for example, has led some lawyers associated with a politics-of-meaning perspective to envision a second stage of trials, in which the adversary system is suspended and the focus is shifted to healing the problems and pain that the initial trial has uncovered in the community.

Anyone who takes these specific examples and regards them as the sum total of the politics of meaning has missed my point. A progressive politics of meaning leads to a rethinking of every aspect of our public and private lives. However, it resists any attempt to impose one particular lifestyle or one particular approach to spirituality.

Most importantly, a politics of meaning is an invitation to transcend all the internalized messages that tell us that it is unrealistic to base our lives on our highest ideals and to fight for their realization. It is no longer appropriate to fight for instrumental goals that do not really express our vision of the kind of world we actually seek. Doing politics in this limited way turns out to be unrealistic and ineffective, because the political Right advocates its full vision enthusiastically. Progressives, to compete successfully, must present their highest, most compelling vision.

THE DESTRUCTIVE WAYS IN WHICH PEOPLE FIND MEANING

The primary dynamic of politics in the twentieth century has been the alternation between repressive communities of meaning and the alienation and loneliness produced by market societies. In reaction to the disintegration of existing communities, and with nothing to protect them from the alienation and loneliness produced by market societies, many people have become so hungry for human connection that they have even been willing to join repressive communities to fill that need. Some have turned back to traditional religious communities. Others have been attracted to the vision of community being offered by xenophobic nationalist movements.

Yet, to the extent that these various communities embody patriarchal privilege and class oppression, they frequently generate their own internal opposition. Just as, in the past, liberalism was created in opposition to the repressive nature of feudal societies, so today we find groups of people within these repressive communities who have begun to question the necessity of sacrificing for the community, once they realize that the sacrifices being sought primarily benefit the interests of small male elites. Disillusioned with such communities, these people turn back to the market, happy to be rid of what they have discovered: that their community is not truly a community, and that it does not really operate according to its proclaimed values. By contrast, in the market there is no similar hypocrisy: everybody really does try to maximize his or her own self-interest, and says so.

However, pursuing the market option and its logic of individualism also proves problematic for many people, since most of them *don't* "make it," and the ethos of selfishness sanctified by market-dominated societies soon yields deep unhappiness. So the cycle continues, and once again, people begin to grasp for communities of meaning, however repressive. [. . .]

ARE PEOPLE READY TO BE PART OF A MEANING-ORIENTED MOVEMENT?

There already are millions of people effectively engaged in challenging the de-meaning of the world. But so far, most of these people do not see themselves as part of a larger meaning-oriented movement.

There already are millions of people involved in the pursuit of meaning and in challenging the atomistic, meaning-denying, economistic, or reductionist accounts of reality that have dominated public discourse. These people recognize that there is something fundamentally wrong with the dominant paradigm in the West, and they are building a more holistic view of our relationships to one another and to the natural world.

Some are involved in alternative approaches to health and healing, to nutrition and diet, to exercise and sports. Some are involved in a worldwide ecological movement which understands that, in order to save the planet, we may need to relate to the world with greater amounts of spiritual energy (awe, wonder, and radical amazement). Some are involved in developing spiritual practices, either in connection with religious communities or in connection with secular approaches to meditation and inner spiritual work.

Many are engaged in creating new relationships between men and women, either through the organized women's movement and the pro-feminist men's movement, or through more personal experiments in creating gentle and caring ways of relating. Many are engaged as teachers, nurses, social workers, and counselors; as rabbis, priests, ministers, and imams; and in other pursuits explicitly dedicated to reconnecting people to one another, to the spiritual energies of our inner selves, and to God's presence in the universe.

Currently, however, most people who are working for meaning operate in relative isolation from one another; they do not share a unified analysis. Some of them are involved in New Age philosophies that I find unpersuasive, or even in right-wing politics. Others resolutely assure one another that they are apolitical and uninterested in social issues. But what all these people do share is an ultimate concern with transcending the mechanistic, atomistic, antispiritual, and nonrelational ways of understanding the universe that have so crippled our thinking in the past.

These people already are challenging the dominant discourse, already are discounting what the official spokespeople have been saying. Some of these challengers turn to self-help and inspirational books, creating best-sellers as they grasp (sometimes indiscriminately) at accounts of reality that affirm our unity and innerconnectedness. Some people, to be sure, want nothing more than a personal solution, and will never be interested in any transformative vision that takes them beyond the confines of their own personal situation. But many more people understand that there is something deeply wrong with our world. Their alienation from the meaning-deadening world in which we live turns them toward new paradigms and new ways of conceptualizing reality. These people, who have faced the spiritual impoverishment of our contemporary world, will form the vanguard of social and political change in the twenty-first century.

One subgroup of this coming vanguard includes the millions of people who were ethically engaged with the social change movements of the 1960s and 1970s, but who today live private lives in part because they can see no political organization that plausibly speaks to their sensibilities. The political struggles of this generation of baby boomers awakened millions of Americans to an ethos of caring for others that was manifested in the civil rights movement, the antiwar movement, the women's movement, and the social justice and environmental movements.

Conservatives have correctly pointed out the ways in which these groups sometimes flirted with a countercultural ethos of individual fulfillment and unchecked self-indulgence that often undermined the moral content of the social change movements. There were moments when "liberation" was construed to mean freedom to do whatever one wanted to do, without regard for the consequences to others. To the extent that countercultural and political movements fell into this way of thinking or acting, they were quintessentially mainstream, providing yet another way for the dominant ethos of the market to permeate and shape mass consciousness. I myself and some of my friends sometimes fell into this self-indulgent definition of liberation, and we were deeply mistaken. I have learned from this experience to be more self-critical and also to listen carefully to the criticisms of people who have conservative politics with which I disagree, since they are sometimes correct and can see things that I cannot see. In fact, I believe that my entire generation has learned to approach politics with the humility that we tragically lacked several decades ago.

Indeed, a politics of meaning runs counter to this earlier tendency toward moral relativism and immediate gratification without moral standards. Yet, I must hasten to add that, even in the 1960s and 1970s, there was within these social change movements a countertendency, often explicitly challenging countercultural indulgence, that emphasized social solidarity and caring for others, that rejected moral relativism, and that articulated a powerful moral critique of the alienation and injustice of the contemporary world. Millions of people who went through that experience remain deeply committed to social justice and to building a more humane and loving society. Many of them have despaired that it ever would be possible to achieve those ends, and have become involved in lifestyles that on the surface seem superficially unconcerned about larger issues. Yet, like so many people in the religious world and in the labor movement, they could be mobilized to a new politics of love and caring were they to learn about it and come to believe that it was possible. Having been burnt by past failures, these former activists will not quickly jump into new political movements. Yet, as a meaning-oriented movement gains momentum, many of them will feel a homecoming that reconnects to their deepest hopes.

These are some of the groups from whom the movement for a politics of meaning will draw its initial support. They will become the transformative agents who move these ideas into the mainstream of American society. These people respond out of a real inner need, not from a commitment to an abstract idea, nor out of a sense that someone else ought to be treated differently. These people know that they cannot secure the kind of life they deeply desire, unless much changes in our society – its structure of values, its relationship to spiritual values, and its opportunities for mutual recognition. These are radical needs. Unlike needs for economic well-being or political rights, these cannot be fulfilled inside our society as it currently is constructed. Nor can these be fulfilled by buying off any one group. In that sense, the condition for the fulfillment of our needs for meaning is the condition for the liberation of our entire society from a materialist and individualist ethos.

Today, people involved in the pursuit of meaning do not yet form a coherent movement, and if they did, it seems unlikely that it would be a politically progressive movement. Nevertheless, the problems that have sensitized them to the crisis of meaning will not be solved by any other contemporary political movements. For this reason, I believe that people who recognize the crisis of meaning will provide the basis for a socially transformative movement in the twenty-first century. This movement will be at the center of creating a different kind of politics in America. That such a movement would be progressive is not guaranteed. It depends on whether the crisis of meaning remains the property of the Right, as it has become in the past decades, or whether liberals and Progressives allow themselves to move beyond the limits of their current conceptual schemes and seriously begin to address meaningful issues.

One function of bringing these people together through the framework of a politics of meaning is to help them recognize their potential power as a transformative and healing force. At this moment, their potential social power appears invisible even to themselves, and their voices remain marginalized.

Nevertheless, their marginality applies only to the conventional political arena, whereas the strategy of social transformation articulated here is not narrowly political in that sense. Changes in consciousness and in the ways people lead their lives will, in the long run, be far more important than who wins this or that election.

Every person engaged in acts of aesthetic and spiritual creativity, every person engaged in acts of mutual recognition that reach beyond the conventions of contemporary isolation, every person engaged in prayer that is spiritually alive, every person who refuses to be cowed by the dominant materialism or ethos of selfishness, every person who rejects technocratic accounts of reality, every person who affirms humor and playfulness and awe and wonder, is herself or himself part of the transformative process that will eventually break through the stranglehold of a meaning-deadening society.

THE IMMEDIATE TASK: SHIFTING THE DOMINANT DISCOURSE

Society's dominant discourse shapes not only its politics but the way people think about their personal lives and choices. Just as John F. Kennedy helped legitimize a discourse of idealism that gave impetus to the social movements of the 1960s, so Ronald Reagan managed to legitimize a discourse of selfishness and insensitivity that has had profound social consequences, far beyond his administration's legislative successes.

Shifting society's discourse – from one of selfishness and cynicism to one of idealism and caring – is the first and most important political goal of a politics of meaning in the next several decades. Long before we can reshape American society in any practical way, we must shift the way we think about our social institutions, politics, and economic practices. Therefore, one of the first priorities of a campaign for a politics of meaning will be to challenge the bottom-line assumption that people must always give priority to looking out for number one.

Changing the dominant discourse will change the messages we give to ourselves and to one another. The more we are able to support the part of ourselves that wishes to commit to a higher vision of who we can be, the likelier we are to take the steps in our personal lives that will make that possible. We are likelier to care for our souls and for our own spiritual and moral development when we live in a society where these kinds of concerns are publicly validated rather than ridiculed or marginalized. We are likelier to find ways to repair the psychic damage done by early childhood mis-recognition and the forced denial of our desire for meaning when we live in a society which publicly values our ability to care as much as it values our ability to dominate or control.

The ultimate test of a politics-of-meaning movement, however, will always be the degree to which it liberates us to engage in small acts of caring and spiritual sensitivity. The more we engage in such acts, the more we are actually building a politics of meaning rather than just talking about it.

THREE

PART FOUR

Sustainable urban development internationally

INTRODUCTION TO PART FOUR

Although the character of cities and towns varies enormously around the world, many urban sustainability issues are similar. Communities almost everywhere these days must decide how to limit or reduce automobile congestion, clean up pollution and contaminated lands, ensure decent, affordable housing for residents, provide infrastructure that enhances rather than degrades natural environments, foster steady local sources of jobs, and promote equity and quality of life. While a few societies remain remote, urban areas throughout Europe, Asia, Africa, and Latin America are experiencing rapid Western-style development. Such transformation is fueled by the expansion of global corporations, ubiquitous technologies such as the automobile, advice from development agencies and First World planning consultants, and the spread of materialist culture generally.

That being said, many differences do exist in urban planning challenges. Third World local governments are frequently overwhelmed with the challenge of meeting demands for basic infrastructure and services. Urban population growth in many developing nations is far more rapid than in the First World. Mexico City's population, for example, grew from 3.5 million people in 1950 to 17.6 million in 2000, and is projected to expand to 19 million by 2015, according to National Intelligence Council estimates. The population in Lagos, Nigeria, grew from 1 million in 1950 to 12.2 million in 2000, and is projected to reach 24.4 million in 2015. That in Bombay grew from 2.8 million in 1950 to 16.9 million in 2000, and is expected to be flooded by 27.4 million people in 2015.[1] The existence of enormous informal settlements (in which residents lack title to land and have usually constructed dwellings themselves out of available materials) also presents unique problems in less-developed countries. At the same time, cities in Europe and elsewhere often benefit from far older urban traditions than in North America, with substantial advantages in terms of compact urban centers, walkable streets, strong public transit systems, and wonderful historic architecture and indigenous design traditions.

The following selections are intended to give a flavor of the issues related to sustainable urban development in different parts of the world. Those wishing more detail on international sustainability planning initiatives are encouraged to visit the websites of the United Nations Division for Sustainable Development (www.un.org/esa/sustdev/), The United Nations Human Settlements Programme (www.unchs.org), the International Council on Local Environmental Initiatives (ICLEI; www.iclei.org), the European Sustainable Cities and Towns Campaign (www.sustainable-cities.org), the International Institute for Sustainable Development (www.iisd.org), the World Wide Web Library on Sustainable Development (www.ulb.ac.be/ceese/meta/sustvl.html), the World Business Council for Sustainable Development (www.wbcsd.ch), and other organizations. Books providing background on international urban sustainability include *Sustainable Cities: Urbanization and the Environment in International Perspective*, edited by Richard Stren and others (Boulder, CO: Westview Press, 1992), *Green Development: Environment and Sustainability in the Third World*, by W.M. Adams (New York: Routledge, 1990), *Environmental Problems in Third World Cities*, by Jorge E. Hardoy and others (London: Earthscan, 1992), *Making Development Sustainable: Redefining Institutions, Policy, and Economics*, edited by Johan Holmberg (Washington, DC: Island Press, 1992), *The Earthscan Reader in Sustainable Cities*, edited by David

Satterthwaite (London: Earthscan, 1999), and *Livable Cities?: Urban Struggles for Livelihood and Sustainability*, edited by Peter Evans (Berkeley: University of California Press, 2002).

NOTE

1 National Intelligence Council. 2000. *Global Trends 2015*. Washington, DC.

"Urban Planning in Curitiba"

from *Scientific American* (1996)

Jonas Rabinovitch and Josef Leitman

Editors' Introduction

More creative or sweeping sustainability-related initiatives can sometimes take place in developing nations than within industrialized countries, because of urgent crises demanding immediate action, centralization of government authority, more dynamic political leadership, or a lack of established bureaucratic tradition. For such reasons the Brazilian city of Curitiba, a metropolis of 1.6 million in the southern part of the country below São Paulo, has emerged as one of the world's leading examples of creative urban development. Although certainly not without problems, Curitiba has managed to innovate consistently over forty years to help manage its extremely rapid growth and provide a more livable and sustainable environment for its residents. It did so largely under the leadership of former mayor Jaime Lerner, an architect who helped found the Research and Urban Planning Institute of Curitiba, a strong independent city planning agency.

This selection from *Scientific American* (March 1996, pp. 46–53) provides an overview of some of Curitiba's achievements. Jonas Rabinovitch is a Brazilian planner who has worked for the city's research and urban planning institute as well as the United Nations. Josef Leitman has worked as a senior urban planner with the World Bank and is the author of *Sustaining Cities: Environmental Planning and Management in Urban Design* (New York: McGraw-Hill, 1999). Additional information about Curitiba is available from the website of the city's urban planning institute at www.ippuc.org.br (English version available).

In contrast to Curitiba's model, which has been led by a strong mayor and planning agency and criticized at times for its top-down nature, the nearby Brazilian city of Porto Alegre has pioneered one of the world's best examples of participatory planning. The latter city has pioneered "participatory budgeting" through which each neighborhood gets to vote on spending priorities for a percentage of the budget. Although Porto Alegre's experiment is more recent than Curitiba's, in the years since the reforms began in 1989 the city has made enormous gains in literacy, sanitation, housing, and other public services. More information on Porto Alegre is available from Rebecca Abers' *Inventing Local Democracy: Grassroots Politics in Brazil* (Boulder, CO: Lynne Rienner Publishers, 2000), and from UN Habitat's Best Practices website at www.sustainabledevelopment.org.

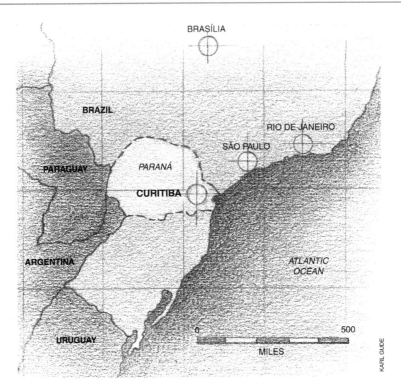

Figure 1. Location map of Curitiba.

Source: Illustration by Karl Gude.

As late as the end of the nineteenth century even a visionary like Jules Verne could not imagine a city with more than a million inhabitants. Yet by the year 2010 over 500 such concentrations will dot the globe, 26 of them with more than 10 million people. Indeed, for the first time in history more people now live in cities than in rural areas.

Most modern cities have developed to meet the demands of the automobile. Private transport has affected the physical layout of cities, the location of housing, commerce and industries, and the patterns of human interaction. Urban planners design around highways, parking structures and rush-hour traffic patterns. And urban engineers attempt to control nature within the confines of the city limits, often at the expense of environmental concerns. Cities traditionally deploy technological solutions to solve a variety of challenges, such as drainage or pollution.

Curitiba, the capital of Paraná state in south-eastern Brazil, has taken a different path. One of the fastest-growing cities in a nation of urban

booms, its metropolitan area mushroomed from 300,000 citizens in 1950 to 2.1 million in 1990. Curitiba's economic base has changed radically during this period: once a center for processing agricultural products, it has become an industrial and commercial powerhouse. The consequences of such rapid change are familiar to students of Third World development: unemployment, squatter settlements, congestion, environmental decay. But Curitiba did not end up like many of its sister cities. Instead, although its poverty and income profile is typical of the region, it has significantly less pollution, a slightly lower crime rate and a higher educational level among its citizens.

DESIGNING WITH NATURE

Why did Curitiba succeed where others have faltered? Progressive city administrations turned Curitiba into a living laboratory for a style of urban development based on a preference for public

Plate 1. *Lakeside parks* serve multiple functions in Curitiba. In addition to providing green space for citizens and forming part of the metropolitan bicycle-path network, they help to control the floods that once plagued the city. The artificial lakes, created during the 1970s, are designed to facilitate drainage and to hold excess rainwater and keep it from inundating low-lying areas.

Source: Photo by Jonas Rabinovitch.

Plate 2. *24-hour street*, an arcade of shops and restaurants that never closes helps to keep Curitiba's downtown area vital. The city has also regulated the locations of banks, insurance companies and other 9 to 5 businesses to prevent the district from becoming a ghost town after working hours.

Source: Photo by Jonas Rabinovitch.

Plate 3. *Historic center* of Curitiba has received special planning protection, including incentives to build elsewhere, that preserves old buildings. Many of the district's streets have been converted to pedestrian use, reducing pollution and fostering a sense of neighborhood. Ceremonial gates mark sections of the central city that were once enclaves for particular immigrant groups (the entrance to the former Italian quarter is shown at the right).

Source: Photo by Jonas Rabinovitch.

transportation over the private automobile, working with the environment instead of against it, appropriate rather than high-technology solutions, and innovation with citizen participation in place of master planning. This philosophy was gradually institutionalized during the late 1960s and officially adopted in 1971 by a visionary mayor, Jaime Lerner, who was also an architect and planner. The past 25 years have shown that it was the right choice; Rafael Greca, the current mayor, has continued the policies of past administrations and built on them.

One of Curitiba's first successes was in controlling the persistent flooding that plagued the city center during the 1950s and early 1960s. Construction of houses and other structures along the banks of streams and rivers had exacerbated the problem. Civil engineers had covered many streams, converting them into underground canals that made drainage even more difficult – additional drainage canals had to be excavated at enormous cost. At the

Plate 4. *Main boulevard* of Curitiba, now devoted to pedestrian traffic, is the site of a weekly celebration of children gathering to paint. The ceremony began more pragmatically in 1972: when motorists threatened to ignore the traffic ban and drive on the street as usual, city workers blocked them by unrolling enormous sheets of paper and inviting children to paint watercolors.

Source: Photo by Jonas Rabinovitch.

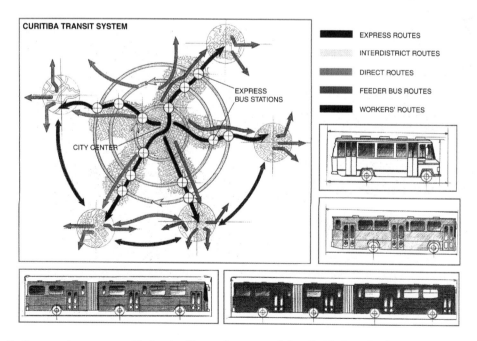

Figure 2. *Bus routes* have grown with the city. Express bus routes define Curitiba's spoke-shaped structural axes; interdistrict and local lines fill in the space between spokes. Each route is serviced by a bus of appropriate scale, from minibuses that carry 40 people on local trips to giant 270-passenger biarticulated vehicles used for express travel.

Source: Illustration by Karl Gude.

Plate 5. *Transport network* includes bicycle paths integrated with streets and the bus network for most efficient travel. The bicycle paths also connect the city's main parks.

Source: Photo by Jonas Rabinovitch.

same time, developers were building new neighborhoods and industrial districts on the periphery of the city without proper attention to drainage.

Beginning in 1966 the city set aside strips of land for drainage and put certain low-lying areas off-limits for building. In 1975 stringent legislation was enacted to protect the remaining natural drainage system. To make use of these areas, Curitiba turned many riverbanks into parks, building artificial lakes to contain floodwaters. The parks have been extensively planted with trees, and disused factories and other streamside buildings have been recycled into sports and leisure facilities. Buses and bicycle paths integrate the parks with the city's transportation system.

This "design with nature" strategy has solved several problems at the same time. It has made the costly flooding a thing of the past even while it allowed the city to forgo substantial new investments in flood control. Perhaps even more important, the use of otherwise treacherous floodplains for parkland has enabled Curitiba to increase the amount of green space per capita from half a square meter in 1970 to 50 today – during a period of rapid population growth.

Plate 6. *Bus tubes.* Most urban bus systems require passengers to pay as they board, slowing loading. Curitiba's raised-tube bus-stops eliminate this step: passengers pay as they enter the tube, and so the bus spends more of its time actually moving people from place to place.

Source: Photo by Jonas Rabinovitch.

Plate 7. *Bus tracks.* Like subways, the buses have a track dedicated entirely to their use. This right-of-way significantly reduces travel time compared with buses that must fight automotive traffic to reach their destinations. By putting concrete and asphalt above the ground instead of excavating to place steel rails underneath it, the city managed to achieve most of the goals that subways strive for at less than 5 per cent of the initial cost.

Source: Photo by Jonas Rabinovitch.

PRIORITY TO PUBLIC TRANSPORT

Perhaps the most obvious sign that Curitiba differs from other cities is the absence of a gridlocked center fed by overcrowded highways. Most cities grow in a concentric fashion, annexing new districts around the outside while progressively increasing the density of the commercial and business districts at their core. Congestion is inevitable, especially if most commuters travel from the periphery to the center in private automobiles. During the 1970s, Curitiba authorities instead emphasized growth

along prescribed structural axes, allowing the city to spread out while developing mass transit that kept shops, workplaces and homes readily accessible to one another. Curitiba's road network and public transport system are probably the most influential elements accounting for the shape of the city.

Each of the five main axes along which the city has grown consists of three parallel roadways. The central road contains two express bus lanes flanked by local roads; one block away to either side run high-capacity one-way streets heading into and out of the central city. Land-use legislation has encouraged high-density occupation, together with services and commerce, in the areas adjacent to each axis.

The city augmented these spatial changes with a bus-based public transportation system designed for convenience and speed. Interdistrict and feeder bus routes complement the express bus lanes along the structural axes. Large bus terminals at the far ends of the five express bus lanes permit transfers from one route to another, as do medium-size terminals located approximately every two kilometers along the express routes. A single fare allows passengers to transfer from the express routes to interdistrict and local buses.

The details of the system are designed for speed and simplicity just as much as the overall architecture. Special raised-tube bus-stops, where passengers pay their fares in advance (as in a subway station), speed boarding, as do the two extra-wide doors on each bus. This combination has cut total travel time by a third. Curitiba also runs double- and triple-length articulated buses that increase the capacity of the express bus lanes.

Ironically, the reasoning behind the choice of transportation technology was not only efficiency but also simple economics: to build a subway system would have cost roughly $60 million to $70 million per kilometer; the express bus highways came in at $200,000 per kilometer including the boarding tubes. Bus operation and maintenance were also familiar tasks that the private sector could carry out. Private companies, following guidance and parameters established by the city administration, are responsible for all mass transit in Curitiba. Bus companies are paid by the number of kilometers that they operate rather than by the number of passengers they transport, allowing a balanced distribution of bus routes and eliminating destructive competition.

As a result of this system, average low-income residents of Curitiba spend only about 10 per cent of their income on transport, which is relatively low for Brazil. Although the city has more than 500,000 private cars (more cars per capita than any Brazilian city except the capital, Brasilia), three quarters of all commuters – more than 1.3 million passengers a day – take the bus. Per capita fuel consumption is 25 per cent lower than in comparable Brazilian cities, and Curitiba has one of the lowest rates of ambient air pollution in the country. Although the buses run on diesel fuel, the number of car trips they eliminate more than makes up for their emissions.

In addition to these benefits, the city has a self-financing public transportation system, instead of being saddled by debt to pay for the construction and operating subsidies that a subway system entails. The savings have been invested in other areas. (Even old buses do not go to waste: they provide transportation to parks or serve as mobile schools.)

The implementation of the public transport system also allowed the development of a low-income housing program that provided some 40,000 new dwellings. Before implementing the public transport system, the city purchased and set aside land for low-income housing near the Curitiba Industrial City, a manufacturing district founded in 1972, located about eight kilometers west of the city center. Because the value of land is largely determined by its proximity to transportation and other facilities, these "land stocks" made it possible for the poor to have homes with ready access to jobs in an area where housing prices would otherwise have been unaffordable. The Curitiba Industrial City now supports 415 companies that directly and indirectly generate one fifth of all jobs in the city; polluting industries are not allowed.

PARTICIPATION THROUGH INCENTIVES

The city managers of Curitiba have learned that good systems and incentives are as important as good plans. The city's master plan helped to forge a vision and strategic principles to guide future developments. The vision was transformed into reality, however, by reliance on the right systems and incentives, not on slavish implementation of a static document.

One such innovative system is the provision of public information about land. City Hall can immediately deliver information to any citizen about the building potential of any plot in the city. Anyone wishing to obtain or renew a business permit must provide information to project impacts on traffic, infrastructure needs, parking requirements and municipal concerns. Ready access to this information helps to avoid land speculation; it has also been essential for budgetary purposes, because property taxes are the city's main source of revenue.

Incentives have been important in reinforcing positive behavior. Owners of land in the city's historic district can transfer the building potential of their plots to another area of the city – a rule that works to preserve historic buildings while fairly compensating their owners. In addition, businesses in specified areas throughout the city can "buy" permission to build up to two extra floors beyond the legal limit. Payment can be made in the form of cash or land that the city then uses to fund low-income housing.

Incentives and systems for encouraging beneficial behavior also work at the individual level. Curitiba's Free University for the Environment offers practical short courses at no cost for homemakers, building superintendents, shopkeepers and others to teach the environmental implications of the daily routines of even the most commonplace jobs. The courses, taught by people who have completed an appropriate training program, are a prerequisite for licenses to work at some jobs, such as taxi driving, but many other people take them voluntarily.

The city also funds a number of important programs for children putting money behind the often empty pronouncements municipalities make about the importance of the next generation. The Paperboy/Papergirl Program gives part-time jobs to schoolchildren from low-income families; municipal day care centers serve four meals a day for some 12,000 children; and SOS Children provides a special telephone number for urgent communications about children under any kind of threat.

Curitiba has repeatedly rejected conventional wisdom that emphasizes technologically sophisticated solutions to urban woes. Many planners have contended, for example, that cities with over a million people must have a subway system to avoid traffic congestion. Prevailing dogma also claims that cities that generate more than 1,000 tons of solid waste a day need expensive mechanical garbage-separation plants. Yet Curitiba has neither.

The city has attacked the solid-waste issue from both the generation and collection sides. Citizens recycle paper equivalent to nearly 1,200 trees each day. The Garbage That Is Not Garbage initiative has drawn more than 70 per cent of households to sort recyclable materials for collection. The Garbage Purchase program, designed specifically for low-income areas, helps to clean up sites that are difficult for the conventional waste-management system to serve. Poor families can exchange filled garbage bags for bus tokens, parcels of surplus food and children's school notebooks. More than 34,000 families in 62 poor neighborhoods have exchanged over 11,000 tons of garbage for nearly a million bus tokens and 1,200 tons of surplus food. During the past three years, students in more than 100 schools have traded nearly 200 tons of garbage for close to 1.9 million notebooks. Another initiative, All Clean, temporarily hires retired and unemployed people to clean up specific areas of the city where litter has accumulated.

These innovations, which rely on public participation and labor-intensive approaches rather than on mechanization and massive capital investment, have reduced the cost and increased the effectiveness of the city's solid-waste management system. They have also conserved resources, beautified the city and provided employment.

LESSONS FOR AN URBANIZING WORLD

No other city has precisely the combination of geographic, economic and political conditions that mark Curitiba. Nevertheless, its successes can serve as lessons for urban planners in both the industrial and the developing worlds.

Perhaps the most important lesson is that top priority should be given to public transport rather than to private cars and to pedestrians rather than to motorized vehicles. Bicycle paths and pedestrian areas should be an integrated part of the road network and public transportation system. Whereas intensive road-building programs elsewhere have led paradoxically to even more congestion, Curitiba's slighting of the needs of private motorized traffic has generated less use of cars and has reduced pollution.

Curitiba's planners have also learned that solutions to urban problems are not specific and isolated but rather interconnected. Any plan should involve partnerships among private-sector entrepreneurs, nongovernmental organizations, municipal agencies, utilities, neighborhood associations, community groups, and individuals. Creative and labor-intensive ideas – especially where unemployment is already a problem – can often substitute for conventional capital-intensive technologies.

We have found that cities can turn traditional sources of problems into resources. For example, public transport, urban solid waste, and unemployment are traditionally considered problems, but they have the potential to become generators of new resources, as they are in Curitiba.

Other cities are beginning to learn some of these lessons. In Brazil and elsewhere in Latin America, the pedestrian streets that Curitiba pioneered have become popular urban fixtures. Cape Town has recently developed a new vision for its metropolitan area that is explicitly based on Curitiba's system of structural axes. Officials and planners from places as diverse as New York City, Toronto, Montreal, Paris, Lyons, Moscow, Prague, Santiago, Buenos Aires and Lagos have visited the city and praised it.

As these planners carry Curitiba's examples back to their homes, they also come away with a crucial principle: there is no time like the present. Rather than trying to revitalize urban centers that have begun falling into decay, planners in already large cities and those that have just started to grow can begin solving problems without waiting for top-down master plans or near fiscal collapse.

BOX 1 Integrated design makes busways work

Curitiba's express bus system is designed as a single entity, rather than as disparate components of buses, stops, and roads. As a result, the busways borrow many features from the subway system that the city might otherwise have built had it a few billion dollars to spare. Most urban bus systems require passengers to pay as they board, thus slowing loading. Curitiba's raised-tube bus-stops eliminate this step: passengers pay as they enter the tube, and so the bus spends more of its time actually moving people from place to place.

Similarly, the city installed wheelchair lifts at bus-stops rather than on board buses (Plate 6), easing weight restrictions and simplifying maintenance – buses with built-in wheelchair lifts are notoriously trouble-prone, as are those that "kneel" to put their boarding steps within reach of the elderly. The tube-stop lifts also speed boarding by bringing disabled passengers to the proper height before the bus arrives.

Like subways, the buses have a track dedicated entirely to their use (Plate 7). This right-of-way significantly reduces travel time compared with buses that must fight automotive traffic to reach their destinations. By putting concrete and asphalt above the ground instead of excavating to place steel rails underneath it, however, the city managed to achieve most of the goals that subways strive for at less than 5 per cent of the initial cost.

Some of the savings have enabled Curitiba to keep its fleet of 2,000 buses – owned by ten private companies under contract to the city – among the newest in the world. The average bus is only three years old. The city pays bus-owners 1 per cent of the value of a bus each month; after ten years it takes possession of retired vehicles and refurbishes them as free park buses or mobile schools.

Companies are paid according to the length of the routes they serve rather than the number of passengers they carry, giving the city a strong incentive to provide a service that increases ridership. Indeed, more than a quarter of Curitiba's automobile owners take the bus to work. In response to increased demand, the city has augmented the capacity of its busways by using extra-long buses – the equivalent of multicar subway trains. The biarticulated bus, in service since 1992, has three sections connected by hinges that allow it to turn corners. At full capacity, these vehicles can carry 270 passengers, more than three times as many as an ordinary bus.

(a)

(b)

Plate 8. (a) *Recycling* in Curitiba takes many forms. As in many other cities, families sort their garbage to ease recovery of glass, metal and plastic. In addition, old buses find second and third careers as free transportation to city parks or as mobile offices and classrooms. (b) Even the city's old electrical utility poles find new life as parts of park buildings and public offices, including the Free University for the Environment.

Source: Photo by Jonas Rabinovitch.

Plate 9. *Botanical gardens* were once a city dump. In addition to providing space for recreation, the gardens serve as a research center for studies of plant compounds.

Source: Photo by Jonas Rabinovitch.

"Planning for Sustainability in European Cities: A Review of Practices in Leading Cities" (2003)

Timothy Beatley

Editors' Introduction

With their compact land development patterns, well-developed transit systems, and pedestrian-oriented mix of small shops and housing, older European cities possess many natural advantages in terms of sustainable urban development. (The same may be said for historic Asian and Latin American cities.) A range of recent innovations by local governments, progressive developers, and civic organizations has also put many of these communities in the forefront of sustainability planning. Yet at the same time European cities and towns face many of the same problems of suburban sprawl, pollution, rising automobile use, persistent inequities, and growing non-local control over their economies that affect communities elsewhere in the world.

In this selection, Tim Beatley, one of the co-editors of this volume, surveys many recent European sustainability planning initiatives. Beatley has studied European sustainability efforts intensively; his other writings include *The Ecology of Place: Planning for Environment, Economy, and Community* (Washington, DC: Island Press, 1996) and *Green Urbanism: Learning from European Cities* (Washington, DC: Island Press, 2000).

INTRODUCTION: LEARNING FROM EUROPEAN CITIES

In few other parts of the world is there as much interest in sustainability as in Europe, especially northern and northwestern Europe, and as much tangible evidence of applying this concept to cities and urban development. For approximately the last six years this author has been researching innovative urban sustainability practice in European cities. The findings from the first phase of this work are presented in the book *Green Urbanism: Learning from European Cities* (Island Press, 2000). What follows is a summary of some of the key themes and most promising ideas and strategies found in the 30 or so cities, in 11 countries, described in this book, as well as more recent case studies and field work.

An initial observation from this work is just how important sustainability is at the municipal level in Europe, especially evident in the cities chosen. "Sustainable cities" resonates well and has important political meaning and significance in these cities, and on the European urban scene generally. One measure of this is the success of the Sustainable Cities and Towns campaign, an EU-funded informal network of communities pursuing sustainability begun in 1994. Participating cities have signed the so-called Aalborg Charter (from Aalborg, Denmark, the site of the first campaign conference), and more than 1,800 cities and towns have done so. Among the activities of this organization are the publication of a newsletter, networking between cities, and initiation of conferences and workshops. The organization has also an European Sustainable City award (with the first of these awards issued in 1996), and it is clear that they have been coveted and highly valued by politicians and city officials.

Many European cities have also gone through, or are currently going through, some form of local Agenda 21 process (including many of the same cities that have signed the Aalborg Charter), and this is another important indicator of the relevance of local sustainability. Indeed, in the countries studied, high percentages of municipal governments are participating (for instance, in Sweden 100 per cent of all local governments are at some stage in the local Agenda 21 process). Often these programs represent tremendous local efforts to engage the community in a dialogue about sustainability, and typically involve the creation of a local sustainability forum, sustainability indicators, local state-of-the-environment reports, and the preparation of comprehensive local sustainability action plans. European cities and towns demonstrate serious commitment to environmental and sustainability values and what follows are a few of the more important ways in which these concerns are being addressed.

Compact cities and regions

Urban form and land use patterns are primary determinants of urban sustainability. While European cities have been experiencing considerable decentralization pressures, they are typically much more compact and dense than American cities. Peter Newman and Jeffrey Kenworthy have monitored and tracked average density in a number of cities throughout the world. Western European cities like Amsterdam and Paris have substantially higher densities, as measured in persons per hectare, than typical American cities. Overall or whole-city densities for European cities are typically in the 40–60 persons per hectare range; American cities are much lower, commonly under 20 persons per hectare (Newman and Kenworthy 1999). Even American cities that we tend to think of as particularly

dense, for example New York, are comparatively less dense when the entire metropolitan wide pattern is considered. Density and compactness directly translate into much lower energy use, per capita, and lower carbon emissions, air and water pollution, and other resource demands compared with less dense, less compact cities.

Many of these European examples, moreover, show that compactness and density need not translate to skyscrapers and excessive high-rise. Density and compactness in cities like Amsterdam happens through a building pattern of predominately low-rise structures. While many sustainability proponents advocate the need for the green high-rise development (e.g. see Ken Yeang's designs for bio-climatic skyscrapers), these European cities demonstrate convincingly that tremendous compactness and density can be accomplished at a clearly human scale. The European model is appealing to many precisely because of its more traditional form of density and compactness, and many believe its more human scale.

These characteristics of urban form make many other dimensions of local sustainability more feasible, of course (e.g. public transit, walkable places, energy efficiency). There are many factors that explain this urban form, including an historic pattern of compact villages and cities, a limited land base in many countries, and different cultural attitudes about land. Nevertheless in the cities studied there are conscious policies aimed at strengthening a tight urban core. Indeed, the major new growth areas in almost every city studied are situated in locations within or adjacent to existing developed areas, and are designed generally at relatively high densities.

Exemplary and for the most part effective efforts at maintaining the traditional tight urban form can be seen in many cities. Cities like Amsterdam are actively promoting urban redevelopment and industrial reuse (e.g. through its eastern docklands redevelopment). Berlin's plan calls for most future growth to be accommodated with its urbanized area through a variety of infill and re-urbanization strategies. Freiburg, Germany, has been able to effectively steer relatively compact, high-density new growth along the main corridors of its tram system, as well as to protect existing housing supply in the center (there is now a prohibition on the conversion of housing to offices and other uses).

European cities are utilizing a variety of planning strategies to promote compactness and to maintain a tight urban form. These include strict limits on building outside of designated development areas, a strong role for municipal governments in designing and developing new growth areas, extensive public acquisition and ownership of land (especially in Scandinavian cities like Stockholm), and a willingness to make significant transportation and other infrastructure investments that facilitate and support compactness.

GREEN URBANISM: COMPACT AND ECOLOGICAL URBAN FORM

Growth areas and redevelopment districts in these European cities are incorporating a wide range of ecological design and planning concepts, from solar energy to natural drainage to community gardens, and effectively demonstrate that *ecological* and *urban* can go together. Good examples of this compact green growth can be seen in the new development districts planned for or recently completed in Utrecht (Leidsche Rijn), Freiburg (Rieselfeld), Amsterdam (e.g. IJburg), Copenhagen (Ørestad), Helsinki (Viikki), and Stockholm (Hammerby Sjöstad).

Leidsche Rijn, for example, is an innovative new growth district in the Dutch city of Utrecht. In addition to incorporating a mixed-use design, and a balance of jobs and housing (30,000 dwelling units and 30,000 new jobs), it will include a number of ecological design features. Much of the area will be heated through district heating supplied from the waste energy of a nearby power plant, a double-water system which will provide recycled water for non-potable uses, and a storm water management through a system of natural swales (what the Dutch call "wadies"). Higher-density uses will be clustered around several new train stations and bicycle-only and bicycle/pedestrian-only bridges will provide fast, direct connections to the city center. Homes and buildings will meet a low-energy standard and only certified sustainably harvested wood will be allowed.

European cities also provide excellent and generally successful examples of redevelopment and adaptive reuse of older, deteriorated areas within the center-city. Good examples include Amsterdam's eastern docklands, where 8,000 new homes have been accommodated on recycled land. In *Java-eiland*, one major piece of this project, an overall plan (prepared by urban designer Sjoerd Soeters) lays out broad density, massing, and circulation for the district. Diversity and distinctiveness in actual design of the buildings, however, was encouraged through a restriction on the number of buildings that could be designed by a single architect. The result is a stimulating community where buildings have been created by scores of different designers. This island district successfully balances connection to the past (a series of canals and building scale reminiscent of historic Amsterdam) with unique modern design (each of the pedestrian bridges crossing the canals offers a distinctive look and design). *Java-eiland* demonstrates that city building can occur in ways that create interesting and organically evolved places, and which also acknowledge and respect history and context, overcoming sameness.

European cities on the whole (and especially the cities examined in this study) have been able to maintain and strengthen their center-cities and urban cores. In no small part this is a function of historic density and compactness, but they are also the result of numerous efforts to maintain and enhance the quality and attractiveness of the city-center. In the cities studied, the center has remained a mixed-use zone, with a significant residential population. *Groningen*, for instance, has undertaken a host of actions to improve its center including the creation of new pedestrian-only shopping areas (creating a system of two linked circles of pedestrian areas), and installation of (yellow) brick surfaces and new street furniture in walking areas, among other actions. Committed to a policy of compact urban form, Groningen has also made a strong effort to keep all major new public buildings and public attractions close-in. As one example, a new modern art museum has been sited and designed to provide an important pedestrian link between the city's main train station and the town center.

SUSTAINABLE MOBILITY

Achieving a more sustainable mix of mobility options is a major challenge, and in almost all of the cities studied in *Green Urbanism* a very high level of priority is given to building and maintaining a relatively fast, comfortable and reliable system of public transport.

There are impressive examples of cities that have been working hard to expand and enhance transit, in the face of rising auto use in many areas. Zürich implements an aggressive set of measures to give priority to its transit on streets. Trams and buses travel on protected, dedicated lanes. A traffic control system gives trams and buses green lights at intersections and numerous changes and improvements have been made to reduce the interference of autos with transit movement (e.g. bans on left turns on tram line roads; prohibiting stopping or parking in certain areas; building pedestrian islands, etc.) A single ticket is good for all modes of transit in the city (including buses, trams, and a new underground regional metro system). The frequency of service is high and there are few areas in the canton that are not within a few hundred meters of a station or stop. Cities like Freiburg and Copenhagen have made similar strides.

In these European cities transit modes are integrated to an impressive degree. This means coordination of investments and routes so that transit modes complement each other. In most of the cities studied, for instance, regional and national train systems are fully integrated with local routes. It is easy, as well, to shift from one mode to another. Local transit centers are viewed in these cities as multi-modal, mixed-use centers of activity. Arnhem's new central train station in the Netherlands is a case in point. It integrates in a single location high-speed and conventional train service, local transit, bicycle parking, rental, and repair, as well as shops, offices and housing. These uses are all within a few hundred meters of the city-center.

The ease of traveling throughout Europe is aided tremendously by the commitment on this continent to high-speed rail. Cross-national movement by high-speed train is increasingly comfortable and easy, and investments in dedicated tracks and infrastructure reflect impressive forward thinking on this issue. And increasingly it is not just the northern and northwestern European nations leading the way. Major new high-speed rail

systems are under construction in Italy and Spain for instance. Overall, plans are on the books to double the length of dedicated high-speed rail track in Europe over the next eight years. And, the newest generation of trains will travel faster – on average 300 kph or higher.

Importantly, investments in transit complement, and are coordinated with, important land use decisions. Virtually all the major new growth areas identified in this study have good public transit service as a basic, underlying design assumption. The cities studied here do not wait until after the housing is built, but rather the lines and investments occur contemporaneously with the projects. The new community growth area *Rieselfeld* in Freiburg, for instance, has a new tram line even before the project has been fully built. In Amsterdam, as a further example, at the new neighborhood of *Nieuw Sloten*, tram service began when the first homes were built. In the new ecological housing district *Kronsberg*, in Hannover, three new tram stops ensure that no resident is further than 600 meters away from a station. There is a recognition in these cities of the importance of providing new residents with options, and establishing mobility patterns early.

Car sharing has become a viable and increasingly popular option in Europe cities. Here, by joining a car sharing company or organization residents have access to neighborhood-based cars, on an hourly or per-kilometer cost. There are now some 100,000 members served by car sharing companies or organizations in 500 European cities. Some of the newest car sharing companies, such as *GreenWheels* in The Netherlands, are also pursuing creative strategies for enticing new customers. This company has been developing strategic alliances, for example with the national train company, to provide packages of benefits at reduced prices. One of the key issues for the success of car sharing is the availability of convenient spaces, and a number of cities, including Amsterdam and Utrecht, have been setting-aside spaces for this purpose. In cities such as Hannover, Germany, the car sharing organization there (a non-profit called Ökostadt) has strategically placed cars at the stations of the Stadtbahn, or city tram, further enhancing their accessibility.

Thinking beyond the automobile

Many of these cities are in the vanguard of new mobility ideas and concepts and are working hard to incorporate them into new development areas. Amsterdam, for example, has taken an important strategy in developing *IJburg*. It is working to develop a comprehensive mobility package that all new residents will be offered and which includes, among other things, a free transit pass (for a certain specified period) and discounted membership in local car sharing companies. Minimizing from the beginning the reliance on automobiles, and giving residents more mobility options, are the goals. Eventually this new area will be served both by an extension of the city's underground metro and fast tram.

An increasing number of carfree housing estates are also being developed in these cities, as a further reflection of the commitment to minimizing auto-dependence. The *GWL-Terrein* project, also in Amsterdam, built on the city's old waterworks site, incorporates only very limited peripheral parking. An on-site car sharing company, in combination with good tram service, are part of what makes this concept work there. The interior of the project incorporates extensive gardens (and 120 community gardens available to residents) and pedestrian environment, with key-lock access for fire and emergency vehicles.

Another carfree experiment is the new ecological district *Vauban*, in Freiburg. Built on the former site of an army barracks, this project is unique because it gives new residents the opportunity to declare their intentions to be carfree, and rewards them financially for doing so. Specifically, if residents choose to have a car, they must pay approximately $13,000 for the cost of a space in the nearby parking garage (a bit less than one-tenth the cost of the housing units). In this way there is a strong financial incentive to choose to be carfree and so far about half the residents have taken the carfree path. Projects like *Vauban* challenge new residents to think and act more sustainably and reward them for doing so.

Bicycles are an impressive mobility option in almost all of the cities studied in *Green Urbanism*, and many of these cities have taken tremendous efforts to expand bicycle facilities and to promote bicycle use. Berlin has 800 km of bike lanes, and Vienna has more than doubled its bicycle network since the late 1980s. Copenhagen now has a policy of installing bike lanes along all major streets, and bicycle use in that city has risen substantially. Few have gone as far, of course, as the Dutch cities, with cities like Groningen, where more than half of the daily trips are made on bicycles. In virtually all new growth areas in the Dutch cities, as well as many Scandinavian and German cities, bicycle mobility is an essential design feature, including providing important connections to existing city bicycle networks.

A number of actions have been taken by these cities to promote bicycle use. These include separated bike lanes with their own signaling, separate signaling and priority at intersections, signage and provision of extensive bicycle parking facilities (e.g. especially at train stations, public buildings), and minimum bicycle storage and parking standards for new development. Many cities are gradually converting spaces for auto parking to spaces for bicycles. Utrecht has discovered that it can fit 6–10 bicycles in the same space it takes to park one automobile. Tilburg, in the Netherlands, has recently built an underground valet bicycle parking facility in the heart of that city's shopping district. Freiburg's mobility center combines two levels of bicycle parking, with car sharing cars on the ground level, a café, travel agency, and office of the Deutsche Bahn (and the structure has a green roof and a photovoltaic array generating electricity!).

These cities are also innovating in the area of public bikes. The most impressive program is Copenhagen's "City Bikes," which now makes available more than 2,000 public bicycles throughout the center of the city. The bikes are brightly painted (companies sponsor and purchase the bikes in exchange for the chance to advertise on their wheels and frames), and can be used by simply inserting a coin as a deposit. The bikes are geared in such a way that the pedaling is difficult enough to discourage their theft. The program has been a success, and a number of bikes has been expanding. These sustainable European cities have discovered that bicycles are an important and legitimate alternative mode of transport to the car and with modest planning and investments substantial ridership can be achieved.

BUILDING PEDESTRIAN CITIES; EXPANDING THE PUBLIC REALM

European cities represent, as well, exemplary efforts at creating walkable, pedestrian urban environments. Relatively compact, dense, and mixed-use urban environments make cities much more walkable, of course. And most European cities and regions benefit from having a compact historic core, designed and evolved around walking and face-to-face commerce. The vitality, beauty, and attraction of European cities is in no small part a function of the impressive public and pedestrian spaces. Cities like Barcelona and Venice remain positive and compelling models of pedestrian urban society. The uses of these spaces are varied and many: they are outdoor stages, the "living rooms" in which citizens socialize, interact, and come together, places where political events occur and democracy plays itself out. These areas are now the social heart of these communities – places where children play, casual conversations and unexpected meetings take place, and people come to watch and be seen.

The overall land use pattern in these cities, and the priority given to maintaining their compact form, certainly make a walking culture more feasible. What is especially impressive, however, is the continued attention given to this issue and the continued expanding of pedestrian areas and the strengthening of the public and pedestrian realm. Cities like Copenhagen have set the stage, beginning in the early 1960s, gradually taking back their urban centers from cars. That city pedestrianized the Ströget, one of its main downtown streets, in 1962. Copenhagen continues this pedestrianizing in a gradual way each year. The city has adopted the policy of converting 2–3 per cent of its downtown parking to pedestrian space each year, to dramatic effect over several decades. Today the amount of pedestrian space is tremendous. Eighteen pedestrian squares have been created in Copenhagen where there was once auto parking – some 100,000 square meters in all. Had proponents of public space in Copenhagen attempted to convert this amount of space all at once it would have been very politically difficult to do so.

Many other cities have followed suit, especially Dutch and German cities, but examples can be found throughout Europe. Cities like Vienna and Groningen have pedestrianized much of their centers, creating delightful, highly functional public spaces. Groningen's compact city policy ensures that major new public buildings and facilities are kept in the center, and accessible through walking – it is a compact city of "short distances." In cities like Leiden, emphasis has been given to installing new pedestrian bridges over canals connecting major streets, and every new residential area is designed to include a grocery, post office, and other shops within an easy walk. The greater mixing of uses means that residents of these cities typically have many shops, services, cafés within a walkable range.

The experience of these European cities in pedestrianizing much of their urban centers has been a positive one, both economically and in terms of quality of life. The spaces created commonly contain fountains, sculptures and public art, extensive seating and, of course, many reasons for being there – restaurants, cafés, shops. Each city has its own unique history and features that can be used to strengthen the unique character of its pedestrian environment. Freiburg's "bächle," or urban streams that run through the streets of its old center, as well as its pebble mosaics are delightful and special and this city has done an excellent job expanding and adding onto to these unique qualities of place.

Good public transit appears a major factor strengthening the pedestrian realm in these cities, as well as commitments to bicycles, as in the case of Copenhagen (Hass-Klau *et al.* 1999). Extensive efforts to calm urban traffic, to restrict auto access, and to raise the cost of parking and auto mobility are also important elements. A number of European cities have experimented with or are anticipating some form of road pricing. The City of London is the most recent notable example, now charging a fee of five pounds for cars wishing to enter central London (and already resulting in a significant reduction in car traffic there). These European experiences support that a pedestrian culture and community life is indeed possible, even where the climate may be harsh and that these spaces serve an incredible range of social, cultural, and economic functions.

GREENING THE URBAN ENVIRONMENT

Ensuring that compact cities are also green cities is a major challenge, and there are a number of impressive greening initiatives among the study cities. First, in many of these cities there is an extensive greenbelt and regional open space structure, with a considerable amount of natural land actually owned by the cities. Extensive tracts of forest and open lands are owned by cities such as Vienna, Berlin, and Graz, among others. Cities such as Helsinki and Copenhagen are spatially structured so that large wedges of green nearly penetrate the center for these cities. Helsinki's large *Keskuspuisto* central park extends in an almost unbroken wedge from the center to an area of old growth forest to the north of the city. It is 1,000 hectares in size and 11 km long.

In Hannover an extensive system of protected green spaces exists, including the *Eilenriede*, a 650-hectare dense forest located in the center of the city. Hannover has also recently completed an 80-kilometer-long *green ring* (der grüne ring) which circles the city, providing a continuous hiking and biking route, and exposing residents to a variety of landscape types, from hilly Borde to the river valleys of the Leineaue river.

There is a trend in the direction of creating and strengthening ecological networks within and between urban centers. This is perhaps most clearly evident in Dutch cities, where extensive attention to ecological networks has occurred at the national and provincial levels. Under the national government's innovative Nature Policy Plan, a national ecological network has been established consisting of core areas, nature development areas, and corridors, which must be more specifically elaborated and delineated at the provincial level. Cities in turn are attempting to tie into this network and build upon it. At a municipal level, such networks can consist of ecological waterways (e.g. canals), tree corridors, and green connections between parks and open space systems. Dutch cities like Groningen, Amsterdam, and Utrecht have full-time urban ecology staff, and are working to create and restore these important ecological connections and corridors.

Many examples exist of efforts to mandate or subsidize the greening of existing urban areas. There is a continuing trend, for instance, towards installation of ecological or green rooftops, especially in German, Austrian, and Dutch cities. Linz, Austria, for instance, has one of the most extensive green roof programs in Europe. Under this program, the city frequently requires building plans to compensate for the loss of green space taken by a building. Creation of green roofs has frequently been the response. Also since the late 1980s the city has subsidized the installation of green roofs – specifically, it will pay up to 35 per cent of the costs. The program has been quite successful and there are now some 300 green roofs scattered around the city. They have been incorporated into many different types of buildings including a hospital, a kindergarten, a hotel, a school, a concert hall, and even the roof of a gas station. Green roofs have been shown to provide a number of important environmental benefits, and to accommodate a surprising amount of biological diversity. Many other innovative urban greening strategies can be found in these cities from green streets, to green bridges, to urban stream daylighting.

RENEWABLE ENERGY AND CLOSED-LOOP CITIES

A number of the cities have taken action to promote more closed-loop urban metabolism, in which, as in nature, wastes represent inputs or "food," for other urban processes (e.g. Girardet 1999). The city of Stockholm has made some of the most impressive progress in this area, and has even administratively reorganized its governmental structure so that the departments of waste, water, and energy are grouped within an ecocycles division. A number of actions in support of ecocycle balancing have already occurred. These include, for instance: the conversion of sewage sludge to fertilizer and its use in food production, and the generation of biogas from sludge. The biogas is used to fuel public vehicles in the city, and to fuel a combined heat and power plant. In this way, wastes are returned to residents in the form of district heating. Another powerful example of the closed-loop concept can be seen in Rotterdam, in the Roca3 power plant supplies district heating and carbon dioxide to 120 greenhouses in the area. A waste product becomes a useful input, and in this case prevents some 130,000 metric tonnes of carbon emissions annually.

Energy is very much on the planning agenda, and these exemplary cities are taking a host of serious measures to conserve energy and to promote renewable sources. The heavy use of combined heat and power (CHP) generation, and district heating, especially in northern European cities, is one reason for typically lower per capita levels of CO_2 production here. Helsinki, for instance, has one of the most extensive district heating systems: more than 91 per cent of the city's buildings are connected to it. The result is a substantial increase in fuel efficiency, and significant reductions in pollution emissions. District heating and decentralized combined heat and power plants are now commonly integrated into new housing districts in these cities. In *Kronsberg*, in Hannover, for instance, heat is provided by two CHP plants, one of which, serving about 600 housing units and a small school, is actually located in the basement of a building of flats.

Many cities, including Heidelberg and Freiburg, have set ambitious maximum energy consumption standards for new construction projects. Heidelberg has recently sponsored a low-energy social housing project, to demonstrate the feasibility of very low-energy designs (specifically a standard of 47 kwh/m^2 per year). The Dutch are promoting the concept of energy-balanced housing – housing that will over the course of a year produce as much energy as it uses – and the first two of these units have been completed in the *Nieuwland* district in Amersfoort.

Many cities such as Heidelberg have undertaken programs to evaluate and reduce energy consumption in schools and other public buildings. Incentive programs have been established which allow schools to keep a certain percentage of the savings from energy conservation and retrofitting investments. Heidelberg has engaged in an innovative system of performance contracts, in which private retrofitting companies get to keep a certain share of the conservation benefits.

There is an explosion of interest in solar and other renewable energy sources in these cities (and countries). Cities like Freiburg and Berlin have been competing for the label "solar city," with each providing significant subsidies for solar installations. In the Netherlands, major new development areas, such as *Nieuwland* in Amersfoort and *Nieuw Sloten* in Amsterdam, are incorporating solar energy, both passive and active, into their designs. In Nieuwland, described as a "solar suburb," there are more than 900 homes with rooftop photovoltaics, 1,100 homes with thermal solar units, and a number of major public buildings producing power from solar (including several schools, a major sports hall, and a childcare facility). What is particularly exciting is to see the effective integration of solar into the architectural design of homes, schools and other buildings.

The degree of public and governmental support in these European cities, financial and technical, for renewable energy developments is truly impressive. Reflecting a generally high overall level of concern for global warming issues and energy self-sufficiency, significant production subsidies and consumer subsidies have both been given. The degree of creativity in incorporating renewable energy ideas and technologies in many of these cities is also quite impressive. Oslo's new international airport, for example, provides heating through a bark/wood bio-energy district heating system. This system provides heat for buildings through 8 km of pipes, as well as the airport's de-icing system. The moist bark fuel is a local product, and costs only one-third as much as fuel oil. In Sundsvall, Sweden, snow is collected, stored, and used as a major cooling source for the city's main hospital. In Copenhagen, twenty 2 MW wind turbines have been installed off-shore which will together generate enough energy for about 30,000 homes.

Green cities, green governance

Many of these cities are taking a hard look at ways their own operations and management can become more environmentally responsible. As a first step, many local governments have undertaken some form of internal environmental audit. Variously called green audits or environmental audits, they represent attempts to study comprehensively the environmental implications of a city's policies and governance structure. A number of local governments are now going through the process of becoming certified (the London Borough of Sutton being the first) under the EU's Eco-Management and Audit Scheme (EMAS), an environmental management system more commonly applied to private companies. Several German cities are preparing environmental budgets, under a pilot program. The cities of Den Haag and London have calculated their ecological footprints and are using these measures as policy guideposts (e.g. see Best Foot Forward Ltd 2002). Albertslund, Denmark, has developed an innovative system of "green accounts," used to track and evaluate key environmental trends at city and district levels, and many of the study cities have developed sustainability indicators (e.g. Leicester, London and Den Haag). Cities like Lahti, Helsinki, and Bologna have gone through extensive in-house education and involvement of city personnel, often as part of the local Agenda 21 process, in examining environmental impacts and in identifying ways that personnel and city departments can reduce waste, energy, and environmental impacts.

Municipal governments have taken a variety of measures to reduce the environmental impacts of their actions. A number of communities have adopted environmental purchasing and procurement policies. Cities like Alberstlund have adopted policies mandating that only organic food can be served in schools and child-care facilities, and restricting use of pesticides in public parks and grounds. Other cities are aggressively promoting the development of environmental vehicles. Stockholm's environmental vehicles program is one of the largest (a pilot program under the EU-funded initiative ZEUS), with over 300 vehicles. A number of cities have sought to modify the mobility patterns of employees, for instance by creating financial incentives for the use of transit or bicycles. Cities like Saarbrucken, Germany, have made great strides in reducing energy, waste, and resource consumption in public buildings.

Communities have also engaged in extensive public involvement and outreach on sustainability matters. A variety of creative approaches have been taken. Leicester, for instance, has developed alliances with the local media and has sponsored a series of educational campaigns on particular community issues.

Understanding European cities: some concluding thoughts

To be sure, many European cities are facing some serious problems and trends working against sustainability, in particular a dramatic rise in automobile ownership and use, and a continuing pattern of de-concentration of people and commerce. And, with their relatively affluent populations consuming substantial amounts of resources, European cities exert a tremendous ecological footprint on the world. Yet, these most exemplary cities provide both tangible examples of sustainable practice, and inspiration that progress can be made in the face of these difficult pressures.

The lessons are several. These cities demonstrate the critical role that municipalities can and must play in addressing serious global environment problems, including reliance on fossil-fuels and global climate change. Innovations in the urban environment offer tremendous potential for dramatically reducing our ecological impacts (European cities produce about half the per capita carbon emissions of American cities), while at the same time enhancing our quality of life (e.g. by expanding personal mobility options with bicycles and transit).

Many, indeed most, of the ideas, initiatives, strategies undertaken in these innovative cities serve, in addition to reducing ecological footprints, to enhance livability and quality of life. Taking back space from the auto and converting it to pedestrian and public space does much to enhance the desirability of these cities. Investments in public transit reduce dramatically energy consumption, CO_2 emissions, and urban air quality problems, but at the same time provide tremendous levels of independence and mobility to the youngest and oldest members of society. Making bicycling safer and easier helps the environment, but also provides a badly needed form of physical exercise.

These experiences demonstrate clearly that it is possible to apply virtually every green or ecological strategy or technique, from solar and wind energy to graywater recycling, in very urban, very compact settings. Green Urbanism is not an oxymoron. Moreover, the lesson of these European cities is that municipal governments can do much to help bring these ideas about, from making parking spaces available for car sharing companies to providing density bonuses for green rooftops, to producing or purchasing green power.

There are also process lessons here. Key among them is an understanding of the great power of partnerships and collaboration between different parties with an interest in sustainability. While not always easy, success at achieving sustainability will depend on them. This means getting different departments to talk to each other and to work together (as in Stockholm), and getting different public and private actors to join together in common initiatives that demonstrate that green urban ideas are possible and desirable.

It is important to recognize, to be sure, the differences in governmental structure. The economic and planning frameworks in place in these countries (compared with, say, the United States) often facilitate many of the exemplary urban sustainability projects described here. The role of economic incentives and the economic incentive structures is critical and undeniable. High prices at the gas pump (typically $4–5 per gallon in Western Europe) have been a conscious policy decision in European countries, and in countries like Germany, have provided essential funding for public transit. Such high prices, relative to countries like the United States, undoubtedly help to encourage more compact land use and personal choices in favor of more sustainable modes of mobility. Also, carbon taxes in countries like Denmark help to substantially level the economic playing field between conventional fossil-fuel energy and more sustainable, renewable forms of energy. Higher energy prices generally help to promote greater conservation and energy efficiency improvements. The important role of adjusting incentives and economic signals is itself a key lesson from the European scene. Rather than seen as a pre-existing background condition, raising gasoline and energy taxes can be seen as an example of an important strategic societal and political choice.

There are other political, social, and cultural conditions, to be sure, that favor many of the exemplary ideas discussed here. Parliamentary governmental structures that give relative voice and power to green party and other social and environmental views (with local representation of these views as well) have been important. Historically stronger planning and land use control systems are helpful also, as well as generally stronger and more proactive roles afforded to government. Many of the important (more activist) urban sustainability activities undertaken in these European cities – as market stimulators, promoters of innovation, and financial underwriters for innovative urban sustainability practices and projects – are common and accepted roles for local governments to play.

But there are also certainly many underlying value differences (compared with the US) that further explain good practice. Prevailing European views of land are less imbued with a sense of personal use and freedom, and there is little expectation, for instance, on the part of a rural landowner or farmer that his or her land will eventually be convertible to urban development.

There are also a number of more regionally unique cultural values and differences, each with significant planning and land use shaping implications. A stronger desire to live within a city or town center clearly exists throughout much of Europe, borne undoubtedly from an older, more developed urban culture. Importance given to strolling, spending time in public places, and to the values of the public realm more generally, in countries like Spain and Italy, certainly help explain the success of pedestrian spaces in these countries. Pace of life, cultural organization of the day, and the number of hours in the work week are also clearly important. In Italy, public and pedestrian spaces are used in part because there is time to use them – the culture organizes its day so as to support the early evening stroll, after the shops close but before the evening meal. To many observers of the European scene there are also lessons to emulate – suggestions and ideas for humanizing cities and strengthening their livability and sociability, as well as their sustainability. The lessons are many and profound on many levels.

REFERENCES

Beatley, Timothy (2000) *Green Urbanism: Learning from European Cities*, Washington, DC: Island Press.

Best Foot Forward Ltd (2002) *City Limits*, London: Best Foot Forward Ltd.

Gehl, Jan and Lars Gemzoe (2000) *New City Spaces*, Copenhagen: The Danish Architectural Press.

Girardet, Herbert (1999) *Creating Sustainable Cities*, Devon, UK: Green Books.

Hass-Klau, Carmen, Graham Crampton, Clare Dowland, and Inge Nold (1999) *Streets as Living Space: Helping Public Places Play Their Proper Role*, London: Lander Publishing Ltd.

Newman, Peter and Jeffrey Kenworthy (1999) *Sustainable Cities: Overcoming Automobile Dependence*, Washington, DC: Island Press.

"Collective Action Toward a Sustainable City: Citizens' Movements and Environmental Politics in Taipei"

from *Livable Cities?: Urban Struggles for Livelihood and Sustainability* (2002)

Hsin-Huang Michael Hsiao and Hwa-Jen Liu[1]

Editors' Introduction

While Curitiba provides an inspiring example of visionary planning in a developing world city, the reality is usually much more mixed. As in wealthier industrialized nations, Third World urban policies often favor wealthy elites and corporations over less affluent residents and natural ecosystems. However, urban social movements have also arisen to fight for a wide variety of environmental and social goals. Urban politics has become increasingly complex in many places; the emerging environmental concerns of middle- and upper-class residents, for example, may run counter to the equity needs of less well-off communities. Great creativity may be required to satisfy both.

In the following selection, Hsin-Huang Michael Hsiao and Hwa-Jen Liu provide a detailed analysis of such dynamics in the context of Taipei, Taiwan's three-million-person capital city. Hsiao is a professor of sociology at the National Taiwan University in Taipei and has written widely on changing Taiwanese social structure. Liu is a doctoral student at the University of California at Berkeley. This selection comes from the excellent collection of articles on Third World cities edited by Peter Evans, *Livable Cities?: Urban Struggles for Livelihood and Sustainability* (Berkeley: University of California Press, 2001).

We begin with two stories of community mobilization in Taipei. The first is about defending living space. It involves a well-to-do middle-class community named Ching-Cheng, which indefatigably fought against the state-owned power company (Taipower), a large department store backed by overseas capital, and finally the city authorities.

From 1988 to 1989, this community faced the imminent construction of a nearby power substation. Residents found out later that the increase in demand for electricity in this district resulted mainly from a newly opened department store, which also planned to rent its basement to a brothel. Unhappy, the community's residents decided to implement a boycott. They fought the department store directly by calling it "a lousy neighbor" in the media. They protested against Taipower for selecting a site without having first informed or consulted them, but they ultimately were unable to overturn the decision. The self-organized Ching-Cheng residents were not defeated, however. In 1991 they successfully saved their community park from becoming a parking garage, and in 1992 they interrupted the municipal government's plan to change the area from a residential zone into a commercial one (Shen 1994). The case of Ching-Cheng, like many others in Taipei, stands in direct opposition to business interests and the top-down decision-making system of government. The statement that the community made was loud and clear: using economic growth to justify abusing citizen rights and the urban environment is no longer acceptable, even if it once was. The mixed success of the Ching-Cheng community over the years is the most frequently cited case of community mobilization among urban reformers and community advocates, and it has also become an exemplar for other mobilized communities in Taiwan.

The second story concerns the birth of a five-hectare city park, Nos. 14–15 park. It illustrates the conflicting interests that can exist between the urban poor and other communities when improving the physical environment of a city as a whole is at stake. The predesignated park site was a Japanese-only cemetery under colonial rule and then was turned into one of the biggest slums in the city, the Kang-Le neighborhood, in the postwar development of Taipei. Kang-Le slum was an enclave surrounded by five-star hotels and skyscrapers in one of Taipei's most prosperous areas. The graphic portrait of urban poverty and the cohabitation of living humans and the dead served only to surprise visitors and offend nearby business owners and middle-class neighbors.

On March 4, 1997, despite strong opposition from urban planners and slum residents, the Taipei municipal authorities sent out bulldozers, escorted by police, and forcibly evicted thousands of slum dwellers, over one-third of whom were handicapped, aged, or extremely poor, and tore down their illegal but long tolerated shacks (Hsiao *et al.* 1997, 3). Overnight, 961 households and the community networks that supported their subsistence were destroyed. After the eviction, the first-ever

freely and directly elected mayor of Taipei, Chen Shui-bian, proudly announced that citizens' urgent need for more green space would soon be satisfied. One year later, the city administration revealed its proposal to name this park "International Plaza" and to use it to host carnivals and celebrations of such holidays as Halloween and St Patrick's Day for foreign residents (Yang 1998). The whole process, as an example of the more vulgar side of "globalization," was praised by bureaucrats and certain media outlets as an effort to create a foreigner-friendly atmosphere and to enhance the visual presentability of the city.

These two vignettes illustrate both the successes and failures of urban environmental activism in Taiwan's capital city, Taipei. On the one hand, in the 1980s and even more in the 1990s the environmental movement epitomized mobilization from below in a society in which top-down politics had predominated since colonial times. On the other hand, community participation in this movement still reflected the inequalities of the society at large. Middle-class communities were often able to protect or even improve their quality of life, even in the face of the ecologically blind developmental programs pushed by the state and private business groups. Poor and slum communities were as likely to be the victims of the environmental movement as they were to be its beneficiaries. . . .

URBANIZATION AND ENVIRONMENTAL CRISES IN TAIPEI

Every city has its own experience of environmental degradation incurred by rapid urban growth. Usually, a city starting its "developmental career" as a manufacturing center accepts the costs of industrial production: noise from mechanization, soil and water pollution, harmful fumes, and damage to the landscape beyond remedy (Mumford 1989, 458–74). Even if the rise of a service economy and the reduction in manufacturing gradually cause change in the industrial structure in the later stages of urban development, early industrial pollution has already left permanent scars on the urban landscape. Classic industrial cities such as Detroit and Manchester have paid for their unsustainable land use by falling in the world city hierarchy and finding themselves incapable of reviving their economic prominence even with enormous reinvestment (Friedmann 1997). But that has not been the case with Taipei.

In contrast to another major metropolis in southern Taiwan, Kaohsiung, Taipei has suffered relatively little harm from extensive industrial pollution because, over the past hundred years, Taipei never developed into a manufacturing city. Industrial products accounted for 88.7 per cent of the total products manufactured in Taipei between 1938 and 1940, but they came from the relatively nonpolluting sectors of food and tea processing and rice milling. In fact, in the 50 years of Japanese colonial rule (1895–1945), aside from its political function as the colony's capital city, Taipei served as only a light industrial center, processing and exporting agricultural products to boost the Japanese economy (Chiang and Hsiao 1985, 194).

When the Kuomintang (KMT) took over Taiwan in 1945, 75 per cent of the industrial infrastructure had been destroyed as a result of World War II, and the engine of industrialization had to be restarted (Huang 1983, 489). In contrast to the exporting of agricultural commodities under colonial rule, the pattern of industrial development was shaped by ISI (import-substituting industrialization) in the 1950s and EOI (export-oriented industrialization) between 1960 and 1972, then shifted to the second ISI and second EOI strategies from 1973 onward (Gereffi 1990, 17–18; Haggard 1990, ch. 4). Along the path of development, the picture of the domestic economy dramatically changed. The drastic decline of the agricultural sector mirrored the significant expansion of the industrial sector, and nationwide environmental degradation has in general been triggered by the growth of certain industries, namely steel and petrochemicals, that were encouraged by state sponsorship during the second ISI phase. While the coastal areas and cities of west and south Taiwan have suffered greatly from these two environmentally devastating industries, however, Taipei has been spared.

If Taipei has benefited from the absence of extensive industrial pollution, then what is damaging Taipei's eco-social system? It is fair to say that the single-minded pursuit of economic prosperity, under pressure from competition to achieve regional economic hegemony during global economic restructuring, and as mediated through local mechanisms of politics, should be held responsible. Since the early 1980s, Taipei has been made into "a real estate profit machine" and is a "world city" candidate in the Asia-Pacific region that, much like its cohort cities (Hong Kong, Singapore, Seoul, Manila, Bangkok, Kuala Lumpur, and Jakarta), strives to attract global capital, expand its hinterland, and invest in vast urban construction projects (Berner and Korff 1995; Douglass 1998a, 1998b; Friedmann 1986; Knight 1993; Boon Thong Lee 1998; Machimura 1992; Yeung 1996).

The prosperity of the real estate market started with Taipei's demographic growth. In 1946, the Taipei metropolitan area (including Taipei city and Taipei county) had a population of less than 800,000 (Tseng 1993, 85), but that figure had jumped to 6.1 million by 1999, amounting to nearly 30 per cent of the total population of Taiwan. With such a concentration of population, the problem of housing and the demand for urban infrastructure soon loomed large. Unfortunately, because Taipei is situated in a basin, the geographical nature of the surrounding environment imposes limits on urban growth (Chiang and Hsiao 1985, 203). The sacrifice of environmentally sensitive areas to unbridled urban growth became "inevitable" (Chang 1993, 450). Though more than 70 per cent of the hillsides around Taipei had been assessed as "improper" places to pursue construction work, a "mountain-removing, then town-building" movement still took place (Sun 1997, 76).

As major developers, the conglomerates and local factions, mainly consisting of political bosses in town, initiated the first wave of purchasing hillside lands around the Taipei basin in the early 1970s. In the next ten years, by working through their political connections, they played with the loose hillside-development regulations to change the designated land use from observed to developable lands, and they acquired permits to build. Though the national government implemented stricter hillside-development regulations in 1983, the new bylaws did not apply to those permits that had already been issued. Since the late 1980s, in response to the booming housing market around the Taipei metropolitan area, developers have sidestepped the new regulations, insisting their old permits still stand, and actually have built large hillside housing projects.

In the Taipei metropolitan area, the total investment from the private sector on hillside development projects in 1997 alone was estimated at more than 360 million US dollars, an amount almost identical to the total expenditure on environmental protection from national revenue in the same year. It also was estimated that solely in one subdistrict of Taipei county, more than 100,000 housing units would be built through the hillside development projects approved in the previous ten years (Chen and Chen 1997). However, this massive amount of investment and development could not guarantee the quality of the housing units being built. Because of the corruption and incompetence of contractors and government officials, many newly built estates became life threatening. As people moved into the new hillside housing in the mid-1990s, tragedies followed. Incidents involving collapsing buildings were widely reported during the wet season each year. The two most notorious tragedies happened in 1997 and 1998. A mudslide in August 1997 took twenty-eight lives, injured fifty people, and completely or partially destroyed two hundred housing units (Wan 1998). As a result, twelve government officials received 5- to 9-year prison sentences for their abuse of power in illegally facilitating improper hillside development (Chen 1998). In October 1998, more than 10,000 households in Taipei county suffered from mudslides and floods and three lives were lost. The cost of urban expansion – land subsidence, deforestation, landslides, soil erosion, disturbance of the watershed, flooding, air pollution – was paid for by the suffering of ordinary citizens.

Unsustainable urban expansion does not happen naturally but is mediated through specific political mechanisms – namely, the unholy alliance between a pro-development state and the agents of the market (conglomerates and local factions). Tracing the history of the strategic coalition between the KMT government and conglomerates and local factions, formerly clients and now competitive partners of the KMT, one finds a clear pattern of "trading political loyalty for economic privileges." In certain economic fields, such as real estate, the government exclusively opened political channels for conglomerates to seek economic advantage in exchange for their political support (Chu 1989, 151; Hsiao and Liu 1997, 51; Chen 1995). Under such circumstances, public goods are inevitably privatized. Without the politically motivated laissez-faire attitude of an authoritarian regime toward land and housing policies, it is hardly possible that the investing of conglomerates and local factions in the real estate market could be so extensive and unconstrained. Taipei is a case in point.

In the last decade, one-fifth of the top 100 conglomerates in Taiwan have invested in the housing market, and local factions in Taipei county, on the outskirts of Taipei city, have owned a total of 161 construction companies and become involved in at least 230 large-scale development projects (Chen 1995, 161–7, 207). In 1998, 42 per cent of Taipei county council members (25 out of 65) were directly engaged in real estate, redevelopment, and constructions. By controlling the urban planning and land use committees, council members affiliated with local factions and conglomerates facilitated unsound development projects, which brought in huge profits for the businesses of politicians and these local factions and conglomerates. The real estate coalition gained huge profit margins by obtaining title at low prices to protected areas not included in urban development plans (hillside or waterfront land, or tillage), then having the land status changed to residential or commercial; bidding for and obtaining newly released public land through political connections; and purchasing land adjacent to designated zones for national construction projects (public transit systems, highways, industrial parks) in advance.

Through manipulating urban planning agencies, conglomerates and local factions have been actively involved in the real estate market since the 1970s (Hsiao and Liu 1997, 46–53). The value of land multiplied by 184.7 times from 1952 to 1975, while the overall value of goods increased by only 4.2 times during the same period (Mi 1988, 112–13). From 1973 to 1981, the index of family income increased by only 3.1 times, but housing prices multiplied by 4 to 5 times, and land value increased by more than 30 times (Hsu 1988, 171). But these figures are not the worst part of the story. The most dramatic instance of speculation in Taiwan's history occurred in the late 1980s. The market price of housing multiplied by 2.5 times in various districts of Taipei between 1987 and 1988 alone. A typical two-bedroom apartment would cost a middle-income white-collar worker more than 20 years' salary (Hsiao and Liu 1997). Constrained by rising real estate prices, urban slums became the solution for low-income families to satisfy their need for affordable housing.

Taipei's land speculation fever in the late 1980s is an excellent example of the maximizing of urban land use at the expense of social and ecological considerations. This value-maximizing ideology is perpetuated in Taiwan's shift from a labor-intensive economy to a capital- and brain-intensive economy. It is also embedded in the concept that, according to the national plan of APROC (the Asia-Pacific Regional Operations Center, a project involving multiple cabinet-level agencies), Taipei is projected to change its role from the leading city in a single country to a regional economic capital whose hinterlands include Taiwan, part of Southeast Asia, and the coastal area of China. It aims to host the headquarters of transnational corporations, provide advanced services such as high-speed information exchange, and be capable of competing with and then superseding Hong Kong and Singapore.

In Taipei, as in other cities that strive to climb the hierarchy of world cities, both national and municipal governments have invested trillions of dollars in transportation infrastructure since the late 1980s: public transit systems, high-speed trains, expressways, connection lines to airports and seaports, and much more. Though environmental impact assessment procedures, required by Taiwanese environmental laws, were conducted in all cases, not one project was turned down. Moreover, the quality of public construction projects was no better than that of private development. Some major projects, such as the public transit system and highways, were filled with rumors of scandals and corruption that disgusted most Taipei residents.

As with profit-oriented hillside development initiated by the private sector, the single-minded ambition of the Taiwan and Taipei governments to gain a bigger share in global prosperity has caused numerous nightmares. Excessive construction has not only heavily changed the physical appearance of the city but also altered its cultural and social landscape. In recent years, several officially designated cultural heritage sites (archaeological sites,

landmarks or architecture with historical significance) have been legally and illegally damaged, if not totally destroyed. For example, the city administration once intended to tear down a first-class national relic, a fortress dating back 100 years, to facilitate the construction of an inner-city highway. After strong protests, the building was finally saved, but a strange, even heartbreaking, juxtaposition of the highway and the relic has resulted – the highway was built directly over the roof of the fortress, and a 50-centimeter distance between the two has been left to safeguard the integrity of the building.

Furthermore, to fulfill the requirements of a world city, regulations on land development and urban zoning have been loosened to create space for high-tech science parks; trade centers; locations for financial, banking, and other service industries; "smart" buildings; and exposition sites for hosting major world events. Not surprisingly, because of the government's redevelopment and rezoning efforts, traditional economic activities and residential areas have been disregarded to make way for developers, transnational corporations, and public agencies. In the process of urban restructuring stimulated by global economic forces, there have been within the last two years numerous public and private development and redevelopment projects, changing large-scale residential zones into commercial zones (Chi 1997).

Despite the increasingly detrimental impact on the environment of traffic congestion, overcrowding, and noise brought about by such changes, the billions of dollars in profits behind them made them attractive to the city government's policy makers. Still, it is because of this very urban restructuring process that the social and economic map of Taipei has been redrawn and that many urban dwellers either have undergone a painful loss of cultural identity or have been totally rubbed off the map. Some development projects commenced with the eviction of a whole village or community that had existed for an extended period of time. In most cases, monetary compensation was provided, but usually without a proper resettlement plan. While the material needs of the city are being met, the social bond, a rare treasure in an overcrowded human setting, and one that residents built over a long period, is in jeopardy and may even be lost forever.

LOCAL RESISTANCE: THE PROFILE OF TAIPEI'S ENVIRONMENTAL MOVEMENT

As a result of urban restructuring, a perception that living in this city has become both difficult and unbearable is widespread among its residents. For those who are affluent enough to own a house, buildings might crumble overnight thanks to wet-season landslides. Among those who have lower incomes, inadequate housing and LULUs (locally unwanted land uses) are prominent issues: the city authorities have threatened to sue the residents of slums for illegally occupying public land, and facilities that have raised health concerns (landfills, incinerators) have been situated near poor communities. Regardless of wealth, air pollution causes respiratory diseases across all generations, and particularly among children. To satisfy the booming demand for office space, recreational area is minimized and skyscrapers are built one after another. Furthermore, within the city proper everything is either overcrowded or under construction. Feeling suffocated by a lack of open space, urbanites desperately look for somewhere with green space and fresh air. Their search usually yields two results: first, such a place is hard to find, since the slopes of the hills surrounding the Taipei basin have been cleared on a massive scale; second, the traffic on the way to assured green spaces is appalling, as reflected in the statistic that the number of vehicles in Taipei increased from 772,297 in 1987 to 1,443,630 in 1996 (Department of Budget, Accounting, and Statistics 1997, 458–9).

Public concern over hillside development was also raised by the frequent mudslides and floods. According to a nationwide poll conducted in 1998, 93 per cent of interviewees considered Taiwan's hillsides in general to be overdeveloped; 67 per cent believed the government and the developers should be held responsible for the detrimental effects of overdevelopment; and 77 per cent supported stricter development regulations. In another poll conducted that same year among Taipei residents only, 68 per cent urged the government to completely prohibit further development on the Taipei metropolitan hillsides, and 70 per cent believed that governmental officials and developers jointly engaged in land speculation (Fei 1998; Tung 1998). All of these factors – hyperurbanization, horrendous air pollution, traffic congestion, housing shortages, inadequate transportation, insecure housing conditions – have led to a diminished quality of urban life in Taipei and increased physical and mental stress for the city's residents.

Alienated and blasé city dwellers . . . have found ways to collectively vent their pain and anger by taking part in nationwide environmental struggles. It is well noted that during the 1980s, Taiwan experienced a surge of massive civil protests in which citizens attempted to gain political rights (freedom of speech, association, and demonstration) repressed under colonial and authoritarian rule for nearly a century and to express their concerns on pressing social issues such as the environment, gender inequalities, labor, welfare, and human rights. Collective action against environmental deterioration is no doubt one of the "early risers" in the process of democratization and has played a critical role in both demonstrating "the vulnerability of authorities" and diffusing "a propensity for collective action" to other social groups with different concerns (Tarrow 1994, 155). . . .

Local environmental protesters have severely questioned the state's pro-development and growth-centered policies and have fully developed a repertoire of collective action, which includes sit-ins, hunger strikes, blockades, religious parades, and theatrical performances. Environmental protests soon became a model for other movements to follow, playing a role similar to that of the civil rights movement in the United States (McAdam 1988, 1999).

As part of the nationwide environmental movement, hundreds of protests took place in the Taipei metropolitan area between 1980 and 1996, and citizens have made explicit demands for a livable environment and a higher-quality urban life. . . . Of the 274 protests in the Taipei metropolitan area between 1980 and 1996, 46 per cent concerned government sponsored development projects; 16.8 per cent opposed development and construction projects proposed by the private sector. Protests against all development projects in Taipei amounted to more than 60 per cent of total protests. . . . It is clear that the evolution of protest in the Taipei metropolitan area tracks the changing macroeconomic context described in the previous section. Despite the nonindustrial characteristic of Taipei's economy, the chief target of environmental protests in the early 1980s (55.6 per cent) was industrial pollution from the private sector, especially the chemical industry. Between 1986 and 1990, the percentage of protests against industrial pollution remained high (60 per cent), but the private business sector no longer was the sole target (32.5 per cent): state-owned enterprises and public construction projects emerged as other important adversaries (27.5 per cent).

In the 1990s, protests began to confront the attempts of real estate developers, politicians, and bureaucrats to make Taipei into a world city. The percentage of protests against specific industrial pollution dropped significantly, but protests against development projects, whether state sponsored or private, increased from around 40 per cent in the late 1980s to nearly 70 per cent between 1991 and 1996.

Though the issues and the strategies of each protest case may have differed, local and community-based mobilization is nevertheless the common characteristic. This confirms the repeatedly highlighted significance of "locality" in environmental movements worldwide (Castells 1997; Diani 1995). The citizens' "war for survival" mainly responds to excessive development and the obvious deterioration of the urban habitat. In the context of Taipei's struggle, the conflicting interest between citizens and global-domestic capital surfaced, and the tension over redistributing spatial resources among diverse social groups exploded. Citizens openly claimed their right to control over their immediate environment and strove to discover and elaborate on the meaning of community.

By means of these protests, Taipei residents have not only expressed their discontent over a deteriorating urban environment, but also challenged the top-down political structure that has long deprived urban dwellers of their right to participate in the decision-making process of public affairs. In most cases, the residents were "informed" by the government about changes in the designation of land use or about the introduction of locally unwanted facilities after the decision had been made. The citizens were outraged not only by the perceived immediate threat brought about by these development projects, but also by the undemocratic, top-down decision-making machinery that had approved them. Thus, these protests also reflect the residents' demands for institutional change toward participatory democracy in the city's public life. In

the process of struggle, local communities have served as agents of environmental action, reflecting not only the immediate interest of their communities, but also the potential for their action to contribute to a more sustainable and just urban environment overall.

COMMUNITY AND GOVERNMENTAL RESPONSIVENESS

In responding to the negative effects caused by urban restructuring, communities rebelled. Ching-Cheng and Kang-Le, both protest cases cited at the beginning of this paper, articulated the idea of "a city for citizens" and aimed to defend their living spaces. Both the middle-class neighborhood and the slum community were pressured by urban redevelopment and the overdue need for public facilities and infrastructure.

Other communities also have been sensitive about hillside development and landscape preservation issues. Wan-Fang, another middle-class neighborhood, campaigned to prevent the Public Housing Office from situating a large-scale housing project on a nearby hillside that experienced landslides during the wet season. Chihshan Yen, with a high proportion of female residents who are members of the Homemakers' Union Environmental Protection Foundation (a female-dominated national environmental NGO), showed strong opposition to the construction of a gas station that, it was feared, might severely damage a historical ruin nearby and lower the quality of life of the community (Yu 1994, 43–56).

The rise of the community-based protests illustrates the efforts of community residents to preserve social bonds and place-specific identity, and to fight against unwanted development projects that usually favor the few over the many. Their demands for stricter environmental sanctions, cultural preservation, livable habitats, and institutional access to decision-making processes have put enormous pressure on government at all levels and finally changed the landscape of local politics. Facing ongoing direct action from these communities and electoral competition among political parties in Taiwan's fledgling democracy, the incumbent parties of both national and municipal governments could not afford to ignore these explicit demands to halt unwelcome urban development and unsound land use on hillsides.

At the national level, island-wide environmental protests have directly contributed to a wave of legislative and institutional reform since the late 1980s. The Pollution Disputes Resolution Law and Environmental Impacts Assessment Act were passed, and the Environmental Protection Basic Law is being reviewed in the legislature (Chi *et al.* 1996). Also, the Environmental Protection Administration (EPA) was upgraded to a cabinet-level agency in 1987. Due to pressure from lobbying by many environmental NGOs, stricter environmental regulations were adopted. In the case of deforestation and landslides, for example, the Ministry of Interior lowered the permitted angle of hillside development twice in one year and thus largely reduced the total area permitted for development.

At the municipal level, party competition in each local election has increased voter sensitivity (note the campaign slogan from various candidates, "citizen first"). This new trend has been more salient since the December 1994 change from appointment by the national government of the Taipei city mayoral post to direct election of this official. Any dubious connection with conglomerates and other local business interests is now viewed as a liability, if not poison, for a candidate in any local election. "Greening" the political platform has become inevitable, and it accurately reflects the rising public demand for a higher quality of urban life. As a result, in the first direct mayoral election, a pro-environment candidate from the Democratic Progressive Party (DPP, the biggest opposition party in Taiwan) was elected. At the same time, in the city council election, quite a number of candidates who had never been involved in plutocracy defeated the longtime representatives of or collaborators with conglomerate interests. The political efficacy of community protests is thus being successfully channeled through the system of representative politics. The question, then, is: What do elected officials do to prove to their voters that they are working toward a livable city? The magic answer is: Build more city parks.

According to a statistic from 1993, the area of park per citizen in Taipei at the time was 2.5 square meters, smaller than in other metropolises such as New York City (13.95), Paris (12.70), Seoul (8.70), and Tokyo (4.70) (Hsiao *et al.* 1997, 12). It

is thus not surprising that in almost all campaign packages in recent elections, city parks have been one of the favorite topics of all candidates. With the urgent need for open space, a formula of "parks equal votes" was adopted. The current and previous mayors have strenuously worked toward building city parks during their tenures, no matter what conflicts the process might attract. The construction of the No. 7 city park in the early 1990s resulted in the eviction of thousands of slum dwellers, as did the Nos. 14–15 park. Between 1991 and 1996, the two mayors built sixty-seven parks, and the total area of parkland increased from 805.8 hectares to 979.4 (Department of Budget, Accounting, and Statistics 1997, 452). The city government was also expected to add at least ten more parks in 1997 (Yang 1997).

In addition to city parks, several pro-community programs were set up. Following his election victory, Chen Shui-bian appointed the first environmentalist with a national reputation to take charge of the local EPA. Also, the city government's Department of Urban Development proposed various redevelopment projects that showed concern for the preservation of traditional landscape and historical architecture. For example, two old and historically significant districts of Taipei city (Tihua Street and the Tatung District) are expected to be economically revitalized and to attract cultural tourism. Each year, ten to fifteen communities are officially sponsored to improve their physical environment and to build stronger community networks. Tougher measures also have been enacted to regulate hillside development at the city government level.

Protest, then, has affected the political arena both by changing the character of political discourse around environmental issues and by forcing modification of the formal rules governing urban development. The municipal government has indeed responded to demands of urban environmental protests, stressing community autonomy and the importance of protecting environmentally sensitive areas by initiating various programs to facilitate community involvement. These effects, however, must be considered only a first step toward making Taipei more livable. Changes in rules and discourse do not necessarily mean changes in practice and behavior "on the ground." Equally important, efforts in the direction of livability continue to have very different implications for poor and middle-class communities.

THE POLITICS OF PROTEST

At first glance, it seems as if citizens' concerns over urban land use were addressed through the tougher measures adopted to inhibit hillside development and by several other governmental initiatives. Tempered by democratization, the top-down nature of urban planning has somehow been softened by the creation of new mechanisms for community participation. The mayor and county magistrate, pressured by the consideration of re-election, must take community voice into account and prevent urban-planning bureaucrats from exhibiting behavior unfriendly to communities.

However, after closer examination, one finds that problems still remain. Unequal treatment of communities prevails, the government's capacity to implement its new policies is highly problematic, the "development first" mentality still lingers, and the vision of becoming a world city has yet to be realized. The positive policy responses toward environmental causes have been tempered by an increasing number of huge infrastructure projects and district redevelopment plans that were approved and quickly and aggressively commenced. . . .

SOCIAL CLASS AND THE EFFICACY OF MOBILIZED COMMUNITIES

Since urban environmental protest is location-specific and community-based, the social attributes of each community to some extent affect its bargaining power with the authorities and adversaries in the course of protest. From various case studies of Taipei's community and environmental movements, it is clear that the class composition of a community is closely associated with the readiness of available resources, the span of information networks, the extent of media coverage, and the level of government responsiveness (Shen 1994; Yu 1994). Middle-class communities are usually in a better strategic position to attract media coverage and to generate public support. They are also more likely than poor communities to receive a fast

government response and to seek assistance from outside NGOs.

In contrast, a poor or slum community is indeed more vulnerable when it encounters powerful social institutions. If poor communities get any attention from the media, it is likely to take the form of distorted and negative coverage, and they are much less successful in dealing with the authorities. Let us take as an example the case of the Kang-Le community. In opposing the eviction of thousands of residents, Kang-Le formed a self-help organization in 1992 and desperately looked for a way to balance the citywide demands for public parks and the needs of slum residents. Before the professors and students from the Graduate School of Architecture and Urban–Rural Planning Studies at National Taiwan University got involved to help organize the anti-bulldozer campaign in early 1997, very few media outlets had paid attention to this story. When the press began to look into the dispute, the government authorities occupied center stage in the news coverage and marginalized the voice of the shantytown residents. Some municipal government officials even used the media to distort the Kang-Le community's image. Slum residents were portrayed as a group of "greedy rich people" who had illegally resided on public land for a long time and who were going to be generously paid up to 52 million US dollars in compensation out of the taxpayers' pockets. They were also criticized for having deprived decent citizens of the right to a park.

Similar problems have been faced by urban Aborigines. Throughout the history of Taiwan, the Aborigine rights to land, natural resources, and cultural heritage have been largely ignored. On the one hand, the officially "reserved" homelands of the Aborigines have borne the weight of public construction projects (mostly dams and, in the case of Orchid Island, a nuclear-waste storage facility). Aboriginal opposition to these projects resulted in very few concessions from the state. On the other hand, having undertaken the process of rural–urban migration in the hope of escaping the troubled mountain economy, urban Aborigines still face the nightmare of eviction. Aboriginal migrants who moved to Taipei illegally built their ethnic settlements either on riverbanks or on hills, where the natural environment was similar to their homeland. These small, scattered ethnic settlements suffered from extremely poor living conditions; some were even without electricity and water, let alone more luxurious amenities. The rationale to evict the Aboriginal shantytown dwellers was quite similar to that which evicted the Kang-Le community: public safety, the retrieval of state-owned property, and the need for public works (in this case, high-speed trains). The Aboriginal struggle with the government received even less media coverage and public attention than did Kang-Le's, and the government response to their concerns was very unfavorable.

In sum, although in principle environmental protests have consistently advocated the right of people to have a say regarding their immediate environment, in practice the claim to that right by different communities is weighed unequally by politicians, the media, and the public. The way each community's requests are dealt with depends very much on the status of the community in the pyramid of power, which is structured around class and ethnicity.

THE MINIMAL INVOLVEMENT OF TRANSLOCAL NGOS

It is intriguing to note that translocal NGOs, whose membership is mainly of the middle class, have played a limited role in the urban struggle, in contrast to the aggressiveness of local communities. According to a survey of environmental organizations conducted in 1997, roughly one-third of Taiwan's environmental NGOs had established their headquarters in Taipei, giving this metropolis the highest density of organizational networks in Taiwan (Hsiao 1997). However, three-quarters of Taipei's NGOs identified themselves as "national organizations" that focus on broader issues, and they are therefore inattentive to local affairs and urban protests in Taipei. Only a few NGOs had half an eye on urban environmental issues. The major interests of these NGOs were recycling, vehicle usage, and hillside development, and furthermore, they unanimously adopted soft-line approaches to change, such as lobbying, instead of actively allying themselves with local communities. Even in the case of the Chihshan Yen community, the involvement of the Homemakers' Union Environmental Protection Foundation

derived more from an overlap in membership than from a primary concern on the part of the NGO for the preservation of Taipei's cultural landscape.

It is not unusual that a division of labor exists between translocal NGOs and local communities within the larger picture of environmental movements. Over the years, Taiwan's NGOs, backed by lobbyists and by think tanks consisting of academic recruits, have played a translating role to reframe local concerns into significant issues in the public sphere. In many cases, combining NGO access to the political system with the pressure derived from residents' direct action has yielded effective and favorable results. For example, communities in southern Taiwan and translocal NGOs have allied with one another to fight successfully against an incoming petrochemical and steel-refining industrial park, to secure the livelihoods of local fish farmers and fishermen, and even to protect endangered wetlands and a threatened species of bird, the blackfaced spoonbill. However, there is no link between NGOs and mobilized communities in Taipei's urban struggle.

Eighteen out of 33 of Taipei's NGOs said that the quality of the urban environment was not even one of their major concerns. Why have Taipei's NGOs shown negligence on urban environmental issues and maintained such a distance from urban communities? What has made Taipei a special case in contrast to the alliances between NGOs and anti-pollution communities elsewhere? To answer these questions we need to consider the following issues: the pro-DPP sentiment among environmental NGOs and these NGOs' perception of "what counts as an environmental issue."

In Taiwan, visible nationwide environmental NGOs fall into roughly two categories: research-oriented groups that specialize in public policy and present their perspective through moderate means such as lobbying and environmental education campaigns, and action-oriented groups that engage in community mobilization and are prone to express their environmental concerns through noninstitutional and confrontational strategies. Both types of group have similarly decent annual budgets and a membership disproportionately made up of academics, but the real difference lies in organizational size. The first type, usually with an elite membership of fewer than 50, acts as a body of consultants for policy makers and never directly participates in local protests. The second type, with a much larger membership pool, tends to bring local concerns into higher-level policy debates and actively engages in protests against large, government-sponsored development projects.

During the struggle against the authoritarian KMT state in the early 1990s, DPP and many action-oriented NGOs developed and cultivated an issue-based comradeship and an overlapping membership. It is well documented that many outspoken leaders of these NGOs are in fact members or supporters of the DPP (Weller and Hsiao 1998). In the 1990s, when the DPP and its incumbents occasionally did not maintain their firm stand on environmental causes, those action-oriented NGOs found it difficult to fight against old friends, except in extreme cases. For example, a DPP county magistrate was strongly attacked because the construction of Taiwan's seventh naphtha cracker (an industrial complex where crude-oil products are refined into the precursors of plastics), a project that environmental NGOs had strongly opposed for years, proceeded partially at his discretion. Besides this extreme case, most NGOs have been friendly toward DPP administrations, and the cooperation between the two has been pervasive. In Taipei's case, instead of working with protesting communities, quite a few environmental NGOs came forward to support DPP mayor Chen Shui-bian's campaign to increase green space. They also endorsed a citywide recycling project and community improvement plans proposed by the city government.

To most environmental NGOs, urban restructuring and changes in urban land use are not really environmental issues but rather technical matters related to urban planning. Compared to the concern of NGOs for forests, rivers, the ocean, wetlands, wildlife, and much more, the urban habitat is rather orphaned – a point that has been nicely elucidated by Harvey (1996, chs 13–14). While making a firm stand against hillside development because of resulting deforestation and soil erosion, NGOs still have difficulty placing other urban issues, such as slum clearance, on their list of priorities. It is thus not surprising that they have remained silent on issues relating to the urban poor. Attentive to a narrowly defined "natural environment," NGOs somehow neglect the "unnatural" urban setting and remain largely uninvolved in many crucial urban

battles. It is rather ironic to note that a movement and numerous NGOs, which "exhort the harmonious coexistence of people and nature, and worry about the continued survival of nature (particularly loss of habitat problems), somehow forget about the survival of humans (especially those who have lost their 'habitats' and 'food sources')" (Harvey 1996, 386).

COMMUNITY INTEREST, SUSTAINABILITY, AND SOCIAL JUSTICE

The third and probably the most important issue in the politics of protest is how Taipei's locally driven protests relate to the long-term sustainability of the city. Is it hypocritical that middle-class communities oppose highway construction and refuse to sacrifice their community parks for parking garages while their residents intensify their use of private automobiles? Is it narrow-minded to boycott the nearby construction of incinerators while the quantity of garbage that each citizen produces increases every year? Is it selfish for slum dwellers to try to save illegal shacks from being torn down regardless of the need for green space? Is it a public-safety problem that Aborigines build their ethnic settlements on the riverbank? In other words, are local communities waving the flag of environmentalism just to protect their own interests? Or can local interest be congruent with ecological concern across a city? On the road toward a more livable city, how is it possible to balance the subsistence of the urban poor and the sustainability of the city?

To praise all community protests as heroic and progressive is as problematic as to blame them for being insular and self-serving. Either romanticism or cynicism may generate an emotional response, but neither provides us with a vision. The motivation behind local protests might be led by self-interest, but once channeled through proper means, these local interests may be translated into larger issues of public concern. The fact that most local protests are interest-driven does not necessarily prevent them from contributing to urban sustainability. Let us take highways and parking garages as examples. As Friedmann has put it (1997, 18): "for an Asian city to replicate Los Angeles' love affair with freeways . . . is to commit collective suicide." In a city like Taipei, with an average density of 9,586 persons per square kilometer (Department of Budget, Accounting, and Statistics 1997), and with some districts even reaching 25,000 persons per square kilometer (Chang 1993, 432), to protest the excessive construction of highways and parking garages is to try to rein in the overuse of private vehicles and encourage the use of more environmentally sound methods of mass transportation. The same reasoning may apply to incinerators. When people refuse to accept an incinerator as their neighbor, we should not immediately denounce them as engaging in NIMBY behavior or as selfish, because their recalcitrance could point to better ways of solving the garbage problem. Instead of endlessly building incinerators and landfills, a task that Taiwan's EPA has been undertaking, one can argue that in the face of insufficient garbage processing facilities, it is more sensible to strengthen the system of recycling and to persuade citizens to voluntarily reduce their garbage output. It is undeniable that public discussion of stringent controls on private vehicles and a citywide recycling plan "was aroused in the wake of these selfish" community protests.

Another issue concerns the conflicting interests of the urban poor and society as a whole. In order to shed light on this sensitive issue, we would first like to go back to the conception of "sustainability." Regardless of its wide variety of connotations, one basic point of consensus on the meaning of sustainability is the concept of "intergenerational justice": in pursuit of a livelihood, the happiness of future generations should not be compromised or sacrificed (Friedmann 1997; Piccolomini 1996). If the pursuit of sustainability should take into account the quality of life of those who are not yet born, then the subsistence and quality of life of the urban poor should not be disregarded in satisfying the needs of other citizens. Like two sides of a coin, in theory or in practice, sustainability should not and cannot be separated from a more just social system. To increase the area of green space might help to mitigate a city's air pollution, but to use that as the justification for dismantling a poor community in its entirety is untenable. If it does not take into account the unequal distribution of economic goods and power among different groups in a society, an ostensible environmental or ecological solution might unintentionally lead to dreadful

social consequences (Beck 1995; Harvey 1992, 1996; criticisms on thoughtless slum demolition projects are also detailed in Gans 1962).

By re-examining the case of the Kang-Le struggle, we discover that the effort to balance both social and environmental demands has failed and that, even more serious, a fast-food approach to improving the urban environment has overwhelmed considerations of social justice, a value that must be maintained on the path toward a sustainable city.

Learning from the classic case of Orcasitas, in Madrid, the urban planners who supported the Kang-Le community's cause made a proposal that would have kept the city park project on course but also saved the community. They said that "the occupants of the illegal shantytown settlements who were . . . responsible for utilizing the area and increasing its value should be the first to benefit once the area had been redeveloped" (Castells 1983, 2–53). In this proposal, 6 per cent of the designated park area would have been used to build affordable government-sponsored housing to relocate the aged and poor residents living on the same site, and the design would have turned the park into an ecomuseum, evoking memories of the area's past by including features of the Japanese cemetery and slum architecture. However, the proposal was turned down by the city's top officials. Their refusal rested on the argument that the resettlement proposal favoring slum dwellers had no legal basis and would only stir up "unrealistic expectations" among slum dwellers in other areas. Immediately after the demolition, there was a 15 to 20 per cent increase in rental and land value surrounding the Nos. 14–15 park (Coalition against the City Government Bulldozers 1975, 11). This further confirms that an ecological vision such as a park system "was easily co-opted and routinized into real-estate development practices for the middle classes" (Harvey 1996, 417).

CONCLUSION

We have tried to set out the central dynamics of the evolution of Taipei's environmental movement, showing both how the movement reflects the changing character of the larger Taiwanese political economy and how the character of the movement itself limits the kinds of changes that it can achieve. The task of enacting change has not been an easy one, since the movement's overall character emerges out of myriad small, heterogeneous actions. Nonetheless, we have tried to show that there is an overall logic to the movement, one whose limitations must be balanced against its undeniably positive potential.

A simplified version of Taipei's struggle can be described as follows: Both the national and municipal governments engaged in a large project of urban space restructuring in order to attract foreign capital, and the KMT government was keen to provide the private sector with incentives to maximize land use on hillsides. Residents later found out that this impressive government-inspired project promised everything but a livable urban environment. Foreign and domestic capital was placed in the "growth machine," leading to assaults on the natural environment, historical buildings, and the living space of various groups of residents. Feeling betrayed and hurt, unhappy residents mobilized neighbors, friends, and relatives to call their local representatives, to vote for other parties, and even to demonstrate in the street. Some translocal environmental NGOs noted this trend and passively participated in a few cases of community mobilization.

Though the governments initially had been unresponsive in meeting resident demands, the voice of the community grew too powerful and the threat of political backlash was serious; thus, several policy revisions were made. Both the national and municipal governments decided to inhibit excessive hillside development, but vast urban infrastructures would proceed as planned. To meet rising social grievances, the municipal government tried hard to build more city parks as an extra bonus for citizens. By doing so, it turned slum dwellers into the victims of urban improvement. . . .

Taipei's experience of urban deterioration is by no means unique. Leading cities in East Asia have been haunted by land speculation and unsound land use. They compete among themselves in the construction of attention-grabbing but wasteful urban infrastructure projects. What makes Taipei special is that civil protests against these macroeconomic processes are facilitated by Taiwan's democratization. Institutionally consolidated party competition provides various communities with more means to address their locally driven interests. Through these means, mobilized communities were finally able to put the long-neglected issue of "urban sustainability" on the national and city agendas. The responsiveness of both the national and municipal governments has proven the political efficacy of community action. In turn, community mobilization has been encouraged by a more responsive and democratic regime – that is why the number of protest cases has increased with time.

One benefit of democracy has been that the voice of Taipei's residents has been heard. However, because of the inequality of power between different social groups, the demands of some communities have been met and those of others blocked. If modern cities are usually portrayed as the center of wealth, power, information, and cultural production, we should not forget that the sharpest disparity between the rich and the poor and between the powerful and the powerless also takes place in urban settings. While demands in the form of environmental protests for a higher quality of urban life have brought positive institutional and legal reforms, poor and slum communities have suffered. The extent to which these proenvironment decrees can restrict excessive public and private development is still unknown; in the meantime, shantytown settlements have already been torn down and traded for city parks.

The dynamics of Taipei's urban environmental struggles have important implications for other Third World urban centers that aspire to be "world cities." In order for their environmental movements to pursue a broader vision of livability, they must structure their actions and their ideology to incorporate more horizontal links between mobilized communities and NGOs. The NGOs themselves need to acknowledge the central importance of dealing with complex urban systems. Most important, the livelihood of poor and slum communities must be incorporated into the concept of sustainability; the broader ecological movement must be more sensitive to equality. The idea that the improvement of the urban environment cannot be divorced from the realization of social justice must become a more integral part of movement ideology. In the end, a city cannot become more sustainable without also being more just.

NOTE

1   The authors' names are in alphabetical order, and each author contributed equally to this collaboration. The authors thank Peter Evans, Mike Douglass, Dungsheng Chen, and two anonymous reviewers for their constructive comments. They are also grateful for editorial assistance from Martin Williams and Fan Chang.

REFERENCES

Beck, Ulrich. 1997. *Ecological Politics in an Age of Risk*. Cambridge: Polity Press.

Berner, Erhard and Rudiger Korff. 1995. "Globalization and Local Resistance: The Creation of Localities in Manila and Bangkok." *International Journal of Urban and Regional Research* 19: 208–22.

Castells, Manuel. 1983. *The City and the Grassroots: A Cross-Cultural Theory of Urban Social Movements*. Berkeley: University of California Press.

—— 1997. *The Power of Identity*. Vol. 2 of *The Information Age: Economy, Society, and Culture*. Oxford: Blackwell.

Chang, Shih-chiao. 1993. "The Experience of and Outlook for Taiwan's Urban Environment and Development" (in Chinese). In *Chinese Cities and Regional Development: Prospect for the Twenty-First Century*. Ed. Yue-man Yeung, 425–52. Hong Kong: Hong Kong Institute of Asia-Pacific Studies, the Chinese University of Hong Kong.

Chen, Cheng-Tzeng and Chun-Kuang Chen. 1997. "The Minister and County Magistrate Calls a Stop to Hillside Development" (in Chinese). *China Times*, August 19.

Chen, Chung-Hsiung. 1998. "Heavy Sentences for County Officials and Developers: The Case of Lincoln Hillside Development" (in Chinese). *China Times*, August 19.

Chen, Dung-Sheng. 1995. *The City of Money and Power: A Sociological Analysis of Local Factions, Conglomerates, and Urban Development in Metropolitan Taipei* (in Chinese). Taipei: Ju-Liu Publishing Co.

Chi, Chih-ko. 1997. "Urban Planning: We Should Listen to the Voice of the People" (in Chinese). *China Times*, July 21.

Chi, Chun-Chieh, H.H. Michael Hsiao, and Juju Chin-Shou Wang. 1996. "Evolution and Conflict of Environmental Discourse in Taiwan." Paper Presented at the Association of Asian Studies Annual Meeting, Honolulu, Hawaii.

Chiang, Nora Huang and Hsin-Huang Michael Hsiao. 1985. "Taipei: History and Problems of Development." In *Chinese Cities: The Growth of the Metropolis since 1949*. Ed. Victor F.S. Sit, 188–209. New York: Oxford University Press.

Chu, Yun-han. 1989. "The Oligarchic Economy and the Authoritarian Political System" (in Chinese). In *Monopoly and Exploitation: The Political Economic Analysis of Authoritarianism*. Ed. Hsin-Huang Michael Hsiao *et al.*, 139–60. Taipei: Taiwan Research Fund.

Coalition against the City Government Bulldozers. 1975. *My Home is at Kang-Le District: The Documents of the Movement against Bulldozers of the City Government* (in Chinese). Taipei: Coalition against the City Government Bulldozers.

Department of Budget, Accounting, and Statistics. 1997. *The Statistical Abstract of Taipei City, 1997*. Taipei: Department of Budget, Accounting, and Statistics, Taipei City Government, Republic of China.

Diani, Mario. 1995. *Green Networks: A Structural Analysis of the Italian Environmental Movement*. Edinburgh: Edinburgh University Press.

Douglass, Mike. 1998a. "World City Formation on the Asia Pacific Rim: Poverty, 'Everyday' Forums of Civil Society, and Environmental Management." In *Cities for Citizens: Planning and the Rise of Civil Society in a Global Age*. Eds Mike Douglass and John Friedmann, 107–37. New York: John Wiley and Sons.

Fei, Kuo-Jen. 1998a. "Seventy Percent of Taipei Residents Urge a Complete Prohibition of Hillside Development" (in Chinese). *China Times*, October 27.

—— 1998b. "East Asian Urbanization – Patterns, Problems, and Prospects." Discussion paper, Institute for International Studies, Asia/Pacific Research Center, Stanford University, CA.

Friedmann, John. 1986. "The World City Hypothesis." *Development and Change* 17: 69–83.

—— 1997. *World City Futures: The Role of Urban and Regional Policies in the Asia-Pacific Region*. Occasional Paper no. 56. Hong Kong: Hong Kong Institute of Asia-Pacific Studies, the Chinese University of Hong Kong.

Gans, Herbert. 1962. *The Urban Villagers: Group and Class in the Life of Italian-Americans.* New York: The Free Press.

Gereffi, Gary. 1990. "Paths of Industrialization: An Overview." In *Manufacturing Miracles: Paths of Industrialization in Latin America and East Asia.* Eds Gary Gereffi and Donald L. Wyman, 3–31. Princeton, NJ: Princeton University Press.

Haggard, Stephan. 1990. *Pathways from the Periphery: The Politics of Growth in the Newly Industrializing Countries.* Ithaca: NY: Cornell University Press.

Harvey, David. 1992. "Social Justice, Postmodernism, and the City." *International Journal of Urban and Regional Research* 16(4): 589–601.

—— 1996. *Justice, Nature, and the Geography of Difference.* Malden, MA: Blackwell.

Hsiao, Hsin-Huang Michael. 1997. *A Symbiotic Relationship with Tension: The Relationship Between EPA and Local Environmental Groups* (in Chinese). Taipei: Environmental Protection Agency.

Hsiao, Hsin-Huang Michael and Hwa-Jen Liu. 1997. "Land-Housing Problems and the Limits of the Non-Homeowners Movement in Taiwan." *Chinese Sociology and Anthropology* 29: 42–65.

Hsiao, Hsin-Huang Michael, *et al.* 1997. *Re-creating the New Functions of Urban Development: An Analysis on the Eviction Case of the Nos. 14–15 Park in Taipei* (in Chinese). Unpublished report sponsored by the Department of Social Affairs, Taipei Municipal Government.

Hsu, K'ung-Jung. 1988. "A Sociological Analysis of the Housing Market in Peripheral Taipei" (in Chinese). *Taiwan: A Radical Quarterly in Social Studies* 1(2–3): 149–210.

Huang, U.-An. 1983. *The History of Taipei's Development* (in Chinese). Taipei: Taipei Archive Committee.

Knight, Richard V. 1993. "Sustainable Development – Sustainable Cities." *International Social Science Journal* 45(1): 35–54.

Lee, Boon Thong. 1998. "Globalization, Tele-revolution and the Urban Space." Paper presented at the Workshop on Southeast Asia under Globalization, the Program for Southeast Asian Area Studies (PROSEA), Academia Sinica, Taipei.

Machimura, Takashi. 1992. "The Urban Restructuring Process in Tokyo in the 1980s: Transforming Tokyo into a World City." *International Journal of Urban and Regional Research* 16: 114–28.

McAdam, Doug. 1988. *Freedom Summer.* New York: Oxford University Press.

—— 1999. *Political Process and the Development of Black Insurgency, 1930–1970.* 2nd ed. Chicago: University of Chicago Press.

Mi, Fu-Kuo. 1988. "Public Housing Policy in Taiwan" (in Chinese). *Taiwan: A Radical Quarterly in Social Studies* 1(2–3): 97–147.

Mumford, Lewis. 1989 [1961]. *The City in History: Its Origins, Its Transformations, and Its Prospects.* New York: MJF Books.

Piccolomini, Michele. 1996. "Sustainable Development, Collective Action, and New Social Movements." *Research in Social Movements, Conflict, and Change* 19: 183–208.

Shen, Yao-pin. 1994. *Community Mobilization and the Transformation of Urban Meaning: The Case Analysis of Taipei's Ching-Cheng Community* (in Chinese). Master's thesis, the Graduate School of Architecture and Urban-Rural Studies, Nation of Taiwan University, Taipei.

Sun, Hsiu-huei. 1997. "They Didn't Know the Buildings Were Dangerous, They Only Saw Them Collapse" (in Chinese). *Global Views Monthly* 135: 76–77.

Tarrow, Sidney. 1994. *Power in Movement: Social Movements, Collective Action, and Politics.* New York: Cambridge University Press.

Tseng, Hsu-Cheng. 1993. "The Formation of the Taipei ren [person]" (in Chinese). In *Research on the Immigration into Taipei County.* Eds Hsin-Huang Michael Hsiao *et al.*, 79–121. Taipei: Taipei County Culture Center.

Tung, Mong-Lung. 1998. "90 Percent of People Interviewed Consider that Hillsides Have Been Overdeveloped" (in Chinese). *China Times,* July 22.

Wan, Zen-Quai. 1998. "No Hope for Monetary Compensation and Resettlement: The Victims of Mudslide Area Trapped" (in Chinese). *China Times,* August 19.

Weller, Robert P. and Hsin-Huang Michael Hsiao. 1998. "Culture, Gender, and Community in Taiwan's Environmental Movement." In *Environmental Movements in Asia.* Eds Arne Kalland and Gerard Persoon, 83–109. Honolulu: University of Hawaii Press.

Yang, Chin-yen. 1997. "Chungshan No. 1 Park – Completed a Year Ahead of Schedule" (in Chinese). *United Daily,* October 23.

FOUR

—— 1998. "Two Parks Developed as Cultural Area, Details to Be Finalized within a Year" (in Chinese). *United Daily*, March 4.

Yeung, Yue-man. 1996. "An Asian Perspective on the Global City." *International Social Science Journal* 48(1): 25–31.

Yu, Tao-ling. 1994. *A Research on the Process of Mobilization of Community Protest: Three Cases of Taipei* (in Chinese). Master's thesis, the Graduate School of Architecture and Urban-Rural Studies, National Taiwan University, Taipei.

PART FIVE

Visions of sustainable community

INTRODUCTION TO PART FIVE

Vision is a key ingredient for long-term change. It can inspire, motivate, spread ideas, and provide guidance for more pragmatic activities. Vision can be developed in many ways, including through community planning documents, blueprints for reform, manifestos, speeches, visual images, films, and a variety of forms of utopian literature. Not for nothing was crusading urban critic Lewis Mumford's first book entitled *The Story of Utopias* (New York: Boni and Liveright, 1922). The very idea of "sustainable communities" represents a long-term vision, albeit one that needs much definition and detail. Documents such as the Brundtland Commission report and *Agenda 21* also fall into the category of vision documents, as do the Charter of the New Urbanism, McDonough's Hannover Principles, and many other writings of authors represented in this book.

In Part 5 we present three fictional visions of more or less sustainable cities, one a well-known historical proposal that has been enormously influential in the actual planning of many new towns, the other two utopias from bestselling popular novels. The philosophy behind Ebenezer Howard's Garden City vision was introduced in the first reading in this volume; here we turn to Howard's description of the garden city itself. Ernest Callenbach's *Ecotopia* achieved a wide distribution in the 1970s, and has staked out a position as the most fully developed ecological utopia of recent decades. Ursula K. Le Guin's science fiction novel *The Dispossessed* probes the political and social dynamics behind the establishment of a utopian civilization on another planet several centuries from now, drawing on anarchist political philosophy as well as environmental sensibilities.

Visions of better places or societies tend to emerge at historical times in which a creative ferment exists and substantial numbers of people are exploring new ideas. The turn of the last century was one such time, with the dawn of the Progressive Era, a growing movement for women's suffrage, and the rapid spread of new technologies (e.g. the electric light, the telephone, the streetcar). This period of rapid change stimulated utopian fiction such as Edward Bellamy's *Looking Backwards* (Boston: Ticknor, 1888), Howard's *Garden Cities of To-morrow* (1898), and Charlotte Perkins Gilman's *Herland* (New York: Pantheon Books, 1915 [1979]). It also gave rise to the organized city planning profession, which began to take shape in the 1900s and 1910s.

The late 1960s and 1970s was another such time when many people's conceptions of possible futures for humanity changed rapidly, with the growth of environmentalism, feminism, civil rights movements, and "small planet" consciousness. Notable utopian writings from this second period include Callenbach's *Ecotopia*, Marge Piercy's *Woman on the Edge of Time* (New York: Knopf, 1976), Dorothy Bryant's *The Kin of Ata Are Waiting For You* (San Francisco: Moon Books/Random House, 1976), and a number of Le Guin's works, especially *Always Coming Home* (New York: Harper & Row, 1984). At the conclusion to *The Granite Garden* (New York: Basic Books, 1984, pp. 268ff.), Spirn presents a vision of "the celestial city" rather similar to Callenbach's, in which nature "is everywhere evident and cultivated in the city." Paul and Percival Goodman's nonfiction *Communitas* (New York: Vintage Books, 1947) presaged these later writings by presenting a very thoughtful consideration of different urban development alternatives. Kevin Lynch's "A Place Utopia" at the end of *Good City Form* (Cambridge, MA: MIT Press, 1981) likewise proposed an ecologically oriented urban utopia, in the form of a decentralized, web-like network of human communities. Such writings coincided with further growth in the agenda of urban planning, especially the rapid development of environmental planning, the emergence of the environmental design field, and growing public participation in planning.

Visionary or utopian writings help expand the framework of permissible ideas for a generation or more. Although mainstream thinkers tend to scorn idealism of all sorts, it has often been extremely influential and successful, as shown by Howard's Garden City vision, the civil rights movement, the environmental movement, several waves of feminism, and recent urban planning movements such as the Congress for the New Urbanism. It is certainly to be hoped that a new wave of visions of healthy and sustainable communities awaits us in the not too distant future, along with pragmatic efforts to actually bring such visions about.

"The Town–Country Magnet"

from *Garden Cities of To-morrow* (1898)

Ebenezer Howard

Editors' Introduction

In this selection from *Garden Cities of To-morrow*, Howard spells out his vision of the Garden City, introduced in this volume's first reading. Although this scheme is highly visionary and quite unlike any previously existing community, Howard does his best to present his idea in a pragmatic, reasonable way that might actually be implemented. He is very specific about the finances, physical extent, design, economic base, and social structure of his proposed community. Howard's efforts paid off, as a Garden Cities Association dedicated to creating such communities was established after the book's appearance, and he himself advised in the creation of the towns of Letchworth (1911) and Welwyn (1926) outside of London.

The reader is asked to imagine an estate embracing an area of 6,000 acres, which is at present purely agricultural, and has been obtained by purchase in the open market at a cost of £40 an acre, or £240,000. The purchase money is supposed to have been raised on mortgage debentures, bearing interest at an average rate not exceeding 4 per cent. The estate is legally vested in the names of four gentlemen of responsible position and of undoubted probity and honour, who hold it in trust, first, as a security for the debenture-holders, and, secondly, in trust for the people of Garden City, the Town–Country magnet, which it is intended to build thereon. One essential feature of the plan is that all ground rents, which are to be based upon the annual value of the land, shall be paid to the trustees, who, after providing for interest and sinking fund, will hand the balance to the Central Council of the new municipality, to be employed by such Council in the creation and maintenance of all necessary public works roads, schools, parks, etc.

The objects of this land purchase may be stated in various ways, but it is sufficient here to say that some of the chief objects are these: To find for our industrial population work at wages of *higher purchasing power*, and to secure healthier surroundings and more regular employment. To enterprising manufacturers, co-operative societies, architects, engineers, builders, and mechanicians of all kinds, as well as to many engaged in various professions, it is intended to offer a means of securing new and better employment for their capital and talents, while to the agriculturists at present on the estate as well as to those who may migrate thither, it is designed to open a new market for their produce close to their doors. Its object is, in short, to raise the standard of health and comfort of all true workers of whatever grade – the means by which these objects are to be achieved being a healthy, natural, and economic combination of town and country life, and this on land owned by the municipality.

Garden City, which is to be built near the center of the 6,000 acres, covers an area of 1,000 acres, or a sixth part of the 6,000 acres, and might be of circular form, 1,240 yards (or nearly three-quarters of a mile) from center to circumference. (Figure 2* is a ground plan of the whole municipal area, showing the town in the center; and Figure 3*, which represents one section or ward of the town, will be useful in following the description of the town itself – *a description which is, however, merely suggestive, and will probably be much departed from.*)

Six magnificent boulevards – each 120 feet wide – traverse the city from center to circumference, dividing it into six equal parts or wards. In the center is a circular space containing about five and a half acres, laid out as a beautiful and well-watered garden; and, surrounding this garden, each standing in its own ample grounds, are the larger public buildings – town hall, principal concert and lecture hall, theatre, library, museum, picture-gallery, and hospital.

The rest of the large space encircled by the "Crystal Palace" is a public park, containing 145 acres, which includes ample recreation grounds within very easy access of all the people.

Running all round the Central Park (except where it is intersected by the boulevards) is a wide glass arcade called the "Crystal Palace," opening on to the park. This building is in wet weather one of the favorite resorts of the people, whilst the knowledge that its bright shelter is ever close at hand tempts people into Central Park, even in the most doubtful of weathers. Here manufactured goods are exposed for sale, and here most of that class of shopping which requires the joy of deliberation and selection is done. The space enclosed by the Crystal Palace is, however, a good deal larger than is required for these purposes, and a considerable part of it is used as a Winter Garden – the whole forming a permanent exhibition of a most attractive character, whilst its circular form brings it near to every dweller in the town – the furthest removed inhabitant being within 600 yards.

Passing out of the Crystal Palace on our way to the outer ring of the town, we cross Fifth Avenue – lined, as are all the roads of the town, with trees – fronting which, and looking on to the Crystal Palace, we find a ring of very excellently built houses, each standing in its own ample grounds; and, as we continue our walk, we observe that the houses are for the most part built either in concentric rings, facing the various avenues (as the circular

roads are termed), or fronting the boulevards and roads which all converge to the center of the town. Asking the friend who accompanies us on our journey what the population of this little city may be, we are told about 30,000 in the city itself, and about 2,000 in the agricultural estate, and that there are in the town 5,500 building lots of an average size of 20 feet × 130 feet – the minimum space allotted for the purpose being 20 × 100. Noticing the very varied architecture and design which the houses and groups of houses display some having common gardens and co-operative kitchens – we learn that general observance of street line or harmonious departure from it are the chief points as to house building, over which the municipal authorities exercise control, for, though proper sanitary arrangements are strictly enforced, the fullest measure of individual taste and preference is encouraged.

Walking still toward the outskirts of the town, we come upon "Grand Avenue." This avenue is fully entitled to the name it bears, for it is 420 feet wide, and, forming a belt of green upwards of three miles long, divides that part of the town which lies outside Central Park into two belts. It really constitutes an additional park of 115 acres – a park which is within 240 yards of the furthest removed inhabitant. In this splendid avenue six sites, each of four acres, are occupied by public schools and their surrounding playgrounds and gardens, while other sites are reserved for churches, of such denominations as the religious beliefs of the people may determine, to be erected and maintained out of the funds of the worshippers and their friends. We observe that the houses fronting on Grand Avenue have departed (at least in one of the wards – that of which Figure 3* is a representation) – from the general plan of concentric rings, and, in order to ensure a longer line of frontage on Grand Avenue, are arranged in crescents – thus also to the eye yet further enlarging the already splendid width of Grand Avenue.

On the outer ring of the town are factories, warehouses, dairies, markets, coal yards, timber yards, etc., all fronting on the circle railway, which encompasses the whole town, and which has sidings connecting it with a main line of railway which passes through the estate. This arrangement enables goods to be loaded direct into trucks from the warehouses and work shops, and so sent by railway to distant markets, or to be taken direct from the trucks into the warehouses or factories; thus not only effecting a very great saving in regard to packing and cartage, and reducing to a minimum loss from breakage, but also, by reducing the traffic on the roads of the town, lessening to a very marked extent the cost of their maintenance. The smoke fiend is kept well within bounds in Garden City; for all machinery is driven by electric energy, with the result that the cost of electricity for lighting and other purposes is greatly reduced.

The refuse of the town is utilized on the agricultural portions of the estate, which are held by various individuals in large farms, small holdings, allotments, cow pastures, etc.; the natural competition of these various methods of agriculture, tested by the willingness of occupiers to offer the highest rent to the municipality, tending to bring about the best system of husbandry, or, what is more probable, the best systems adapted for various purposes. Thus it is easily conceivable that it may prove advantageous to grow wheat in very large fields, involving united action under a capitalist farmer, or by a body of co-operators; while the cultivation of vegetables, fruits, and flowers, which requires closer and more personal care, and more of the artistic and inventive faculty, may possibly be best dealt with by individuals, or by small groups of individuals having a common belief in the efficacy and value of certain dressings, methods of culture, or artificial and natural surroundings.

This plan, or, if the reader be pleased to so term it, this absence of plan, avoids the dangers of stagnation or dead level, and, though encouraging individual initiative, permits of the fullest co-operation, while the increased rents which follow from this form of competition are common or municipal property, and by far the larger part of them are expended in permanent improvements.

While the town proper, with its population engaged in various trades, callings, and professions, and with a store or depot in each ward, offers the most natural market to the people engaged on the agricultural estate, inasmuch as to the extent to which the townspeople demand their produce they escape altogether any railway rates and charges; yet the farmers and others are not by any

means limited to the town as their only market, but have the fullest right to dispose of their produce to whomsoever they please. Here, as in every feature of the experiment, it will be seen that it is not the area of rights which is contracted, but the area of choice which is enlarged.

The principle of freedom holds good with regard to manufacturers and others who have established themselves in the town. These manage their affairs in their own way, subject, of course, to the general law of the land, and subject to the provision of sufficient space for workmen and reasonable sanitary conditions. Even in regard to such matters as water, lighting, and telephonic communication – which a municipality, if efficient and honest, is certainly the best and most natural body to supply – no rigid or absolute monopoly is sought; and if any private corporation or any body of individuals proved itself capable of supplying on more advantageous terms, either the whole town or a section of it, with these or any commodities the supply of which was taken up by the corporation, this would be allowed. No really sound system of action is in more need of artificial support than is any sound system of thought. The area of municipal and corporate action is probably destined to become greatly enlarged; but if it is to be so, it will be because the people possess faith in such action, and that faith can be best shown by a wide extension of the area of freedom.

Dotted about the estate are seen various charitable and philanthropic institutions. These are not under the control of the municipality, but are supported and managed by various public-spirited people who have been invited by the municipality to establish these institutions in an open healthy district, and on land let to them at a pepper-corn rent, it occurring to the authorities that they can the better afford to be thus generous, as the spending power of these institutions greatly benefits the whole community. Besides, as those persons who migrate to the town are among its most energetic and resourceful members, it is but just and right that their more helpless brethren should be able to enjoy the benefits of an experiment which is designed for humanity at large.

* Figures not shown in this Reader.

"The Streets of Ecotopia's Capital" and "Car-Less Living in Ecotopia's New Towns"

from *Ecotopia* (1975)

Ernest Callenbach

Editors' Introduction

In retrospect, given the enormous growth of environmentalism in the 1960s and 1970s, it seems inevitable that someone would write an ecological utopia and call it "Ecotopia." But it was a University of California Press editor named Ernest Callenbach who made the idea reality. Published in 1975, *Ecotopia* (New York: Bantam Books, 1975) sold half a million copies and was translated into eight languages. It told the story of a journalist named William Weston who visited the new nation of Ecotopia – northern California, Oregon, and Washington – some years after it had seceded from the USA and sealed its borders. The changes were, of course, astonishing.

Callenbach followed up *Ecotopia* with a prequel, *Ecotopia Emerging* (Berkeley, CA: Banyon Tree Books, 1981), which told the story of the ecotopian revolution. This event began, appropriately enough, in the anarchic coastal hamlet of Bolinas on a peninsula in Marin County north of San Francisco, when local residents rolled stones across the only access road and declared themselves independent from the USA. The revolution spread, the rest of the country being distracted by economic crisis, toxic contamination, nuclear meltdowns, and dysfunctional politics. The resulting grassroots democracy in Ecotopia featured a strong female president who unlike other politicians took ideas seriously and spoke directly and honestly to the public. And with such leadership, people from all walks of life joined into the challenge of creating an ecological society.

THE STREETS OF ECOTOPIA'S CAPITAL

San Francisco, May 5. As I emerged from the train terminal into streets, I had little idea what to expect from this city – which had once proudly boasted of rising from its own ashes after a terrible earthquake and fire. San Francisco was once known as "America's favorite city" and had an immense appeal to tourists. Its dramatic hills and bridges, its picturesque cable cars, and its sophisticated yet relaxed people had drawn visitors who returned again and again. Would I find that it still deserves its reputation as an elegant and civilized place?

I checked my bag and set out to explore a bit. The first shock hit me at the moment I stepped onto the street. There was a strange hush over everything. I expected to encounter something at least a little like the exciting bustle of our cities – cars honking, taxis swooping, clots of people pushing about in the hurry of urban life. What I found, when

I had gotten over my surprise at the quiet, was that Market Street, once a mighty boulevard striking through the city down to the waterfront, has become a mall planted with thousands of trees. The "street" itself, on which electric taxis, minibuses, and delivery carts purr along, has shrunk to a two-lane affair. The remaining space, which is huge, is occupied by bicycle lanes, fountains, sculptures, kiosks, and absurd little gardens surrounded by benches. Over it all hangs the almost sinister quiet, punctuated by the whirr of bicycles and cries of children. There is even the occasional song of a bird, unbelievable as that may seem on a capital city's crowded main street.

Scattered here and there are large conical-roofed pavilions, with a kiosk in the center selling papers, comic books, magazines, fruit juices, and snacks. (Also cigarettes – the Ecotopians have not managed to stamp out smoking!) The pavilions turn out to be stops on the minibus system, and people wait there out of the rain. These buses are comical battery-driven contraptions, resembling the antique cable cars that San Franciscans were once so fond of. They are driverless, and are steered and stopped by an electronic gadget that follows wires buried in the street. (A safety bumper stops them in case someone fails to get out of the way.) To enable people to get on and off quickly, during the fifteen seconds the bus stops, the floor is only a few inches above ground level; the wheels are at the extreme ends of the vehicle. Rows of seats face outward, so on a short trip you simply sit down momentarily, or stand and hang onto one of the hand grips. In bad weather fringed fabric roofs can be extended outward to provide more shelter.

These buses creep along at about ten miles an hour, but they come every five minutes or so. They charge no fare. When I took an experimental ride on one, I asked a fellow passenger about this, and he said the minibuses are paid for in the same way as streets – out of general tax funds. Smiling, he added that to have a driver on board to collect fares would cost more than the fares could produce. Like many Ecotopians, he tended to babble, and spelled out the entire economic rationale for the minibus system, almost as if he was trying to sell it to me. I thanked him, and after a few blocks jumped off.

The bucolic atmosphere of the new San Francisco can perhaps best be seen in the fact that, down Market Street and some other streets, creeks now run. These had earlier, at great expense, been put into huge culverts underground, as is usual in cities. The Ecotopians spent even more to bring them up to ground level again. So now on this major boulevard you may see a charming series of little falls, with water gurgling and splashing, and channels lined with rocks, trees, bamboos, ferns. There even seem to be minnows in the water – though how they are kept safe from marauding children and cats, I cannot guess.

Despite the quiet, the streets are full of people, though not in Manhattan densities. (Some foot traffic has been displaced to lacy bridges which connect one skyscraper to another, sometimes fifteen or twenty stories up.) Since practically the whole street area is "sidewalk," nobody worries about obstructions – or about the potholes which, as they develop in the pavement, are planted with flowers. I came across a group of street musicians playing Bach, with a harpsichord and a half dozen other instruments. There are vendors of food pushing gaily colored carts that offer hot snacks, chestnuts, ice cream. Once I even saw a juggler and magician team, working a crowd of children – it reminded me of some medieval movie. And there are many strollers, gawkers, and loiterers – people without visible business who simply take the street for granted as an extension of their living rooms. Yet, despite so many unoccupied people, the Ecotopian streets seem ridiculously lacking in security gates, doormen, guards, or other precautions against crime. And no one seems to feel our need for automobiles to provide protection in moving from place to place. . . .

Ecotopians setting out to go more than a block or two usually pick up one of the sturdy white-painted bicycles that lie about the streets by the hundreds and are available free to all. Dispersed by the movements of citizens during the day and evening, they are returned by night crews to the places where they will be needed the next day. When I remarked to a friendly pedestrian that this system must be a joy to thieves and vandals, he denied it heatedly. He then put a case, which may not be totally farfetched, that it is cheaper to lose a few bicycles than to provide more taxis or minibuses.

Ecotopians, I am discovering, spout statistics on such questions with reckless abandon. They have a way of introducing "social costs" into their calculations which inevitably involves a certain amount of optimistic guesswork. It would be interesting to confront such informants with one of the hard-headed experts from our auto or highway industries – who would, of course, be horrified by the Ecotopians' abolition of cars.

As I walked about, I noticed that the downtown area was strangely overpopulated with children and their parents, besides people who apparently worked in the offices and shops. My questions to passersby (which were answered with surprising patience) revealed what is perhaps the most astonishing fact I have yet encountered in Ecotopia: the great downtown skyscrapers, once the headquarters of far-flung corporations, have been turned into apartments! Further inquiries will be needed to get a clear picture of this development, but the story I heard repeatedly on the streets today is that the former outlying residential areas have largely been abandoned. Many three-story buildings had in any case been heavily damaged by the earthquake nine years ago. Thousands of cheaply built houses in newer districts (scornfully labeled "ticky-tacky boxes" by my informants) have been sacked of their wiring, glass, and hardware, and bulldozed away. The residents now live downtown, in buildings that contain not only apartments but also nurseries, grocery stores, and restaurants, as well as the shops and offices on the ground floor.

Although the streets still have an American look, it is annoyingly difficult to identify things in Ecotopia. Only very small signs are permitted on the fronts of buildings; street signs are few and hard to spot, mainly attached to buildings on corners. Finally, however, I navigated my way back to the station, retrieved my suitcase, and located a nearby hotel which had been recommended as suitable for an American, but still likely to give me "some taste of how Ecotopians live." This worthy establishment lived up to its reputation by being almost impossible to find. But it is comfortable enough, and will serve as my survival base here.

Like everything in Ecotopia, my room is full of contradictions. It is comfortable, if a little old-fashioned by our standards. The bed is atrocious – it lacks springs, being just foam rubber over board – yet it is covered with a luxurious down comforter. There is a large work table equipped with a hot plate and teapot. Its surface is plain, unvarnished wood with many mysterious stains – but on it stands a small, sleek picturephone. (Despite their aversion to many modern devices, the Ecotopians have some that are even better than ours. Their picturephones, for instance, though they have to be connected to a television screen, are far easier to use than ours, and have much better picture quality.) My toilet has a tank high overhead, of a type that went out in the United States around 1945, operated by a pull chain with a quaint carved handle; the toilet paper must be some ecological abomination – it is coarse and plain. But the bathtub is unusually large and deep. Like the tubs still used in deluxe Japanese inns, it is made of slightly aromatic wood.

I used the picturephone to confirm advance arrangements for a visit tomorrow with the Minister of Food, with whom I shall begin to investigate the Ecotopian claims of "stable-state" ecological systems, about which so much controversy has raged.

[. . .]

CAR-LESS LIVING IN ECOTOPIA'S NEW TOWNS

San Francisco, May 7. Under the new regime, the established cities of Ecotopia have to some extent been broken up into neighborhoods or communities, but they are still considered to be somewhat outside the ideal long-term line of development of Ecotopian living patterns. I have just had the opportunity to visit one of the strange new minicities that are arising to carry out the more extreme urban vision of this decentralized society. Once a sleepy village, it is called Alviso, and is located on the southern shores of San Francisco Bay.

You get there on the interurban train, which drops you off in the basement of a large complex of buildings. The main structure, it turns out, is not the city hall or courthouse, but a factory. It produces the electric traction units – they hardly qualify as cars or trucks in our terms – that are used for transporting people and goods in Ecotopian cities and for general transportation in the countryside. (Individually owned vehicles were prohibited in "carfree" zones soon after Independence. These zones at first covered only downtown areas where pollution and congestion were most severe. As the minibus service was extended, these zones expanded, and now cover all densely settled city areas.)

Around the factory, where we would have a huge parking lot, Alviso has a cluttered collection of buildings, with trees everywhere. There are restaurants, a library, bakeries, a "core store" selling groceries and clothes, small shops, even factories and workshops – all jumbled amid apartment buildings. These are generally of three or four stories, arranged around a central courtyard of the type that used to be common in Paris. They are built almost entirely of wood, which has become the predominant building material in Ecotopia, due to the reforestation program. Though these structures are old-fashioned looking, they have pleasant small balconies, roof gardens, and verandas – often covered with plants, or even small trees. The apartments themselves are very large by our standards – with ten or fifteen rooms, to accommodate their communal living groups.

Alviso streets are named, not numbered, and they are almost as narrow and winding as those of medieval cities – not easy for a stranger to get around in. They are hardly wide enough for two cars to pass; but then of course there are no cars, so that is no problem. Pedestrians and bicyclists meander along. Once in a while you see a delivery truck hauling a piece of furniture or some other large object, but the Ecotopians bring their groceries home in string bags or large bicycle baskets. Supplies for the shops, like most goods in Ecotopia, are moved in containers. These are much smaller than our cargo containers, and proportioned to fit into Ecotopian freight cars and onto their electric trucks. Farm produce, for instance, is loaded into such containers either at the farms or at the container terminal located on the edge of each minicity. From the terminal an underground conveyor belt system connects to all the shops and factories in the minicity, each of which has a kind of siding where the containers are shunted off. This idea was probably lifted from our automated warehouses, but turned backwards. It seems to work very well, though there must be a terrible mess if there is some kind of jam-up underground.

My guides on this expedition were two young students who have just finished an apprenticeship year in the factory. They're full of information and observations. It seems that the entire population of Alviso, about 9,000 people, lives within a radius of a half mile from the transit station. But even this density allows for many small park-like places: sometimes merely widenings of the streets, sometimes planted gardens. Trees are everywhere – there are no large paved areas exposed to the sun. Around the edges of town are the schools and various recreation grounds. At the northeast corner of town you meet the marshes and sloughs and saltflats of the Bay. A harbor has been dredged for small craft; this opens onto the ship channel through which a freighter can move right up to the factory dock. My informants admitted rather uncomfortably that there is a modest export trade in electric vehicles – the Ecotopians allow themselves to import just enough metal to replace what is used in the exported electric motors and other metal parts.

Kids fish off the factory dock; the water is clear. Ecotopians love the water, and the boats in the harbor are a beautiful collection of both traditional and highly unorthodox designs. From this harbor, my enthusiastic guides tell me, they often sail up the Bay and into the Delta, and even out to sea through the Golden Gate, then down the coast to Monterey. Their boat is a lovely though heavy-looking craft, and they proudly offered to take me out on it if I have time.

We toured the factory, which is a confusing place. Like other Ecotopian workplaces, I am told, it is not organized on the assembly line principles generally thought essential to really efficient mass production. Certain aspects are automated: the production of the electric motors, suspension frames, and other major elements. However, the assembly of these items is done by groups of workers who actually fasten the parts together one by one, taking them from supply bins kept full by the automated machines. The plant is quiet and pleasant compared to the crashing racket of a Detroit plant, and the workers do not seem to be under Detroit's high output pressures. Of course the extreme simplification of Ecotopian vehicles must make the manufacturing process much easier to plan and manage – indeed there seems little reason why it could not be automated entirely.

Also, I discovered, much of the factory's output does not consist of finished vehicles at all. Following the mania for "doing it yourself" which is such a basic part of Ecotopian life, this plant chiefly turns out "front ends," "rear ends," and battery units.

Individuals and organizations then connect these to bodies of their own design. Many of them are weird enough to make San Francisco minibuses look quite ordinary. I have seen, for instance, a truck built of driftwood, almost every square foot of it decorated with abalone shells – it belonged to a fishery commune along the coast.

The "front end" consists of two wheels, each driven by an electric motor and supplied with a brake. A frame attaches them to a steering and suspension unit, together with a simple steering wheel, accelerator, brake, instrument panel, and a pair of headlights. The motor drives are capable of no more than thirty miles per hour (on the level!) so their engineering requirements must be modest – though my guides told me the suspension is innovative, using a clever hydraulic load-leveling device which in addition needs very little metal. The "rear end" is even simpler, since it doesn't have to steer. The battery units, which seem to be smaller and lighter than even our best Japanese imports, are designed for use in vehicles of various configurations. Each comes with a long reel-in extension cord to plug into recharging outlets.

The factory does produce several types of standard bodies, to which the propulsion units can be attached with only four bolts at each end. (They are always removed for repair.) The smallest and commonest body is a shrunken version of our pick-up truck. It has a tiny cab that seats only two people, and a low, square, open box in back. The rear of the cab can be swung upward to make a roof, and sometimes canvas sides are rigged to close in the box entirely.

A taxi-type body is still manufactured in small numbers. Many of these were used in the cities after Independence as a stop-gap measure while minibus and transit systems were developing. These bodies are molded from heavy plastic in one huge mold.

These primitive and underpowered vehicles obviously cannot satisfy the urge for speed and freedom which has been so well met by the American auto industry and our aggressive highway program. My guides and I got into a hot debate on this question, in which I must admit they proved uncomfortably knowledgeable about the conditions that sometimes prevail on our urban throughways – where movement at any speed can become impossible. When I asked, however, why Ecotopia did not build speedy cars for its thousands of miles of rural highways – which are now totally uncongested even if their rights of way have partly been taken over for trains – they were left speechless. I attempted to sow a few seeds of doubt in their minds: no one can be utterly insensitive to the pleasures of the open road, I told them, and I related how it feels to roll along in one of our powerful, comfortable cars, a girl's hair blowing in the wind. . . .

We had lunch in one of the restaurants near the factory, amid a cheery, noisy crowd of citizens and workers. I noticed that they drank a fair amount of the excellent local wine with their soups and sandwiches. Afterward we visited the town hall, a modest wood structure indistinguishable from the apartment buildings. There I was shown a map on which adjacent new towns are drawn, each centered on its own rapid-transit stop. It appears that a ring of such new towns is being built to surround the Bay, each one a self-contained community, but linked to its neighbors by train so that the entire necklace of towns will constitute one city. It is promised that you can, for instance, walk five minutes to your transit station, take a train within five minutes to a town ten stops away, and then walk another five minutes to your destination. My informants are convinced that this represents a halving of the time we would spend on a similar trip, not to mention problems of parking, traffic, and of course the pollution.

What will be the fate of the existing cities as these new minicities come into existence? They will gradually be razed, although a few districts will be preserved as living museum displays (of "our barbarian past," as the boys jokingly phrased it). The land will be returned to grassland, forest, orchards, or gardens – often, it appears, groups from the city own plots of land outside in the country, where they probably have a small shack and perhaps grow vegetables, or just go for a change of scene.

After leaving Alviso we took the train to Redwood City, where the reversion process can be seen in action. Three new towns have sprung up there along the Bay, separated by a half mile or so of open country, and two more are under construction as part of another string several miles back from the Bay, in the foothills. In between, part of the former suburban residential area has already been turned into alternating woods and grassland. The scene reminded me a little of my boyhood country summers in Pennsylvania. Wooded strips follow the winding lines of creeks. Hawks circle lazily. Boys out hunting with bows and arrows wave to the train as it zips by. The signs of a once busy civilization – streets, cars, service stations, supermarkets – have been entirely obliterated, as if they never existed. The scene was sobering, and made me wonder what a Carthaginian might have felt after ancient Carthage was destroyed and plowed under by the conquering Romans.

"Description of Abbenay"
from *The Dispossessed* (1974)

Ursula K. Le Guin

Editors' Introduction

A number of leftist thinkers have historically been attracted to a decentralized, anarchist model of society in which very little centralized power exists and there is a strong emphasis on equality and individual responsibility toward others. In theory such a social structure avoids the oppressive concentration of wealth and power found in other societies. Thus it would emphasize the equity goals of sustainable development, as well as potentially the environmental and economic goals. Although commentators frequently dismiss anarchism by equating it with acts of terrorism against the state, it has a rich heritage as a political philosophy, having attracted thinkers such as Peter Kropotkin, Pierre-Joseph Proudhon, Bertrand Russell, Emma Goldman, and Paul Goodman.

In her science fiction novel *The Dispossessed* (New York: Avon Books, 1974), Ursula K. Le Guin describes a utopian, quasi-anarchist society that has left behind a corrupt and oppressive world of warring nation-states not unlike our own to colonize the inhabitable moon of the parent planet. On the home world, Urras, a wealthy class lives in luxury while its servants inhabit vast slums. Women are denied rights and professional roles, social conformity is extreme, and free thought is discouraged. On the renegade world, Annares, a new social structure has been established in which men and women are completely equal, no one is rich, decisions are made by committee, and everyone partakes in the labor of providing the essentials of life. This new society follows the principles of a female philosopher, Odo, who had led the rebellion on the mother planet. Its culture is somewhat reminiscent of the kibbutz model followed in the early Israeli state.

Le Guin goes beyond simply describing this utopia to explore the problems it faces in living sustainably on a dry, barren terrain and the inherent tensions between anarchic decentralization and the need for some sort of organizational structure. The society has difficulty being completely self-sufficient, and by cutting itself off from other worlds it has lost much of its own heritage and the benefits of trade and communication. In the course of the book a pioneering physicist, Shevek, makes the first voyage back to Urras and seeks to rebuild ties between the two worlds in ways that can have benefits for both.

This science fiction vision portrays a society struggling for many qualities that we might equate with sustainability. Though not perfect by any means, the residents of Anarres have sought to live harmoniously on their planet, to minimize depletion of natural resources, to promote equality between individuals, and to foster a sense of mutual responsibility. Their philosophy is also based on principles of organic unity that reflect those of many current earthly environmentalists. Although set in a far distant time and place, Le Guin's beautifully written novel describes many dynamics that our own human communities will need to contend with if they are to become more sustainable in the long run.

Decentralization had been an essential element in Odo's plans for the society she did not live to see founded. She had no intention of trying to de-urbanize civilization. Though she suggested that the natural limit to the size of a community lay in its dependence on its own immediate region for essential food and power, she intended that all communities be connected by communication and transportation networks, so that goods and ideas could get where they were wanted, and the administration of things might work with speed and ease, and no community should be cut off from change and interchange. But the network was not to be run from the top down. There was to be no controlling center, no establishment for the self-perpetuating machinery of bureaucracy and the dominance drive of individuals seeking to become captains, bosses, chiefs of state.

Her plans, however, had been based on the generous ground of Urras. On arid Anarres, the communities had to scatter widely in search of resources, and few of them could be self-supporting, no matter how they cut back their notions of what is needed for support. They cut back very hard indeed, but to a minimum beneath which they would not go; they would not regress to pre-urban, pretechnological tribalism. They knew that their anarchism was the product of a very high civilization, of a complex diversified culture, of a stable economy and a highly industrialized technology that could maintain high production and rapid transportation of goods. However vast the distances separating settlements, they held to the ideal of complex organicism. They built the roads first, the houses second. The special resources and products of each region were interchanged continually with those of others, in an intricate process of balance: that balance of diversity which is the characteristic of life, of natural and social ecology.

But, as they said in the analogic mode, you can't have a nervous system without at least a ganglion, and preferably a brain. There had to be a center. The computers that coordinated the administration of things, the division of labor, and the distribution of goods, and the central federatives of most of the work syndicates, were in Abbenay, right from the start. And from the start the Settlers were aware that that unavoidable centralization was a lasting threat, to be countered by lasting vigilance.

O child Anarchia, infinite promise
infinite carefulness
I listen, listen in the night
by the cradle deep as the night
is it well with the child

Pio Atean, who took the Pravic name Tober, wrote that in the fourteenth year of the Settlement. The Odonians' first efforts to make their new language, their new world, into poetry, were stiff, ungainly, moving.

Abbenay, the mind and center of Anarres, was there, now, ahead of the dirigible, on the great green plain.

That brilliant, deep green of the fields was unmistakable: a color not native to Anarres. Only here and on the warm shores of the Keran Sea did the Old World grains flourish. Elsewhere the staple grain crops were ground-holum and pale mene-grass.

When Shevek was nine his afternoon schoolwork for several months had been caring for the ornamental plants in Wide Plains community – delicate exotics, that had to be fed and sunned like babies. He had assisted an old man in the peaceful and exacting task, had liked him and liked the plants, and the dirt, and the work. When he saw the color of the Plain of Abbenay he remembered the old man, and the smell of fish-oil manure, and the color of the first leafbuds on small bare branches, that clear vigorous green.

He saw in the distance among the vivid fields a long smudge of white, which broke into cubes, like spilt salt, as the dirigible came over.

A cluster of dazzling flashes at the east edge of the city made him wink and see dark spots for a moment: the big parabolic mirrors that provided solar heat for Abbenay's refineries.

The dirigible came down at a cargo depot at the south end of town, and Shevek set off into the streets of the biggest city in the world.

They were wide, clean streets. They were shadowless, for Abbenay lay less than thirty degrees north of the equator, and all the buildings were low, except the strong, spare towers of the wind turbines. The sun shone white in a hard, dark, blue-violet sky. The air was clear and clean, without smoke or moisture. There was a vividness to things, a hardness of edge and corner, a clarity. Everything stood out separate, itself.

The elements that made up Abbenay were the same as in any other Odonian community, repeated many times: workshops, factories, domiciles, dormitories, learning-centers, meeting halls, distributaries, depots, refectories. The bigger buildings were most often grouped around open squares, giving the city a basic cellular texture: it was one subcommunity or neighborhood after another. Heavy industry and food-processing plants tended to cluster on the city's outskirts, and the cellular pattern was repeated in that related industries often stood side by side on a certain square or street. The first such that Shevek walked through was a series of squares, the textile district, full of holum-fiber processing plants, spinning and weaving mills, dye factories, and cloth and clothing distributories; the center of each square was planted with a little forest of poles strung from top to bottom with banners and pennants of all the colors of the dyer's art, proudly proclaiming the local industry. Most of the city's buildings were pretty much alike, plain, soundly built of stone or cast foamstone. Some of them looked very large to Shevek's eyes, but they were almost all of one storey only, because of the frequency of earthquake. For the same reason windows were small, and of a tough silicon plastic that did not shatter. They were small, but there were a lot of them, for there was no artificial lighting provided from an hour before sunrise to an hour after sunset. No heat was furnished when the outside temperature went above 55 degrees Fahrenheit. It was not that Abbenay was short of power, not with her wind turbines and the earth temperature-differential generators used for heating; but the principle of organic economy was too essential to the functioning of the society not to affect ethics and aesthetics profoundly. "Excess is excrement," Odo wrote in the *Analogy*. "Excrement retained in the body is a poison."

Abbenay was poisonless: a bare city, bright, the colors light and hard, the air pure. It was quiet. You could see it all, laid out as plain as spilt salt.

Nothing was hidden.

The squares, the austere streets, the low buildings, the unwalled workyards, were charged with vitality and activity. As Shevek walked he was constantly aware of other people walking, working, talking, faces passing, voices calling, gossiping, singing, people alive, people doing things, people afoot. Workshops and factories fronted on squares or on their open yards, and their doors were open. He passed a glassworks, the workman dipping up a great molten blob as casually as a cook serves soup. Next to it was a busy yard where foamstone was cast for construction. The gang foreman, a big woman in a smock white with dust, was supervising the pouring of a cast with a loud and splendid flow of language. After that came a small wire factory, a district laundry, a luthier's where musical instruments were made and repaired, the district small-goods distributory, a theater, a tile works. The activity going on in each place was fascinating, and mostly out in full view. Children were around, some involved in the work with the adults, some underfoot making mud pies, some busy with games in the street, one sitting perched up on the roof of the learning center with her nose deep in a book. The wiremaker had decorated the shopfront with patterns of vines worked in painted wire, cheerful and ornate. The blast of steam and conversation from the wide-open doors of the laundry was overwhelming. No doors were locked, few shut. There were no disguises and no advertisements. It was all there, all the work, all the life of the city, open to the eye and to the hand. And every now and then down Depot Street a thing came careering by clanging a bell, a vehicle crammed full of people and with people festooned on stanchions all over the outside, old women cursing heartily as it failed to slow down at their stop so they could scramble off, a little boy on a homemade tricycle pursuing it madly, electric sparks showering blue from the overhead wires at crossings: as if that quiet intense vitality of the streets built up every now and then to discharge point, and leapt the gap with a crash and a blue crackle and the smell of ozone. These were the Abbenay omnibuses, and as they passed one felt like cheering.

Depot Street ended in a large airy place where five other streets rayed in to a triangular park of grass and trees. Most parks on Anarres were playgrounds of dirt or sand, with a stand of shrub and tree holums. This one was different. Shevek crossed the trafficless pavement and entered the park, drawn to it because he had seen it often in pictures, and because he wanted to see alien trees, Urrasti trees, from close up, to experience the greenness of those multitudinous leaves. The sun was setting, the sky was wide and clear, darkening to purple at the zenith, the dark of space showing through the thin atmosphere. He entered under the trees, alert, wary. Were they not wasteful, those crowding leaves? The tree holum got along very efficiently with spines and needles, and no excess of those. Wasn't all this extravagant foliage mere excess, excrement? Such trees couldn't thrive without a rich soil, constant watering, much care. He disapproved of their lavishness, their thriftlessness. He walked under them, among them. The alien grass was soft underfoot. It was like walking on living flesh. He shied back onto the path. The dark limbs of the trees reached out over his head, holding their many wide green hands above him. Awe came into him. He knew himself blessed though he had not asked for blessing.

Some way before him, down the darkening path, a person sat reading on a stone bench.

Shevek went forward slowly. He came to the bench and stood looking at the figure who sat with head bowed over the book in the green-gold dusk under the trees. It was a woman of fifty or sixty, strangely dressed, her hair pulled back in a knot. Her left hand on her chin nearly hid the stern mouth, her right held the papers on her knee. They were heavy, those papers; the cold hand on them was heavy. The light was dying fast but she never looked up. She went on reading the proof sheets of *The Social Organism*.

Shevek looked at Odo for a while, and then he sat down on the bench beside her.

He had no concept of status at all, and there was plenty of room on the bench. He was moved by a pure impulse of companionship.

He looked at the strong, sad profile, and at the hands, an old woman's hands. He looked up into the shadowy branches. For the first time in his life he comprehended that Odo, whose face he had known since his infancy, whose ideas were central and abiding in his mind and the mind of everyone he knew, that Odo had never set foot on Anarres: that she had lived, and died, and was buried, in the shadow of green-leaved trees, in unimaginable cities, among people speaking unknown languages, on another world. Odo was an alien: an exile.

The young man sat beside the statue in the twilight, one almost as quiet as the other.

At last, realizing it was getting dark, he got up and made off into the streets again, asking directions to the Central Institute of the Sciences.

PART SIX

Case studies of urban sustainability

INTRODUCTION TO PART SIX

What follow are brief case studies – in text and images – that illustrate the reality and practicality of many of the urban sustainability ideas discussed in this book. This list is not exhaustive, to be sure, but includes many inspiring recent examples of urban sustainability practice. Following each brief description are sources and websites for learning more about these positive examples.

URBAN SUSTAINABILITY AT
THE BUILDING AND SITE SCALE

Commerzbank Headquarters, Frankfurt

Designed by Norman Foster architects, this commercial office building has become the tallest building in Europe (See plate CS1). This 53-story, nearly 300-meters-tall building, completed in 1997, is designed to incorporate creatively a number of environmental and sustainability features. The building takes a triangular form, with a continuous atrium in its center, extending the entire length of the structure.

A major design element is a series of sky gardens (See plate CS2). The triangular building incorporates these gardens every four floors, with a larger garden open to the elements on the 43rd floor. Ten gardens in all are provided, each with trees and vegetation, and serving both as places for employees to visit and relax, and also for climate control regulators for the building. Convection in summertime draws air through the sky gardens and the atrium, providing natural cooling and serving as a "natural chimney."

Other ecological elements of the structure include reliance on natural ventilation (operable windows in all offices), extensive daylighting, sensors that automatically adjust artificial lighting inside, and low-emissity windows (windows coated to allow in short wavelength solar energy but reduce radiant loss of interior heat). The building is enclosed in a high-tech double skin with the space between acting as a thermal buffer. The city's district heating system provides the building with heat. The building is predicted to use about one-third less energy than a conventional office building.

For more information, see: www.fosterandpartners.com.

Menara Mesiniaga bio-climatic skyscraper, Kuala Lumpur, Malaysia

Designed by architect Ken Yeang, the IBM headquarters building in Kuala Lumpur, Malaysia represents an important built example of what Yeang calls "bio-climatic" skyscrapers (See plates CS3 and CS4). Such buildings are designed from the start to take full advantage of local climate, to incorporate plants and vegetation, and to substantially reduce their energy and resource consumption levels compared with typical high-rise structures. The exterior of the Menara Mesiniaga tower serves as an "environmental filter" rather than a hard façade; a permeable membrane that allows movement of air and natural ventilation and breaks up the visual monotony of the exterior. Extensive exterior louvers provide shading on the east and west sides of the building, also adding to the building visual distinctiveness. Perhaps most impressively is the "vertical landscaping," as Yeang calls it, that spirals up and around the structure and connects with a series of recessed "sky courts." These sky courts facilitate ventilation and act as thermal buffers. A sun-shaded roof is designed as important habitable space, and includes a gym and pool. A partially louvered sunroof also acts as a wind scoop, directing air back into the interior of the structure. Other elements of the design include placement of the elevators and core services on the hottest side of the structure. The building rises out of a green terraced base, and is intended to connect the fifteen-story building to earth and land.

Completed in 1992, this is perhaps Ken Yeang's best example of a completed bio-climatic skyscraper design. Other important examples include the UMNO Tower in Penang, the planned EDITT Tower in Singapore, and a new design for a ecotower as part of a comprehensive redevelopment scheme for the Elephant and Castle area of London.

For more information, see: Ken Yeang, *Bioclimatic Skyscrapers*, revised edn (London: Ellipsis London Press, 2000); *The Green Skyscraper: The Basis for Designing Sustainable Intensive Buildings* (New York: Prestel, 2000).

Adelaide EcoVillage (Christie Walk)

A new ecological co-housing project is under construction in the Southern Australian city of Adelaide, with the first five units occupied in 2002. Green city ideas have a long history in Adelaide, under the advocacy of Paul Downton, and the organization Urban Ecology Australia. This project, known as Christie Walk (See plates CS5 and CS6), is an example of ecological infill. When completed it will include four townhouses, six apartments and four straw-bale cottages. A community house is also included. Ecological features include onsite sewage treatment and graywater recycling. Stormwater is retained onsite and used for toilet flushing.

The homes are designed to be very energy-efficient, and include both active and passive solar. The building designs take advantage of high thermal mass, extensive insulation, and a natural ventilation system. Stairwells act as ventilation flues. Vegetation and landscaping using native plants cools the air.

Extensive use has been made of recycled materials (e.g., flyash in concrete, recycled timber in windows, reuse of brick and stone from demolished buildings), as well as non-toxic paints and finishes. The outer shell of the building has been designed to last longer than 100 years, with interior doors and walls made from renewable resources. A rooftop garden is included, as well as a community garden where food will be produced for the neighborhood.

Siting this development on an L-shaped parcel in the heart of Adelaide reflects its sustainability values as well. Its urban location will permit living with little or no dependence on cars. Public transit and shopping are nearby. In recognition of the project's location, some relief from the city's parking requirement was given – only ten parking spaces were required for these fourteen units.

For more information, see: "Urban Ecology Australia – Christie Walk," at www.urbanecology.org.au/christiewalk/main.html

Condé Nast building (4 Times Square), New York

Nicknamed the "green giant," the Condé Nast building, also known as 4 Times Square, is the first major office structure in New York City designed and built around sustainability principles (See plates CS7 and CS8). Completed in 1999, the building is forty-eight stories in height and includes 1.6 million square feet of space. Designed by the architectural firm Fox and Fowle, the structure incorporates many impressive sustainability features. These include a very energy-efficient building design, utilizing large low-emissity windows (that capture sunlight and retain heat) that provide extensive daylight to the building, natural gas absorption chiller/heaters, added insulation, thin film photovoltaic panels (on the south and east façades, on the top nine floors of the building, producing about 15 kW at peak), and two 200-kW fuel cells that produce enough energy to operate the building in the evening. The building's unique ventilation system delivers much more fresh air to building occupants than a typical building – five times the amount required by code.

Careful planning of construction deliveries (reducing engine idling), and management and recycling of construction waste were also important elements. Other green elements of the building include non-CFC air-conditioning, use of energy-efficient, variable speed motors and pumps, use of low-water-use fixtures, and extensive use of recycled materials in its construction. All floors are equipped with waste recycling chutes. A set of tenant guidelines has been prepared to suggest ways in which tenants can reduce their environmental impacts (e.g., by selecting environmentally friendly furniture).

An important urban sustainability dimension of the building is its location in the center of Manhattan. Built on the foundations of a former building, this is an urban infill project, embedded in a very pedestrian urban environment, with great access to transit. In fact, the building provides no parking.

For further information, see: USDOE, Office of Energy Efficiency and Renewable Energy, undated. Case Study: The Condé Nast building at 4 Times Square, found at www.eere.energy.gov/buildings/documents/pdfs/29940.pdf

URBAN SUSTAINABILITY AT THE NEIGHBORHOOD OR DISTRICT SCALE

Kronsberg Ecological District, Hannover, Germany

This model ecological housing district is Hannover's newest growth area. Designed and built as a model development for the 2000 World Expo, it incorporates almost every urban sustainability or ecological design element imaginable. The sustainability dimensions begin with its basic form: relatively high-density, multi-family housing, sited along a new line of the city's tram system (with three very accessible new tram stops), and with a car-minimal grid street pattern. The entire district is a traffic-calmed (30-km restricted) zone, with extensive bike lanes and onsite car sharing providing additional alternatives to the automobile. The district's new town hall takes sustainability as a key theme. This building is constructed from sustainable materials, with PVs on its rooftop, and houses social service offices, meeting space, and a library specializing in the environment.

The district captures and contains all stormwater onsite through an innovative system of treatment bioswales that feed into two long stormwater retention boulevards, serving also as important green features and delightful pathways (See plate CS9). The Germans refer to this as the Mulden-Rigolen-System, or gulley and trench system. Through this stormwater collection system and a number of other water design elements, water is present and made visible in this community. Other sustainability features include extensive use of green rooftops, green courtyards and water features, and community gardens.

This model district demonstrates a number of important energy features. Homes are designed to meet an impressive low-energy standard, and two very efficient combined heat and power plants provide heat for about 3,000 units in the district's first phase of development. (One of these power plants is actually in the basement of a building of flats!) Three wind turbines have also been built, including one large 1.8-MW turbine, and all are but a few hundred meters away from the housing. A number of solar energy technologies and design ideas are being tried, including a centralized solar hot water heating system (plate CS10) which serves one portion of the district (and stores hot water in a partially underground 2,800-square meter tank, which doubles as a children's play area). Other sustainability elements include a demonstration ecological farm, a sustainable landscape management plan, and a green elementary school.

For more information, see: KUKA, *Living in Kronsberg* (Hannover: Kronsberg-Umwelt-Kommunikations-Agentur GmbH (KUKA), 2000).

Beddington Zero Energy Development (BedZED), London

Beddington Zero Energy Development (BedZED) is an impressive new ecological housing project in the Hackbridge neighborhood in South London. Designed by Bill Dunster architects, it is billed as the first carbon-neutral development in the UK (See plate CS11). The project includes a mix of housing and workspace (eighty-two homes, including fourteen home/work units), on a reclaimed sewage works. These three-story buildings are oriented with living space to the south to capture the sun, with most units provided with rooftop sky gardens. Well insulated, and with floors and walls providing extensive thermal mass, these buildings provide a unique system of natural ventilation is provided through a visually distinctive set of windcowls (plate CS12) that rotate into the wind, capturing fresh air but also extracting heat from outgoing air. Most construction materials were obtained within a thirty-five-mile radius of the site, and wood both for construction and to burn as fuel in the new combined heat and power plant comes from local forests. Other elements include photovoltaic panels to provide enough electricity to recharge a fleet of electric car-sharing vehicles.

Much of the credit for this bold project goes to the Peabody Trust, a London housing association known for its support of innovative and sustainable design. A new, larger extension of the BedZED ideas is now in the works also in London: Ladbroke Green is a mixed-use project located on a former gasworks, where 308 units of ecological housing and 16,000 square meters of commercial space will be built on about ten acres. With photovoltaic panels covering the rooftops of this project, this mixed-use development will be the largest application of solar panels in the UK.

For more information, see: Beddington Zero Energy Development at: www.bedzed.org.uk/

Greenwich Millennium Village, London

This new ecological neighborhood in the heart of the city of London is built on a former industrial site. It will eventually consist of 1,377 homes and 5,000 m² of office space on about twenty-nine hectares. Also included will be office and commercial space, a community center, cafés, and a large village green. Residents will be served by good transit (the Jubilee Line of the London Tube), and extensive bicycle and pedestrian routes have been included in the design (See plate CS13).

Buildings in the village, including the Ralph Erskine-designed phase one housing, will incorporate many green features, including sustainably harvested wood and climate-sensitive designs to maximize use of the sun and minimize effects of prevailing winds (See plate CS14). The goal of an 80 per cent reduction in energy in comparison with conventional development has been set. The Village will incorporate a combined heat and power plant (the first in Great Britain as part of a private housing development).

An interesting feature of the houses is the design of flexible living spaces. The units use sliding interior walls which allow adjustment and reorganization of the spaces as family needs change. Structural systems are designed, as well, to accommodate upgrades and modifications later (e.g., the addition of balconies).

For more information, see: "Greenwich Millennium Village" at: www.greenwich-village.co.uk

Nieuwland (solar suburb), Amersfoort, Netherlands

One of the largest demonstrations of a community organized around solar energy, this new district in the Dutch city of Amersfoort has been nicknamed the "solar suburb" (See plates CS15 and CS16). The sustainability of this unique town begins with orientation – some 85 per cent of the buildings are oriented to the south in this otherwise unique circular urban form containing a town center (with grocery, post office, shopping) at the core. Everything is within walking distance, and the community is served by a good bus service. Both photovoltaic and solar hot water panels are used throughout. PVs on building rooftops and façades generate 1.35 megawatts. Major institutional buildings, including two elementary schools, a sports complex, a kindergarten, and multi-family social housing, all have photovoltaics on their rooftops. Two pilot energy-balanced homes have also been constructed here. These are grid-connected homes that generate as much energy as they need over the course of a year.

The unique energy features of this community have benefited from substantial technical and financial support from the regional energy company REMU (Regionale Energiemaatschappij Utrecht), as well as from the Dutch Energy Agency (Novem) and the European Union. The positive environmental impacts of this project are tremendous, with an estimated reduction in carbon dioxide of almost 89,000 kilograms/year (approximately 98 tons).

For more information, see: REMU, *Building Solar Suburbs: Renewable Energy in a Sustainable City* (Utrecht: REMU, 1999).

Village Homes, Davis, California

The brainchild and creation of husband and wife team Judy and Michael Corbett, this sustainable neighborhood in Davis, California, has been an inspiring model since the late 1970s (See plates CS17 and CS18). This mostly single-family community (240 units on 60 acres) is organized around a series of interior green fingers, which collect the development's stormwater (there is no conventional stormwater collection system), and provides a beautiful network of community green spaces and walkways. Relatively small homes on small lots are grouped in clusters of eight, with small green spaces connecting to larger spaces. Narrower than typical streets provide access to the rear of the homes (homes front on the connected green spaces), and the east–west orientation allows for solar access which most of the homes take advantage of. Fruit trees and edible landscaping pervade the neighborhood, and there is a vineyard, orchard, and community gardens. Three hundred almond trees are harvested commercially, providing a portion of the funds needed to pay for the three full-time gardeners.

While the Corbetts faced significant obstacles in building Village Homes (such as objections to the natural drainage system by the city's public works department and the fire department's opposition to the narrow streets), they were able to creatively overcome them. For example, a compromise on street width was reached, providing for a three-foot clear zone on each side, to allow emergency vehicle movement. The project has been a success on virtually every measure: the homes are highly sought after, residents tend to know each other, and crime in the neighborhood is very low compared with nearby conventional suburban-style development. The natural stormwater system has worked well, and incidentally resulted in a $600-per home saving, providing much of the funding for the neighborhood's impressive landscaping and green features.

For more information, see: Michael Corbett and Judy Corbett, *Toward Sustainable Communities: Learning from Village Homes* (Washington, DC: Island Press, 1996).

Los Angeles Eco-Village

Los Angeles Eco-Village is located three miles west of downtown in an ethnically diverse working-class neighborhood. Its focus is on transforming and retrofitting an existing urban neighborhood, rather than building a new community from scratch. Under the inspiration and guidance of Lois Arkin, the village, which consists of two city blocks, was initiated in 1993. About 500 residents are involved in the Eco-Village, and its primary goal is "to reduce our environmental impacts while raising the quality of neighborhood life."

A number of ecological retrofitting activities have occurred in the neighborhood, including the purchasing and renovation of two apartment buildings, the planting of fruit trees and community gardens, traffic calming (including neighborhood meals where tables are placed in the streets), ecological demonstration projects (e.g., graywater reclamation), and neighborhood dinners and meals and other gatherings (See plates CS19 and CS20). An Ecological Revolving Loan Fund, operated by Arkin's non-profit Cooperative Resources and Services Project, has provided the funds for purchasing and retrofitting the apartment buildings.

The LA Eco-Village project demonstrates clearly the potential of transforming older urban neighborhoods into more sustainable places. Many of the basic building blocks of urban sustainability are already in place – a more compact, walkable urban form, access to transit, stores, and so on. On top of these elements are then layered new forms of participation and new strategies for greening, together creating urban communities that exert a small ecological footprint and at the same time build community.

For more information, see: "Los Angeles Eco-Village Overview" at www.ic.org/laev

Civano and Armory Park del Sol, Tucson, Arizona

Civano is a relatively large new development on the outskirts of Tucson, a productive combination of ecological design and new urbanist thinking. Growing out of an early effort to build solar homes (See plate CS21) (and originally known as the Tucson Solar Village), the development will eventually contain 2,600 homes on 1,200 acres. The community's master plan reflects the desire to create a complete community, with housing, schools, parks, and employment within walking distance (See plate CS22). Much of the site is to be protected in a nature preserve. Amenities include a solar-heated pool, and extensive hiking and biking trails.

Sustainability is reflected in this development in other ways as well: The homes are built to a more stringent energy standard. Homes are estimated to use 50 per cent less energy than homes typical for the area. A double-water system is used (one line for potable water, another for less clean reclaimed water), and the houses and buildings take advantage of solar orientation. Civano has been substantially underwritten by Fannie Mae's American Communities Fund.

The emphasis on natural plants and vegetation carries over to Civano's commercial nursery, which specializes in such plants. Among other things, the nursery provides workshops on xeriscaping and native planting, and has salvaged some 6,000 plants and trees from Civano development sites.

One goal is to provide at least one job for every two households, as a way to minimize automobile commuting. Already a solar panel company, Global Solar, has located in Civano's commercial center.

Despite its many positive sustainability features, Civano's location on a greenfield location, at the edge of the city of Tucson, raises serious questions about its ultimate value as a model. Another green development in Tucson, much smaller in size, is Armory Park del Sol. In contrast, Armory Park del Sol is being build at a much higher density (99 units on about 15 acres) within an existing historic neighborhood, within Tucson city. The brainchild of John Wesley Miller, who originally conceived of the Tucson Solar Village, each of these homes will be Energy Star rated and will provide 1 kW of power from rooftop photovoltaic panels.

For more information, see: www.civano.com

EcoCity Cleveland and Cleveland EcoVillage

Cleveland is emerging as a testing ground for many new urban sustainability ideas. In the heart of the Midwest rustbelt, much of the growing interest in sustainability there has been the result of the creative advocacy of a small non-profit organization named EcoCity Cleveland. With a full-time staff of about three, the organization has accomplished much in terms of promoting new ideas and thinking, building a new awareness about the ecology of the city and region, and showing the potential benefits of a more sustainable urban direction. EcoCity Cleveland has been a tireless campaigner, publishing an award-winning newsletter, issuing policy papers, networking and partnering with other groups. It has published a sort of biological owner's manual for the region as well as bioregional maps, and each year gives out its Bioregional Hero awards.

Much success has been seen, and a considerable shift in thinking experienced in that city. Accomplishments have included new commitments to bikes and bicycling in the region, a new Green Environmental Center that will house several local environmental groups, new and renovated buildings incorporating green features, and fresh attention being given to urban sustainability in a number of recent and ongoing planning initiatives in the city (e.g., a new waterfront development initiative).

The Cleveland EcoVillage is perhaps the most tangible new reflection of these efforts. Here, a west end neighborhood, near a transit station slated for redevelopment, has been reconceptualized as a green urban village. As David Beach, founder of EcoCity Cleveland says, the goal is "to develop a model urban village that will realize the potential of urban life in the most ecological way possible" (Scott, 2002). Community meetings have been held, and a conceptual plan (See plate CS23) for the neighborhood has been prepared. Important elements of the vision include new infill housing, built with ecological design features, a renovated neighborhood park, community gardens, and the rapid transit station, itself to reflect a green theme (incorporating passive and active solar features, a roof from recycled materials, native landscaping), and within a short walk for residents of the neighborhood.

The first of twenty new green townhouses have been completed (See plate CS24), homes that incorporate a variety of green features (solar energy, high energy-efficiency, non-toxic materials, recycled and flyash in concrete, and use of ISC-certified wood, among others).

For more information, see the EcoCity Cleveland website: www.ecocitycleveland.org

REFERENCE

Scott, M. Robert (2002) Gettin' Easier Bein' Green: Eco-city Thrives in Cleveland. *Ohio Realtor*, September.

URBAN SUSTAINABILITY AT THE CITY AND REGIONAL SCALE

Vancouver, British Columbia

Vancouver, BC, with a population of a little over two million, has been a model of a relatively compact, sustainable North American city and region. It has been able, especially in recent years, to guide much of its growth into compact, dense, walkable, urban neighborhoods. The Vancouver development style has largely been one of accommodating much growth in tall, thin skyscrapers, benefiting, certainly, from spectacular mountain views there.

Growth in the city is governed by a comprehensive set of urban design guidelines that, among other things, stipulate a minimum number of affordable units, and a minimum number of family-friendly units (e.g., units where day care and schools are within a certain walking distance). Buildings are oriented toward the street, and great importance is given to promoting a vibrant street life and amenity-rich urban environment. High-rise buildings must be flanked by four-to-six-story structures that serve to soften the visual effects of the skyscrapers, and advance a human street scale (See plate CS25).

A regional livability strategy has been in place for fifteen years, administered by the Greater Vancouver Regional District (GVRD). The plan calls for steering development in the region into designated town centers (one of these being Vancouver), lying along the route of the SkyTrain, an elevated rail system. Much rural land in the region is contained in an Agricultural Land Reserve (ALR) and is thus off-limits to suburban development.

There are many explanations for Vancouver's success, including its relatively strong planning framework. The fact that Vancouver essentially has no major highways (Canada has never had an interstate highways program like the US's) is also helpful, and residents there must choose to live closer in if they wish to have access to cultural amenities, schools, housing, and so on.

The success of Vancouver's land use policies has been demonstrated in a recent analysis of land use trends there between 1986 and 2001. While the region grew in population during that period by almost 50 per cent, the portion of the city living in more compact, walkable neighborhoods actually increased. A joint study by Northwest Environment Watch and Smart Growth BC concluded that the percentage of residents living in compact neighborhoods (defined as neighborhoods with twelve or more residents) rose impressively from 42 per cent to 62 per cent during this period.

Building on these planning successes, much attention in recent years has been given to pursuing a broader sustainability agenda. The GVRD, for instance, has been developing a Sustainable Region Initiative. This agency is serving as a catalyst to promote sustainability initiatives and projects throughout Vancouver (e.g., by sponsoring regional workshops and publishing case studies), by taking action directly (e.g., through employee trip reduction programs), and by looking for ways to implement its regional programs more sustainably (e.g., combined greenway and utility corridors) (See plate CS26).

For more information, see: Northwest Environment Watch and Smart Growth BC. 2002. *Sprawl and Smart Growth in Greater Vancouver*, September 12, found at: www.northwestwatch.org/press/Vancouvergrowth.html. See also www.gvrd.bc.ca/sustainability/index.html

Bogotá, Colombia

It is remarkable what can be accomplished in a city like Bogotá, in a country, Colombia, that is essentially at war. Bogotá, a city of seven million people, has emerged as a place of innovation and inspiration, forging ahead with an impressive agenda to make the city more livable and sustainable. Largely through the leadership of former Mayor Enrique Peñalosa, the city has undertaken a number of creative transportation strategies. Like Curitiba, it is now using a bus-only transit system, with articulated buses operating like a subway (See plate CS27). Called TransMilenio, it operates along two main corridors, but plans are that twenty-two corridors will be in operation by 2015 (80 per cent of the city's population will be within 500 meters of a bus-stop).

Other accomplishments include 300 km of new bicycle paths, hundreds of new neighborhood parks, and a new greenway. One of the boldest actions was to convert a major street in the downtown to pedestrians only. This 17-kilometer-long street, which includes pedestrian amenities, has become cherished public space. Bogotá has also started something it calls its Ciclovia – the closing every Sunday of the city's main road arteries, some 120 km in all, in order to make them available to bicycles, walking, and socializing (See plate CS28). Former major Peñalosa, a champion of this idea, calls it a "marvelous community building celebration," and it attracts an incredible 1.5 million residents each week. Even more impressive is the number of people who participate when the city closes those same streets on a Christmas evening – some three million residents participate.

For more information, see: Enrique Peñalosa, "Urban Transport and Urban Development" at: www.worldbank.org/html/fpd/urban/forum2002/docs/peñalosa-pres.pdf

Gaviotas, Colombia

This small community, a sixteen-hour jeep ride from Bogotá, Colombia, has been called Colombia's model ecological city. Really a village more than a city, about 200 people live here. Its founder, Paolo Lugari, thinks of it not as a utopia, but a real place. Gaviotas and its residents have found many creative ways to live more sustainably. The emphasis here has been on low-tech ideas and tools, and on designing and building and creating those things uniquely suited to this wet savanna environment. Groundwater for the village is extracted through a specially designed hand pump, so easy to operate that it has been fashioned into a children's see-saw (See plate CS29). There is extensive use of solar energy – for providing hot water, for generating electricity, and for purifying drinking-water. Specially designed small windmills capture the modest wind energy. Travel is by way of a specially designed bicycle. Food is produced through a hydroponics system. Gaviotas is, essentially, a solar-powered, self-sufficient ecological village.

It is also an ecologically restorative village. One important economy-building step was the planting of some 20,000 acres of new forest. Caribbean pines were planted specifically because they were discovered to be unusually suited to the region's acidic soil. Turpentine is distilled from the bark resin of these trees, and has become a major economic product for the village. In addition, a rich undergrowth has become re-established in this new forest, with a flourishing of plants and wildlife. Eventually, the pines will be replaced by this regenerated rainforest.

The village's hospital, the only one within a half a day's car ride, is also designed to function ecologically. Its unique ventilation system brings cool air from a nearby hillside, its electricity is generated from PV panels, and hot water and distilled drinking-water are provided from solar hot-water panels. Methane is extracted from livestock dung and sent to the hospital (See plate CS30).

Gaviotas has become a model of sustainable village building, in Colombia as well as in other parts of Latin America. The water pump, for instance, is now in use in some 700 other Colombian villages, and the windmills, solar water-heating systems, and other technical innovations have been replicated elsewhere also. Gaviotas shows the power of creative innovation and place-based sustainable living.

For more information, see: Alan Weisman, *Gaviotas: A Village to Reinvent the World* (Post Mills, VT: Chelsea Green, 1999).

Auroville, India

This ecological, utopian new town is located in southeast India, about 160 kilometers south of Madras. Planned to eventually have a population of 50,000, there are now about 1,600 residents, representing about thirty different countries. Auroville has been conceived as a "universal township," inspired by the thinking of philosopher and yogi Sri Aurobindo, as a place where "men and women of all countries are able to live in peace and progressive harmony, above all creeds, all politics and all nationalities." Human unity is a central ideal behind the town. The town's physical form is in the shape of a galaxy (See plate CS31), with a matrimandir, or large golden temple, in the very center. This spiritual center of the community, or Peace Area (See plate CS32), also contains a Banyan tree and an amphitheater for community events. The community's more elaborated master plan sets out a series of zones, essentially concentric circles, that extend out from the town's center. These include residential, international, industrial, and cultural zones, as well as a large Green Belt. The nearly 1500-hectare Green Belt is home to agriculture, forests, and biodiversity.

Auroville's many other ecological elements include extensive efforts at environmental regeneration of the landscape (planting of more than two million trees, extensive soil and water conservation, training programs in sustainable farming), the use of earth construction techniques in building (use of onsite, locally produced compressed earth blocks), and extensive use of renewable energy (e.g., thirty windmills, and a solar community kitchen producing 900 meals a day with 50 per cent of the energy coming from a "giant mirrored solar dish"), among others.

The region around Auroville contains thirteen villages and 40,000 people. Much of the work of Auroville has focused on improving the quality of life and sustainable practices in these surrounding villages. This has happened through training and workshops, for instance, and through co-operative projects.

For more information, see: www.auroville.org

IBA Emscher Park, Germany

The Ruhr Valley in northwestern Germany is a landscape heavily scarred by industrialization and littered with the remnants of coal-mining and steel production. In 1989 the state of North-Rhine Westphalia along with the federal government embarked on an ambitious project to promote long-term rehabilitation and reuse of this immense industrial area covering 800 square kilometers (307 square miles) along a 70-kilometer (43-mile) urban corridor. Called the International Building Exhibition (IBA) Emscher Park, more than a hundred demonstration projects have been funded over a ten-year period, with a remarkable impact on the cities and landscape, and a new appreciation and pride for an industrial past that most in the region sought to hide or forget. An impressive array of projects and creative examples of land and landscape recycling has resulted. Projects have included the establishment of eleven technology centers, new ecological housing on reclaimed brownfield sites, and the conversion of industrial buildings to cultural uses (e.g., conversion of the gasometer in Oberhausen to an exhibition hall). One of the most creative projects is the Landscape Park in Duisberg-North (See plate CS33). Here, a former steel mill has been converted into a regional park and industrial monument, where formal gardens have been created and trees planted amidst the remnant buildings, and massive foundation pillars have been converted into areas for climbing and repelling (See plate CS34).

Mont-Cenis Academy, in Herne-Sodingen, Germany, is one of the most impressive buildings sponsored through IBA. This government training center, designed by Jourda and Perrandin, takes the form of a large glass building, a "vast timber framed hanger" that contains offices, seminar spaces, a library, town hall, cafés, restaurant, and hotel. Nine wooden buildings lie protected under this "glass weather shield." A series of computer-controlled flaps and louvers open and close, depending on outside weather patterns, to ensure comfortable temperature and ventilation inside the glass shield. The structure permits use much of the year with minimal heating, and allows natural cross-ventilation during the summer. Most dramatic of all are the photovoltaic (PV) cells, 10,000 square meters in all, on the building's rooftop. More power is produced than is needed by the structure. Careful thought was given (and simulation done) on how to configure or array the PV panels. In the end they were grouped in a way that they "form cloud-like patterns that magically diffuse light into the great greenhouse" (Kugel 1999). It is the largest solar roof in the world.

Sited on former coalfields, the project builds on this history in several ways. The architecture evokes its mining and industrial past; according to one observer, "its monumental scale and repetitive structure evoke the big sheds, furnaces and factories of the Ruhr's old industrial landscape" (Kugel 1999). A combined heat and power plant, moreover, takes advantage of this site in even more tangible ways, as it utilizes escaping methane to produce heat and electricity.

For more information, see: Claudia Kugel. 1999. "Green Academy." *Architectural Review*. October.

London, England

With the re-establishment of city-wide governance in London (the new Greater London Authority), an impressive new emphasis has been given to sustainability in this metropolitan area of 7.5 million people (See plates CS35 and CS36). The stated vision is to "develop London as an exemplary, sustainable world city." Already, a number of specific sustainability plans and actions have been developed in the city. Especially impressive is the city's newly prepared Biodiversity Strategy and draft Energy Plan (including greenhouse gas emission targets).

Under Mayor Ken Livingstone's leadership, new institutional structures have been formed to promote and consider sustainability, including the creation of the London Sustainable Development Commission. The key charge of the Commission is to establish the London Sustainable Development Framework. Elements of this framework include the development of a comprehensive set of sustainable indicators (a draft of these has been issued), and review of the proposed spatial plan for London. The city is making dramatic efforts to reduce automobile traffic in its center, through a bold and controversial road pricing scheme that now charges £5 per car entering central London. Further initiatives include a new spatial plan for the region and a new green procurement code.

A new city-commissioned study entitled *City Limits*, commissioned by the city, presents a detailed analysis of the material flows required by the city and calculates its ecological footprint. The first such detailed study for a city of this size, the report concludes that London's footprint is extremely large, nearly 300 times its own land area. The study documents a tremendous material flow of inputs and outputs. Londoners consume, for instance, 154,407 gigawatts of energy each year, two million tons of wood products, and 730,000 tons of vegetables. At the same time, they emit fifty million tons of carbon dioxide and produce eight million tons of sewage sludge.

For more information, see: Greater London Regional Authority, "Sustainable Development," at www.london.gov.uk/londonissues/sustainability.jsp

Chicago, Illinois

Popular five-term mayor Richard Daley has declared that he intends for Chicago to be the "greenest city in the world." This is not just political rhetoric, but has already been translated into a host of exemplary urban sustainability ideas and initiatives in this region of about eight million people. Under Daley's leadership the city has undertaken a number of greening programs, including programs for encouraging and subsidizing green rooftops (and the dramatic retrofitting of City Hall with a green rooftop (See plate CS37)), extensive tree planting, and conversion of former gas stations and brownfield sites into neighborhood parks. An ambitious new energy plan promotes renewable energy (photovoltaic panels have now been installed on the rooftops of many city buildings), and establishes the goal of providing at least 20 per cent of the city's energy needs through renewable energy. Through the city's brightfields initiative, a new Center for Green Technology has been built on a former industrial site. The city is also moving forward to reclaim and restore a large former industrial area, Lake Calumet, for purposes such as sustainable energy production. A large solar energy plant and methane recovery facility are envisioned.

A number of green pilot programs have been initiated, including a competition to design and build prototype green homes (five designs were selected and are being built), and a green bungalow retrofit program. Most recently, the city has issued a set of sustainability principles, called the "Chicago Principles," which encapsulate its urban sustainability values and goals, and which will serve to guide the city into the future.

At a regional level, 160 organizations have joined forces in a unique coalition to create Chicago Wilderness. This is both an umbrella organization and a vision for a 200,000-acre system of protected lands and landscapes in the region. Chicago Wilderness has already accomplished much, including preparation of a regional biodiversity atlas, extensive habitat restoration work, and perhaps the first urban biodiversity conservation plan. Chicago Wilderness has also done much to raise awareness about biodiversity, and has sponsored numerous public workshops and educational programs throughout the region. Community-based conservation work has already taken place through this coalition.

For more information, see: City of Chicago, Department of Environment, www.ci.chi.il.us/Environment/

Austin, Texas

Austin, Texas began its Community Sustainability Initiative (CSI) in 1997 and has undertaken a number of key projects. The position of Sustainability Office was created to oversee these initiatives. Important projects and activities within CSI's first years have included a capital improvements program matrix used by city departments in evaluating proposed capital projects, the development of a set of sustainability indicators, and preparation of a set of sustainable building guidelines for city buildings and facilities.

A number of other initiatives in the areas of urban planning and environmental management have been undertaken in Austin, though outside the direct auspices of the CSI. Especially notable are Austin's Smart Growth Initiative and Green Builder Program. Under its Smart Growth Initiative, Austin prepared a Smart Growth Map (See plate CS38) indicating "desired development zones" where future growth is encouraged through a system of incentives. The goal is to discourage sprawl and to promote more compact, higher-density development within the existing urbanized areas of the city. Development proposals are evaluated through the City's Smart Growth Matrix and the resulting score (based primarily on the proposed location) determines the extent of financial incentives available in the form of infrastructure fee reductions or waivers. Fee waivers and expedited permit review are also available under the city's SMART housing incentives for the provision of affordable housing.

Austin's Green Builder Program is one of the oldest such efforts in the United States. Under this program, residential and commercial buildings are rated according to the extent and number of green features (receiving from one to five stars). The city provides technical assistance, convenes workshops and training for builders and designers, helps market green homes, and provides a variety of rebates and subsidies for green construction and rehabilitation (e.g., energy efficiency loans).

For more information, see: City of Austin, "The City of Austin Sustainable Communities Initiative," www.ci.austin.tx.us/sustainable.html

Portland, Oregon

Few American cities have accomplished as much as Portland, Oregon in the area of urban sustainability. This region of about two million people has developed a deserved reputation as an American testing ground for innovative urban planning ideas.

Promoting a more compact urban form is one of this region's key accomplishments. In part this was a response to the passage of Senate Bill 100 in Oregon in 1973. This law mandated, among other things, that all cities adopt urban growth boundaries (UGBs) and that land use plans and regulations protect productive forest and farmland. UGBs are intended to delineate between developable or urbanizable lands and areas that are to be conserved and protected from development. Around Portland this UGB has been delineated on a regional basis, encompassing twenty-three smaller cities and portions of three counties. This regional scale growth management has been aided substantially through the establishment of a strong regional government, the Portland Metropolitan Services District, or "Metro" for short. The only popularly elected regional government in the USA, it has real powers especially in the areas of land use, transportation, parks and greenspaces, and solid waste management.

The City of Portland has itself taken many actions to reduce the reliance on private automobiles, to promote a more walkable city, and to protect its environment. A series of exemplary downtown plans have sought to increase the amount of housing in the center, to reconnect the city with its riverfront, and to create a highly attractive urban environment. A percent-for-art program has resulted in delightful street sculptures and public art. The city, moreover, has a long history of recapturing space from the automobile and giving it back to pedestrians. Dramatic examples include tearing out a riverfront highway and putting a park in its place – Tom McCall Waterfront Park – and removing a downtown parking garage to create Pioneer Courthouse Square, scene of many concerts, festivals, and other public events. A highly popular regional light rail system, the Metropolitan Area Express, or "MAX," (See plate CS39) connects much of the region in addition to an exemplary bus system, and much regional planning has occurred to guide growth along transit corridors. The city and region have also increased significantly the number of bikeways, now extending for more than 240 kilometers (149 miles).

Together these planning tools have led to a more compact regional development pattern than most other American cities with a much more vibrant urban core and relatively effective public transit. While the UGB has been extended several times, the overall pattern of development is gradually becoming denser and more pedestrian-friendly. A recently adopted, long-range regional plan, Region 2040, developed through an extensive public process, envisions maintaining the region's tight Metro urban form and steering much new growth into a series of centers along the spine of the MAX system (See plate CS40). The goal is that 85 per cent of new residents will be within a five-minute walk of a transit station.

While Portland has secured its reputation on these successful land use and transportation achievements, in recent years it has embraced a broader, more expansive urban sustainability agenda. It now boasts a city Office of Sustainable Development (formerly Portland Energy Office), housed in the Jean Vollum Natural Capital Center. The city has adopted a set of sustainable city principles and has been promoting a variety of other green city ideas. These have included green building, sustainable technologies and practices, energy conservation and efficiency, and solid waste and recycling. The City's Green Investment Fund has provided important funding to support green building, and a green building policy mandating minimum standards for city-funded facilities was adopted in 2001.

Portland has not been without its critics. Sharp rises in the price of housing in recent years have been blamed on its UGB and growth controls, for instance (though it is far from clear that this has been the cause). Others have been critical that while greater density and compactness have been achieved, the overall form this development takes is not dramatically different than other places; much of it still very suburban and car-dependent, critics such as new urbanist architect Andres Duany charge.

For more information, see: Portland Office of Sustainable Development, www.sustainable Portland.org

Burlington, Vermont

This progressive city of about 40,000 people has made a name for itself for embracing and implementing a host of community sustainability measures. The city has adopted the Earth Charter, an international declaration of environmental principles, and has made significant commitments to sustainability. It boasts one of the few successful pedestrian malls in the nation, and its town center is vibrant, walkable, and mixed-use.

Burlington is the center of many creative urban sustainability ideas. A non-profit organization called The Good News Garage, for instance, repairs donated automobiles, supplies them to low-income families at cost, and also provides job training. The Burlington Community Land Trust, formed in 1984 with seed funding from the City, is an innovative model for both conserving land resources and providing affordable housing. The first of its kind in the nation, the Trust buys land, builds housing, and then rents or sells the homes to low- and moderate-income families. The Trust has also been instrumental in securing new parks, healthcare, and other community facilities, and in restoring and recycling older buildings in the city, for instance, recently converting old bus and trolley barns into apartments.

Burlington residents give strong support to local organic agriculture. The Intervale Farm Area has become an important national model. Consisting of about 800 acres of floodplain land along the Winooski River, the Intervale is now home to two community supported agriculture (CSA) operations and several other commercial organic farms. Managed now by a non-profit foundation, the Intervale is also home to the Burlington Compost Program, a living machine (a biological system for purifying wastewater) producing a Tilapia fish, a gardeners' supply store, and a small farms incubator program. A Food Enterprise Center is currently under development.

Energy conservation and renewable energy are also important topics in Burlington. The McNeil Generating Station (See plate CS41), run by the city's Electric Department, generates 50 MW of electricity from burning wood from local private woodlands and mill wood waste. This plant generates nearly enough electricity for the city.

Perhaps most impressive is the Burlington Legacy Project, an extensive community-based process created in 1999 that developed a vision and plan for the city for the year 2030. The Burlington Legacy document lays out a common future vision, identifies and discusses key themes (economy, neighborhoods, governance, youth and life skills, and environment), sets goals and priority actions, and provides profiles of projects already underway. The bottom-up planning process involved unusually extensive opportunities for community participation and some unique elements, including community surveys, focus groups convened at the neighborhood level, active involvement of the city's youth (e.g., poster and essay contests, school focus groups), public hearings, and a "Summit on the City's Future," held in March 2000 (See plate CS42).

For more information, see: The Burlington Legacy Project," at www.cedo.ci.burlington.vt.us/legacy/strategies/index.html

San José, California

San José adopted its "sustainable city major strategy" in 1994, as an element of its new General Plan, San José 2020. But concern about sustainability in that city extends back at least to 1980, when an influential report was prepared for the city entitled *Toward A Sustainable City*. Since then San José has done many things already to become more sustainable. It has adopted an ambitious recycling program, including the provision of financial incentives, following its "close the loop" philosophy for the reuse of recycled materials. After decades of sprawl in the 1950s, 1960s, and 1970s, it has imposed relatively stringent growth management provisions, including an urban growth boundary, has invested in light rail transit, has adopted zoning changes and other policies to promote infill development, and has placed new importance on strengthening its downtown (See plates CS43 and CS44).

Under the city's energy management program, staff have converted its 48,000 incandescent and mercury vapor streetlights to use sodium lamps (incidentally saving $1.5 million per year in the process), and have prepared a set of solar design guidelines. San José is also one of a growing number of communities promoting green building practices, and is participating in a program to reduce the urban heat island effect ("cool roofs"). Other environmental accomplishments have included water conservation and efficiency improvements, and a policy for purchasing recycled products.

For more information, see: Smart Growth Network, "San José, The Sustainable City Project" at www.sustainable.doe.gov/success/sust_city.shtml

Santa Monica, California

Santa Monica, a city adjacent to Los Angeles known for its progressive policies, initiated its sustainable city program in 1994. In 2001 a Sustainable City Working Group was formed to evaluate the program and recommend future directions. This led to the preparation of an updated Santa Monica Sustainable City Plan (and a change in name from "project" to "plan"). The new plan lays out a comprehensive vision of a sustainable city, delineating nine Guiding Principles, as well as goals, indicators, and targets. A new Sustainable City Steering Committee is envisioned, as well as an interdepartmental Sustainable City Implementation Group.

Santa Monica has already taken many actions to advance urban sustainability, and has a number of new initiatives in progress (See plate CS45). The municipality, for instance, has adopted a set of mandatory Green Building Design and Construction Guidelines, as well as environmental purchasing and toxic reduction policies. The city has added parks and planted trees, constructed the innovative Santa Monica Urban Runoff Recycling Facility (SMURRF) (See plate CS49), and planted organic gardens at all of its public schools.

Many impressive steps have also been taken in Santa Monica to support renewable energy. The municipality has undertaken extensive energy-efficiency improvements in government buildings, and boasts that it is the first city (perhaps in the world?) to purchase 100 per cent of its municipal power (the electricity needed to power street lights, city buildings and facilities) from green, renewable sources. The city was even willing to pay a premium to ensure that the sources of this power are renewable.

Santa Monica has developed a set of Sustainability Indicators and is using these to judge overall city progress. These indicators evaluate the city's success in several areas, including tons of waste sent to the landfill, water and energy usage, transit ridership, number of community gardens in the city, and the amount of public open space.

For more information, see: City of Santa Monica. 2003. *Santa Monica Sustainable City Plan*. Adopted February 11.

Plate CS2: Sky garden in Commerzbank, Frankfurt, Germany.

Source: Norman Foster and Partners.

Plate CS1: The exterior of Commerzbank, Frankfurt, Germany (Norman Foster and Partners).

Source: Photograph by Timothy Beatley.

Plate CS3: Menara Mesiniaga, a bio-climatic skyscraper designed by architect Ken Yeang.

Source: Photograph by K.L. Ng, from ArchNet.

Plate CS4: Menara Mesiniaga, garden terraces.

Source: Photograph by K.L. Ng, from ArchNet.

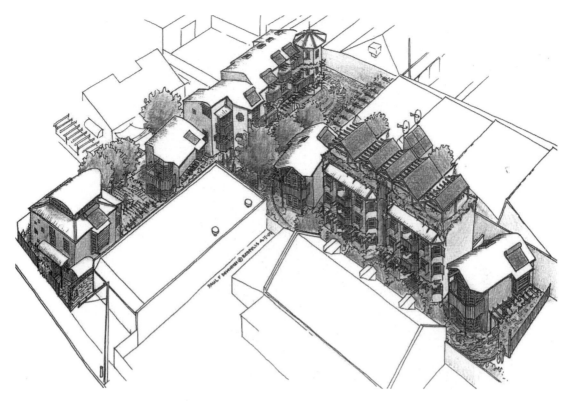

Plate CS5: Artist's rendering of Christie Walk, Adelaide, Australia.

Source: Courtesy Paul Downton, Urban Ecology Australia.

Plate CS6: Christie Walk, Adelaide, Australia.

Source: Photograph by Paul Downton.

Plate CS8: Condé Nast building, neon frontage on Times Square, sign powered by photovoltaic production.

Source: Photograph by Timothy Beatley.

Plate CS7: Condé Nast building, 4 Times Square, New York.

Source: Photograph by Timothy Beatley.

Plate CS9: Natural stormwater collection system at the Kronsberg Ecological District, in Hannover, Germany.

Source: Photograph by Timothy Beatley.

Plate CS10: Centralized solar hot water heating system, Kronsberg, Hannover, Germany.

Source: Photograph by Timothy Beatley.

Plate CS11: Beddington Zero Energy Development, Hackbridge, London.

Source: Photograph by Timothy Beatley.

Plate CS12: Dramatic windcowls that scoop up air ventilating BedZED buildings and giving the neighborhood a distinctive and colorful look.

Source: Photograph by Timothy Beatley.

Plate CS13: Greenwich Millennium Village ecological housing.

Source: Photograph by Timothy Beatley.

Plate CS14: Greenwich Millennium Village ecological housing.

Source: Photograph by Timothy Beatley.

Plate CS15: Nieuwland, solar community, Amersfoort, The Netherlands.

Source: Photograph by Timothy Beatley.

Plate CS16: Solar homes at Nieuwland.

Source: Photograph by Timothy Beatley.

Plate CS17: Village Homes, Davis, California.

Source: Photograph by Timothy Beatley.

Plate CS18: Green stormwater swales, Village Homes.

Source: Photograph by Timothy Beatley.

Plate CS19: Traffic calming brunch at the Los Angeles Eco-Village.

Source: Photograph by Jesse Moorman.

Plate CS20: Tree planting at the Los Angeles Eco-Village.

Source: Photograph by Jesse Moorman.

Plate CS21: Civano Community, Tucson, Arizona solar house.

Source: GEO Advertising and Marketing.

Plate CS22: Civano streetscape, Tucson.

Source: GEO Advertising and Marketing.

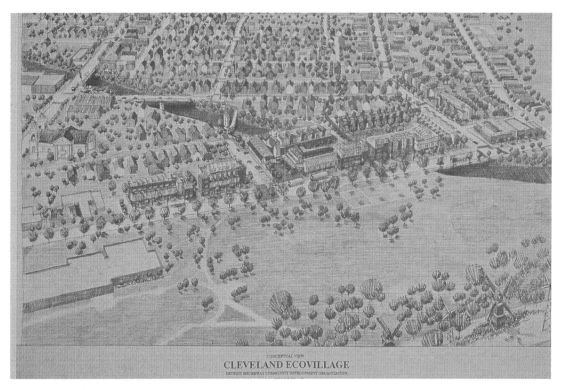

Plate CS23: Cleveland EcoVillage, EcoCity Cleveland, Conceptual Master Plan.

Source: Courtesy of David Beach, EcoCity Cleaveland.

Plate CS24: Cleveland EcoVillage, EcoCity Cleveland, new green townhouses.

Source: Courtesy of David Beach, EcoCity Cleaveland.

Plate CS25: New high-rise development along the waterfront of False Creek, Downtown Vancouver, British Columbia.

Source: Photograph by Timothy Beatley.

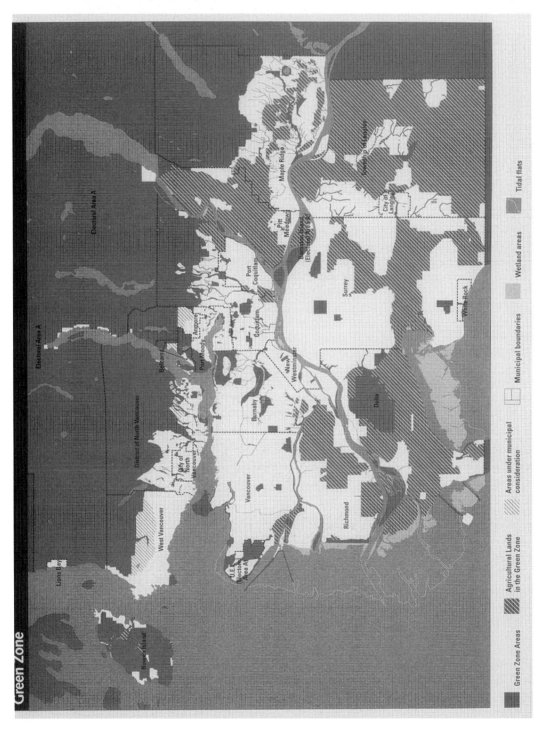

Plate CS26: Map of Vancouver, BC, region, showing protected Green Zone.

Source: Courtesy Creation Vancover Regional District.

Plate CS27: Bogotá TransMilenio, transit system.

Source: TransMilenio S.A.

Plate CS28: Bogotá Ciclovia.

Source: Instituto Distrital de Recreación y Deporte de Bogotá.

Plate CS29: Paolo Lugari sitting on one of Gaviotas' famous water pumps, Gaviotas, Colombia.

Source: ZERI Foundation, Gunter Pauli.

Plate CS30: Self-sufficient hospital, Gaviotas, Colombia.

Source: ZERI Foundation, Gunter Pauli.

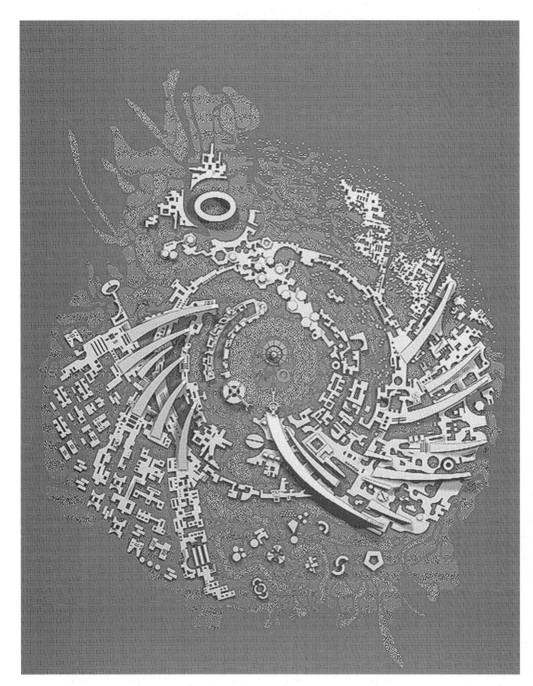

Plate CS31: Auroville Galaxy Plan, Auroville, India.

Source: Photograph by Pino Marchese.

Plate CS32: Matrimandir (temple) and Peace Area, Auroville, India.

Source: Photograph by Pino Marchese.

Plate CS34: Major foundation pillars used for climbing and repelling. Landscape Park, Duisburg-North.

Source: Photograph by Timothy Beatley.

Plate CS33: Landscape Park, Duisburg-North, former steel mill, design by Pater Latz.

Source: Photograph by Timothy Beatley.

Plate CS36: New ecological building in Central London. Swiss Reinsurance building, also a Norman Foster design.

Source: Photograph by Timothy Beatley.

Plate CS35: New London City Hall, a green building designed by Norman Foster.

Source: Photograph by Timothy Beatley.

Plate CS37: Green roof retrofit of Chicago City Hall.

Source: Photograph by Timothy Beatley.

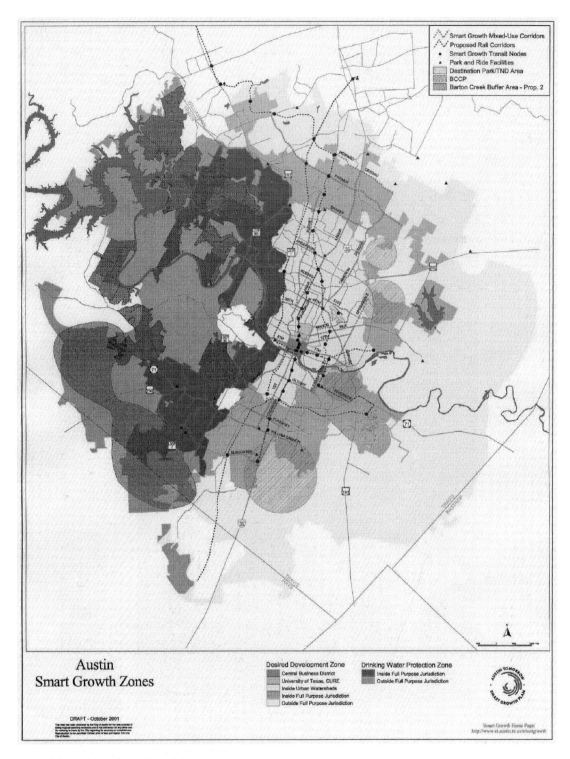

Plate CS38: Austin Smart Growth Map.

Source: Courtesy of City of Austin, Department of Growth Management.

Plate CS39: Portland MAX light rail system.

Source: Photograph by Timothy Beatley.

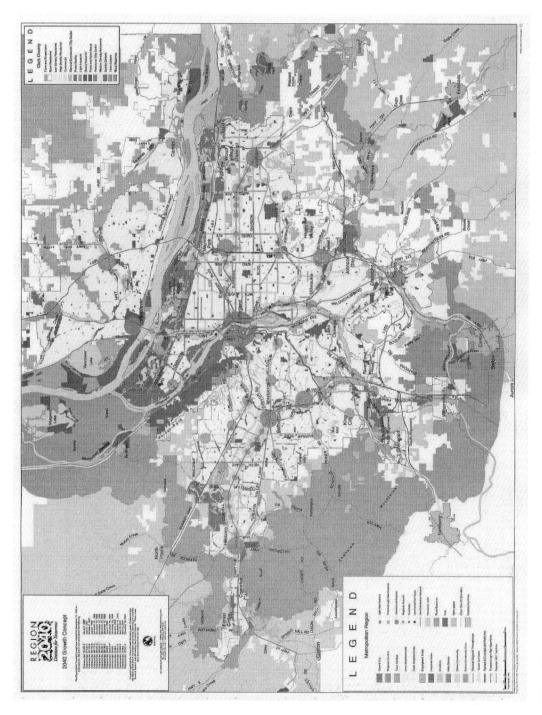

Plate CS40: Portland Metro 2040 Concept Map.

Source: Courtesy of Portland Metropolitan Services District.

Plate CS41: McNeil Generating Plant, Burlington, Vermont.

Source: Burlington Electric Department.

Plate CS42: Young and old together at the McClure Intergenerational Center, Burlington, Vermont.

Source: Photograph by Nicole Craft.

Plate CS43: The Ohlone Chynoweth apartments, an example of transit-orientated development in San José, California.

Source: Photograph by Jeffrey Peters.

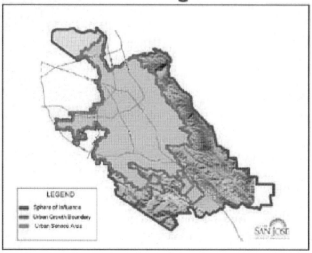

Plate CS44: San José, California, map of urban growth boundary.

Source: Courtesy City of San José.

Plate CS45: Colorado Court green and affordable housing project, Santa Monica, California.

Source: Courtesy City of Santa Monica.

Plate CS46: Santa Monica Urban Runoff and Recycling Facility, Santa Monica, California.

Source: Courtesy City of Santa Monica.

PART SEVEN

Sustainability planning exercises

INTRODUCTION TO PART SEVEN

Many of us learn best "by doing," that is, by working actively to find solutions to various challenges in which we must process information and apply concepts. Consequently, hands-on exercises are a good way to understand what sustainability planning might involve. The exercises in this section are appropriate either for classroom use or for individuals or groups to complete on their own. They have been developed by one of us (Wheeler) in conjunction with courses taught at the University of California at Berkeley. An entire undergraduate or graduate class can even be structured around a series of such exercises, which examine sustainability planning at different scales and in different contexts. (Bringing in guest speakers undertaking similar projects in the real world can help supplement the class material.)

Interested readers are encouraged to modify or expand these teaching exercises to fit their own needs. Generally these tasks are done in groups, on the theory that we can all learn enormously from each other and from the process of testing our ideas against those of others. Many of these exercises require some graphic representation – simple diagrams, plans, or maps to convey ideas. Although many of us are not used to presenting ideas visually, and may not have drawn anything since kindergarten, graphic work has been very much a part of the urban planning profession from the start. It is of course integral to architecture and landscape architecture as well. What matters in this case are not fine works of art but relatively simple diagrams, maps, or plans that can convey concepts effectively, especially for presentation to an audience. Clear labels that can be read at a distance are important for these poster-size graphics. Even students or community members with no background in graphic representation should be encouraged to experiment with different ways of illustrating their sustainable urban planning ideas.

Cognitive Mapping Exercise

The purpose of this exercise is to gain understanding into the nature of the environments we grew up in, how they affected us, and how they might relate to our views of sustainable urban development. This exercise draws upon cognitive mapping techniques developed by Kevin Lynch at MIT and others in the field of environmental design research. Such researchers have systematically asked particular groups to draw maps of their home environments to see what may be learned about how people perceive neighborhoods, cities, or the region as a whole.

INSTRUCTIONS TO PARTICIPANTS

Close your eyes and imagine yourself as a child at an age that was particularly important to your development. Put yourself in a location that you most identified with at that age and picture the landscape around you, including any features that were important to you such as your family's house and yard, locations where your friends lived, places where you played or worked, and the surrounding city, town, or countryside. What elements in the human or natural environment were important to your daily life? Did any parts of the landscape feel pleasant or unpleasant, safe or threatening, nurturing or draining? With whom did you interact on a daily basis? Where could you go? How did you get around?

Using colored markers on a large sheet of paper, map this environment as it appeared to you at that time (not as it really is). Put yourself in the center (stick figures are fine) and work outwards, adding those features that were most influential in your daily life. These elements can be represented symbolically in whatever way works for you and do not have to be to scale (i.e., in realistic spatial relationship to one another). Dramatize them as much as possible. Express how things felt to you then, rather than trying to reconstruct a realistic map. Fill the paper. Use bright color and thick lines so that someone else can understand this map from a distance. Label your drawing clearly so that others can understand it. Drawing skill does not matter.

After fifteen or twenty minutes get together with several others and take turns explaining what you have drawn. The others can ask clarifying questions but should not debate anything you have shown them. Focus on how this environment affected you and your family personally, and what may be learned from it regarding how this place functioned, how different people or cultures related to one another, and how they related to the natural world. If you are doing this exercise individually at home, you might try writing up a couple of pages analyzing this environment and its social, environmental, and economic implications.

Future Visions Exercise

This second exercise may be combined with the first to analyze how formative urban environments from our past influence our views on how cities should be developed in the future. In this case, participants can draw both maps one after the other, and then discuss them in groups. For some, ideal environments may be rather similar to childhood environments. For others, they will be quite different, perhaps in order to respond to some of the problems experienced while growing up.

INSTRUCTIONS TO PARTICIPANTS

Close your eyes again and imagine your ideal community – a place you would really like to live in if you could, and that is sustainable in whatever way you want to define that term. How is this place laid out in terms of streets, buildings, public spaces, parks, homes, workplaces, and shopping areas? How do people get around (what transportation networks exist)? What types of people live there? How do they interact or relate? How does the built environment relate to the natural landscape?

On a large sheet of paper draw a plan of this place. Do it at whatever scale seems appropriate to you. Include symbols and labels to identify key elements. Again, fill the paper and use lots of color, so that someone can read this map from a distance. Drawing quality does not matter. At one side list four to five key characteristics of this place that will help convey its unique qualities.

As before, discuss your vision in small groups. Each person should explain their image in a few minutes, with others asking only clarifying questions, not debating issues or components. Consider whether or how this ideal relates to the formative environment(s) you experienced when you were young. If desired, write up a couple of pages on this relationship.

Definitions of Sustainable Development

> This is a brief exercise to stimulate discussion of different perspectives toward sustainability. In three to five minutes have participants write down their own personal definition of sustainable development. They should keep this definition as simple as possible, something that they could use to explain the concept to a parent or grandparent. Then ask participants to repeat their formulations for the others. Write down key themes from each friend. Then go through the different definitional approaches in a group discussion, highlighting the pros and cons of each approach, and discussing the perspective or worldview that might most endorse it.

For comparison, here are some definitions from various international sources.

Theme: Meeting the needs of future generations

"Sustainable development is development that meets the needs of the present without compromising the ability of future generations to meet their own needs" *(Brundtland Commission, 1987)*.

Theme: Carrying capacity of ecosystems

Sustainable development means "improving the quality of human life while living within the carrying capacity of supporting ecosystems" *(World Conservation Union, 1991)*.

Theme: Maintain natural capital

"Sustainability requires at least a constant stock of natural capital, construed as the set of all environmental assets" *(British environmental economist David Pearce, 1988)*.

Theme: Maintenance and improvement of systems

"Sustainability . . . implies that the overall level of diversity and overall productivity of components and relations in systems are maintained or enhanced" *(American ecological economist Richard Norgaard, 1988).*

Theme: Positive change

Sustainable development is "any form of positive change which does not erode the ecological, social, or political systems upon which society is dependent" *(Canadian planner William Rees, 1988).*

Theme: Sustaining human livelihood

Sustainability is "the ability of a system to sustain the livelihood of the people who depend on that system for an indefinite period" *(Indonesian economist Otto Soemarwoto, 1991).*

Theme: Protecting and restoring the environment

"Sustainability equals conservation plus stewardship plus restoration" *(Ecological architect Sim Van der Ryn, 1994).*

Theme: Oppose exponential growth

"Sustainability is the fundamental root metaphor that can oppose the notion of continued exponential material growth" *(*Ecotopia *author Ernest Callenbach, 1992).*

Theme: Composite approach

"Sustainable development seeks . . . to respond to five broad requirements: (1) integration of conservation and development, (2) satisfaction of basic human needs, (3) achievement of equity and social justice, (4) provision of social self-determination and cultural diversity, and (5) maintenance of ecological integrity" *(International Union for the Conservation of Nature, 1986).*

Analyzing the Three Es in an Urban Planning Debate

In this more complicated exercise, the objective is to re-enact a recent urban planning decision in your community and to analyze how the Three Es of sustainable development (environment, equity, and economy) play out within this debate. Usually the best subject of this role-play is a controversial development proposal such as a large subdivision at the edge of the city. The students or group participants will play roles of developers, city council members, environmentalists, neighbors, and other constituencies. They should pay particular attention to how environmental, equity, and economic concerns are expressed during this debate and by whom. They should also analyze how the values, worldviews, and assumptions of each party affect their views of the debate.

Provide the participants with a few pages of background on the selected project (from press clippings, local government materials, or web sources) as well as a map or site plan. Make up as many roles as needed (this exercise may be done with groups as large as thirty people). Prepare a few sentences describing the background and views of each participant, and give a sheet with this description to the individual assigned this role. Often it works best to cast participants against type, e.g. to give those participants who consider themselves the strongest environmentalists the role of developers, and vice versa.

Roles might include the following:

- City Council or County Board of Supervisors (usually five to nine members; one is Mayor or Chair).
- County Staff (Planning or Community Development Director, Transportation Planner, Housing Planner).
- Development Team (Landowner, Principal of Development Firm, Architect, Landscape Architect or Environmental Consultant).
- Public Commenters (Representatives of environmental groups such as the Sierra Club, Audubon Society, and local "Friends of" groups; representatives of a local Citizens Transportation Alliance; local affordable housing advocates; Chamber of Commerce or local Business Development Association representatives; owners of nearby stores; the President of the regional Homebuilders Association; local union representatives; landowners in or near the project area; other nearby residents or landowners concerned about noise, views, traffic, loss of open space, effects on property values, and so on).

INSTRUCTIONS TO PARTICIPANTS

This exercise will re-create a development approval decision by a local city council or board of supervisors. In real life, this hearing would probably be on an appeal of an earlier decision by a planning commission or zoning board. In our simplified version of this process, the Council members or supervisors will hear from the project applicants, municipal staff, and the public in turn, and will then vote on whether to approve the project and if so what conditions to attach. You will each be assigned a role. Feel free to make up additional details that you think would fit with your character, and play the role as much as possible. A prize will be given for best actor/actress, as voted by your classmates.

Exercise format

(adapt this format to reflect local procedures):

- Ten minutes – Participants review roles and meet with allies.
- Ten minutes – The Mayor or Chair of the Board of Supervisors opens the hearing. Project backers present their proposal and describe its benefits. Council members or supervisors may ask brief clarifying questions.
- Five minutes – The city or county's planning staff presents its recommendations, usually to approve the project with mitigations to reduce its environmental impacts.
- Fifteen minutes – Public hearing. Any member of the public wishing to make a statement must fill out a speaker card (distribute index cards) and hand it to the Chair before the hearing begins. The Chair calls these cards in random order. Speakers identify themselves and their organizations, if any, and have two or three minutes to make a statement.
- Fifteen minutes – Council members or supervisors state their own positions when recognized by the Chair. Then any member may make a motion, in this case to approve the project, approve it with conditions, deny it, or postpone action pending some further process. If the motion is seconded by another supervisor, discussion on it takes place. Other supervisors can offer amendments to the motion, which must be accepted by the primary sponsor to be incorporated. After discussion on the motion has taken place the Chair calls for a vote. Motions must be approved by majority vote.

Afterwards, have a group discussion and/or prepare a short write-up analyzing how the Three Es were represented within this planning debate. Who speaks for each perspective? Which perspectives are underrepresented, if any? How might alternative decisions come about that better meet goals such as the Three Es in the long run? Which individuals or groups might develop such alternatives, and how?

Sustainability Indicators Exercise

This exercise explores how progress toward sustainable development might be measured in your particular community or region. First, review the factors making for good sustainability indicators, as discussed in the Maclaren reading or other sources.

Have small groups brainstorm six key indicators of urban sustainability for your area. These indicators should integrate a range of themes such as environment, economy, and equity, reflect meaningful trends, targets, or goals, reflect specific regional conditions, use readily available and updatable information, and be easily understandable to the general public. You may need to advise the groups on what types of information are available from which agencies or other sources.

The groups should take between thirty and forty minutes to come up with their lists, writing them on large poster paper. Then each team should present their recommendations to the rest of the participants in five or ten minutes. Follow-up discussion can focus on what types of public policies might affect these indicators, what agencies or levels of government have authority over each area, and how significant change might come about.

Personal Ecological Footprints/Household Sustainability Audit

Sustainability begins at home, so this exercise asks participants to evaluate their own lifestyle and household, and to come up with recommendations for changes. The exercise is probably best completed individually by participants or housemates and then presented in the form of a short paper or class presentation.

First, have participants complete the personal ecological footprint analysis online at www.rprogress.org. This will give them a figure for the number of acres that would be required to offset their personal resource consumption, and the number of earths that would be required if everyone in the world lived at their consumption level. Share and discuss these findings. Are there ways that these individual footprints might be reduced?

Next, participants should prepare an audit (a careful and systematic survey) of their own households to determine what improvements could be made to reduce resource usage and otherwise improve sustainability. Background readings on home energy and resource conservation measures may be helpful. Your local utility company's website may have extensive information on these, as well as websites of non-profit organizations such the American Council for an Energy Efficient Economy (www.aceee.org).

INSTRUCTIONS TO PARTICIPANTS

Systematically examine your home (both the building and lot) to determine how your use of energy and resources might be reduced, and sustainability otherwise improved. Review data such as utility bills (for electricity, gas, and water consumption) and if possible compare these with past bills or averages for your community to see how your consumption varies and what factors might be affecting it. Tabulate all uses of energy and water in your home, and list what conservation measures have been applied or might be applied to each. Examine also the landscaping of your lot, calculate the amount of paved surfaces, and investigate possibilities for changes in either. You may want to consider the following questions in your analysis:

- Where and how extensively have energy and water conservation measures been used?
- What other conservation measures are possible?
- Are there ways to improve the recycling of various materials?
- Is solar hot water or electricity a possibility? Where might such devices go?
- Is use of graywater a possibility? If so, how might that be done?
- How might landscaping be improved in terms of water consumption, use of native or drought-tolerant species, and/or creation of habitat?
- Is urban agriculture a possibility on your lot? How might space for that be created?
- How much of the site is covered by impermeable surfaces? Where does the runoff go? How might runoff be reduced, made cleaner, or made less severe after storms?
- Does this site contribute to an urban heat island effect? How might that be reduced?
- Might the planting of trees or changes to the building reduce air-conditioning needs in the summer?
- Does the existing development on this parcel best contribute to the city and region's overall needs? Might additional housing units be created on this site? If so, how? Is the site zoned in the best way to meet urban sustainability goals? Should the zoning be changed? (Zoning codes are often online these days at your city's website.)
- Should measures be adopted to ensure the long-term affordability of this housing?
- Is there anything else about your lifestyle that you would change to promote sustainability?

Write a short analysis of key steps to improve the sustainability of your home. Be comprehensive in your analysis, but focus on what you feel are the most promising strategies. Include any graphics or photos you feel are necessary.

Firsthand Analysis of Urban Environments

In line with Allan Jacobs' piece "Seeing Change" from *Looking at Cities*, a walking tour is an indispensible laboratory for teaching sustainability planning. It is an opportunity for students to practice firsthand analysis of urban environments, seeing how particular places have evolved in the past and how they might change in more sustainable directions in the future. A tour probably works best if it is in a relatively dense urban area where pedestrian travel is relatively easy and a number of interesting sites may be viewed within a three- to four-hour stretch. Bike tours to cover larger distances are also possible. It is a good idea to choose an area with some historical sites or contexts, some diversity of building types and land uses, some parks or remnants of natural ecosystems, and some new building or streetscape projects that the students can evaluate.

Prepare a map of the tour route, and if desired structure in some mini-exercises such as calculating residential densities in several sites (by pacing off lot or block sizes, determining numbers of units, and calculating the net density figure), brainstorming traffic-calming strategies at dangerous streets and intersections, and making observations of public squares or parks. Read local histories beforehand to gain historical information that can be passed on to students. Much of the value of a walking tour depends on how good the tour leader is at pointing out detail and history, and at raising questions for the participants.

INSTRUCTIONS TO PARTICIPANTS

On this walking tour we will practice observing urban environments firsthand, to see how they have evolved in the past and how they might become more sustainable in the future. Walk slowly and see how much information about the urban environment you can gather through careful observation. See if you can identify opportunities for infill development, ecological restoration, and improved street design. Note also how people use particular places, and what groups will feel welcome or not in different contexts. Try to imagine how particular blocks and streetscapes may look in the future, and think about what urban planning strategies might help these areas change.

Regional Vision Exercise

The purpose of this exercise is to have participants think about physical planning on a regional scale – not a level most citizens are used to considering. In particular, they should focus on how land use, transportation, environmental, and equity planning at this scale can help improve urban sustainability. Participants will consider how they would structure development in your region if they could start over from scratch. This exercise is not as far-fetched as it may sound; many metropolitan regions these days are developing Smart Growth plans which seek to shape regional form in such ways, and new towns or large subdivisions not infrequently undertake physical planning on very large scales.

This exercise requires obtaining or developing a large-format base map of the region, preferably containing only natural landscape features (e.g., rivers, streams, shorelines, hills, mountains, wetlands). A simple but workable base map may be created by tracing these features onto vellum or mylar from any good existing regional map, then printing copies on to bond paper for the students to use (one per group). Alternatively, student teams can create their plans by placing tracing paper on top of any decent large-format regional map, ignoring existing roads and urbanization.

INSTRUCTIONS TO PARTICIPANTS

In this exercise your team will determine a sustainable development scenario for your region. Assume that you can design the region from scratch (i.e., there is no existing development). What would be the form of communities? What mixture of land uses and urban densities would you suggest? What types of transportation? What kind of street network? What sort of park and open space system? How would the physical development reflect economic and equity goals?

Be aware of the natural landscape – which parts of it you might recommend for development, and which you would not. You will be given a large base map to use for this exercise. Illustrate your concept on this map. Don't try to be extremely detailed; you probably won't be able to show every street, for example (you may want to show a sample neighborhood at a larger scale off to the side to give viewers an idea of what details you would incorporate). The overall concepts are what is important.

First, decide as a group on an overall theme and a short list of physical design strategies that will carry out this theme. Write these in large print off to the side of your map. Next, map basic land uses and transportation systems with standard city planning land use colors. Either colored pencils or markers are fine, although the former may look better. Show the following:

- Developed areas. (Use several shades of yellow and brown to show different residential densities, with brown being the densest. Use red for mixed-use development or centers of activity, blue or purple for heavy industry, and gray for large institutional land uses such as airports and universities.)
- Parkland and agricultural land. (Use two shades of green.)
- Transportation systems. (Show major transit routes and highways in black).

Label as much as you can on the map, including types of economic activity you want to promote in particular places. Include a key showing which color corresponds to which type of land use.

You should plan your vision to accommodate the current population of the region. This means that you will have to make some trade-offs with density. If you plan a low-density environment with single family homes and big yards, you will probably have to cover most of the map with development. If you are planning medium-density communities with a mixture of housing types (including townhouses and small apartment buildings), you will have to cover only about one-third as much land area. If you are planning a higher density environment (with some high-rise apartment buildings, many townhouses and three- to five-story apartment buildings, and a few single family homes), you will need to cover only about one-sixth of the area. But then be able to say how you are going to make this higher density environment livable for people who don't currently like density.

Make sure everyone on your team participates in each part of the project. Be prepared to describe your vision to the class in a few minutes.

Economic Development Exercise

After covering economic development topics in lectures and readings, prepare brief case studies of four different cities or towns in your area, including mention of historical, cultural, demographic, and economic circumstances. These towns should illustrate different scenarios such as:

- a wealthy, fast-growing new suburb with both jobs and housing
- a declining central city area with falling tax base
- a stable, built-out inner-ring suburb with concerns about future stagnation
- a far-flung, working-class bedroom community which has allowed rapid housing development but has little employment base.

Next, divide the class into four teams and give each the following instructions, along with a map of the region.

INSTRUCTIONS TO PARTICIPANTS

Congratulations! Your consulting firm has been hired to produce an economic development strategy for a local city that has become interested in sustainable development. The background and challenges facing your particular city will be handed out separately. Based on this information and what you know of the region and sustainability planning, your firm must come up with a carefully thought out strategy for the mayor and city council.

This strategy should include:

- An overall theme that will be catchy and particularly appropriate for the unique context of this city.
- A recommendation on particular types of business or industry the city should focus on (the history, culture, and location of the city should come into play here, as well as your judgment about what might be the most dependable and ecologically appropriate sources of jobs in the future).
- Particular incentives, programs, or regulatory changes by which the city might support these.
- Land use, infrastructure, and urban design changes that can complement these forms of economic development.

You will have thirty minutes to develop your recommendations. Then your team will make a brief presentation to the city council and planning staff (the rest of the class). Short, punchy concepts will be easy for the city council staff to understand and will prevent them from going to sleep. Be as visionary as you like, but back up your recommendations with pragmatic suggestions for how these may be implemented. You will also need to keep your city's fiscal situation in mind. A particularly important question is: How can this community improve its fiscal, social, and ecological health in the long run without seeking continual growth in its land area?

Some potential economic development strategies

- *Business recruitment*: City tax breaks and subsidized infrastructure or land to lure businesses from other cities (a traditional strategy).
- *Business retention*: Support for existing businesses in the form of loans, technical assistance, and efforts to help them find larger spaces within the city.
- *Microenterprise development*: A special focus on small start-up companies to help them get going, often through loans, technical assistance, and subsidized office space and services within a business incubator (a building with flexible office space and shared facilities).
- *Eco-business development*: A focus on environmentally protective and restorative types of businesses (recycled products, pollution control and clean-up, alternative energy, alternative building materials).
- *Eco-industrial parks*: A strategy to link industrial and manufacturing businesses so that they can use one another's waste products as inputs.
- *Sustainable agriculture*: If your city or town still has farmland, a strategy to preserve this while encouraging ecologically appropriate production for local markets.
- *Local self-sufficiency*: A strategy emphasizing locally owned businesses using local labor and providing products and services for local markets. This might imply policies to keep out chain stores and big-box retailers.
- *Co-operative businesses*: Support for worker- or consumer-owned businesses.
- *Conservation-based development*: Efforts in rural areas to support sustainable resource production and agriculture.
- *Tourism*: A leading economic development strategy in many places, taking advantage of environmental amenities and cultural heritage as economic assets (implies support for historic preservation).

▪ *Place-oriented strategies*: Focus on creating particular centers or hubs of economic activity that will then generate further economic growth, for example, through efforts to create twenty-four-hour downtowns that attract entertainment-related businesses.

Creek Mapping Exercise

This exercise is appropriate for participants with some background in environmental issues, or for classes that include background material on creek restoration, biodiversity, landscape ecology, or related topics.

Divide the group into small teams. Assign each team one creek or other watershed feature within your community. Provide each team with the best possible map of this feature, preferably one showing the original route of the waterway in cases where it is now culverted, and/or locations of buried culverts. Such maps are often available from local environmental organizations or city public works departments.

INSTRUCTIONS TO PARTICIPANTS

1 Study your creek's route on the map provided, and if desired enlarge portions of the map to assist in field observation and note-taking. You may also find additional information about this creek or local creek restoration activities on various local websites.
2 Follow the current or original route of the creek as best you can (preferably by foot or bicycle) and make systematic notes of present conditions and restoration opportunities. Where the creek is underground, follow the approximate route of the culvert or original channel, and identify locations where the creek might potentially be unearthed. Where the creek is above ground, follow the channel as best you can, observing the condition of the stream (flow, erosion, pollution, trash, form of the channel, vegetation, wildlife, and human use). Map these conditions as accurately as you can, and take photos if possible. Don't try to get through brushy parts of the creek corridor or trespass on to private property without permission. Often creek channels can be observed where they cross streets.
3 Write up a concise set of recommendations for restoring the creek, including near-term activities (trash clean-up, removal of non-native vegetation, revegetation), and longer term possibilities (channel restoration, unearthing from culverts). Add photos of existing conditions if possible. Present these to the larger group, and share copies with local environmental groups and city planners.

Potential opportunities to look for:

▪ Clean-up of trash or debris.
▪ Removal of invasive or non-native species and replanting with native riparian species.
▪ Restocking with native fish and creation of good habitat for these.
▪ Replacement of concrete channel stabilization devices with more naturalistic measures.
▪ Unearthing of culverted creeks.
▪ Creation of parks, greenways, ecological educational facilities, or other community amenities in conjunction with creek restoration.

Neighborhood Planning Exercise

In this role-play, teams of students meet in a charrette format to develop sustainable development proposals for a particular neighborhood. Ideally this workshop can replicate a public meeting held during a recent local Specific Plan, Area Plan, or Neighborhood Plan process, using the same background materials acquired from the city. Or it can focus on another neighborhood likely to be well-known to participants. To start with, the instructor should prepare a packet of background information and maps, potentially including a printout of the relevant zoning for the area, traffic statistics, and demographic data from the census bureau. A walking tour of the neighborhood also provides a good starting point.

Divide students into groups of five or six and prepare this number of roles, with a short description of each character. Roles might include a city planner (facilitating the workshop), several residents of differing backgrounds, the owner of a local store or automobile dealership, a representative of an environmental group, and a local developer or community development corporation staff person.

INSTRUCTIONS TO PARTICIPANTS

During the next thirty minutes, city staff will lead a public workshop in which members of the public will sit down to develop a potential plan of action for this neighborhood. Informational material and maps of the area are provided. Each group should come up with a list of proposed strategies for sustainable development of this area, and show these graphically on the maps or flip-chart paper provided. You will be assigned a role for this exercise. Feel free to improvise based on what you think this character might be concerned about.

Issues the city suggests groups might focus on include the following:

- Ways to promote reuse of vacant or underutilized sites.
- The lack of a pedestrian-friendly street environment.
- Heavy traffic and dangerous street crossings.
- Attracting new locally owned and local serving businesses to the area.
- Restoration of deteriorated housing in nearby blocks.
- The city's general need for more housing, particularly affordable housing.
- Creating more pocket parks and attractive public spaces, including some sort of focal point or community center.
- Ecological restoration of parks, creeks, median strips, and other public green space.

Possible strategies you might consider include the following:

- Changing zoning, for example, to require taller or higher density development, to reduce setbacks from the street, or to require mixed-use buildings (housing above store fronts).
- Trying to establish a new neighborhood center by zoning so as to create a dense cluster of buildings with public amenities and services, and/or by having the city acquire land and carry out improvements.

- Providing grants or loans to affordable housing providers if they will build on sites in the area.
- Purchasing or acquiring through eminent domain certain sites that would be used for public purposes such as parks, civic facilities, or affordable housing.
- Requiring owners to fix up properties with building or housing code violations, and/or providing loan funds for them to do this.
- Prohibiting certain types of businesses from locating in the area through zoning changes (e.g., drive-through fast-food restaurants, liquor stores, or automobile dealerships).
- Daylighting parts of creeks which run through this area in culverts, or restoring native habitat to city parks.
- Adding street trees and widening sidewalks to improve the pedestrian environment.
- Adding stop signs or lights to allow pedestrians to cross (you will need to balance this with motorists' desire not to stop too often!).
- Adding or removing other types of traffic calming anywhere in the neighborhood (the fire and police departments will still need reasonably fast access everywhere).
- Adding attractive lighting, planters, banners, or public art to improve the streetscape.
- Establishing urban design guidelines or design review standards for new buildings.
- Designating the area a redevelopment district to be able to issue bonds (to be repaid by the projected increase in tax revenues) to raise money for improvements.
- Creating ongoing public–private partnerships or organizations to improve the area.
- Subsidizing new initiatives such as a car-sharing co-op.

The city does not have unlimited resources, however, and your proposals should take this into account. If possible, propose funding mechanisms through which your recommendations may be implemented.

An Ecological Site Plan

This exercise allows participants to investigate sustainable development possibilities for a particular site. Choose a vacant lot, parking lot, or other redevelopable site relatively close by, in a location where participants can visit it easily for firsthand observation. Obtain a base map for this parcel from the city or other sources and make copies for each team. The exercise is probably best done in small groups of two or three.

INSTRUCTIONS TO PARTICIPANTS

Your consulting firm has been hired by the property owner to come up with sustainable development ideas for this site. This owner is interested in any development scenarios that would generate an economic return and enhance local sustainability. Key planning challenges include:

- Appropriate intensity and mix of uses for this site. (Check the city's zoning for the parcel, usually available online through the municipal website. You can, however, recommend that the developer seek a variance from local zoning requirements if appropriate.)
- Ways that development can help create more pedestrian-friendly streets.
- Orientation and design of buildings so as to reduce energy and resource consumption.

- Appropriate landscaping as well as preservation or restoration of any natural landscape features on the site.
- Location of walkways, open spaces, plazas, or mini-parks.
- Ways that people will enter the site, pass through it, park, and otherwise circulate.
- Meeting overall city needs, especially for housing and reduced automobile use.

Procedure

1 Decide on a "program": a set of uses that you think is most appropriate for this site.
2 Decide on a set of sustainable design principles to help implement your program, related to the general size and mass of buildings, their relation to the natural landscape and climate, their relation to streets and the surrounding context, and their use of materials and construction techniques.
3 Using tracing paper or copies of the base map, sketch out several different possible building layouts using these strategies.
4 Decide on one and depict it on the base map provided. Don't worry about architectural detail. Simply show building outlines (indicate the height and uses of each structure), open spaces, walkways, landscaping, parking (if any), and street design features. Use several different colors of pencil or marker. Add arrows or other symbols to show how sunlight hits the site, how people will pass through the spaces, key entrances or gathering places, flows of water and drainage, and other important characteristics. Label your site plan clearly.
5 Be prepared to present your proposal in a space of five to ten minutes.

International Development Exercise

Although perhaps the most difficult exercise to do, an international case study can help students explore how sustainability principles might apply on a large scale to a developing country. While in the real world sustainability policies should be grounded in a detailed understanding of a given culture and place, it is possible for students to explore many issues in two to three class periods with background reading beforehand.

Choose a country about which you have firsthand knowledge, or a place with an interesting and timely range of development problems, such as Afghanistan, Iraq, Colombia, or Thailand. Prepare a background reading packet for the students, and if possible show a video or slides to help them gain a feeling for the country (Bullfrog Films at www.bullfrogfilms.com has many excellent videos on developing world issues). An introductory lecture can then briefly outline the nation's history, geography, and development issues, as well as discussing some historical strategies of development taken by the World Bank, other international institutions, and national governments.

For the exercise itself, split the students up into teams and have each develop a sustainable development strategy for the nation. This can be done either as a role-play, with students assigned roles representing a cross-section of major national constituencies, or with students simply acting as foreign advisors or consultants. The teams can meet outside of class, or they can be given part of several class periods to caucus and develop their recommendations. During the final class period they will then make presentations, outlining key recommendations and displaying their scenario on large sheets of paper. These recommendations might include an overall theme or set of long-term goals for the country, a policy paradigm that can best achieve those goals (see below), a handful of specific strategies or technologies that should be promoted, and a half-dozen key indicators to assess progress toward sustainability.

If desired, teams may be assigned four scenarios:

1 A relatively stable political situation, substantial international assistance.
2 A relatively stable political situation, little international assistance.
3 Unrest and instability, substantial international assistance.
4 Unrest and instability, little international assistance.

INSTRUCTIONS TO PARTICIPANTS (ONE VERSION)

Your team has been appointed as a special advisor to the national government on sustainable development. Your job is to recommend an appropriate development strategy for the country, given the nation's context and the resources and political situation determined in your scenario. You should focus primarily on government policies, programs, and investment strategies that can meet long-term economic, environmental, and equity needs. Key questions might include the following:

- How should the government assist people in obtaining long-term shelter?
- How can the nation combat environmental problems such as soil degradation, overgrazing, deforestation, and desertification?
- How should the country generate electric power?
- What industries or types of agriculture should the country concentrate on?
- How can the nation best provide jobs for those currently unemployed?
- What strategies will ensure that the country's population can feed itself?
- Should the country aggressively develop its gas, oil, coal, or mineral reserves?
- How might the country promote equity, in particular the rights of minorities and women?
- How can the nation effectively make use of returning refugees and émigrés from abroad?
- Should the country aim for a strategy of self-sufficiency in basic products, an export-oriented economy tied in closely with the global economic and financial system, or some other system?
- Should the country take extensive loans from international institutions such as the World Bank in order to develop rapidly, even though it will then have to repay the loans with interest and may be subject to pressure to develop natural resources, allow unrestricted operations by multinational corporations, and reduce government spending and social programs?

Each group should appoint a facilitator to co-ordinate the discussion, a note-taker to write down strategies on large sheets of paper, and several spokespersons to make its final presentation (or you can all do this together).

Your group should decide on the following:

- An overall theme or set of long-term goals for the country.
- A policy paradigm that can best achieve those goals (see below).
- A handful of specific strategies or technologies that are most appropriate given the context.
- Priorities for government investment.
- A handful of key indicators to assess progress toward sustainability.

If the country in question is basically going through nation-building from scratch, you should decide on a broad policy paradigm that makes most sense for the government. Would you recommend a free-market approach with minimal government intervention in the economy and privatized public services? A social democratic approach with a strong government role in directly building infrastructure, providing services, and regulating the economy? A decentralized model in which the government seeks to empower local or regional communities and the non-profit sector to do the work? Some other paradigm?

Outline your recommendations on a large sheet of poster paper and be prepared to present them to the class in ten minutes or so.

FURTHER READING

Adams, W.M. (1990) *Green Development: Environment and Sustainability in the Third World*, New York: Routledge.

Audirac, Ivonne (1997) *Rural Sustainable Development in America*, New York: John Wiley & Sons.

Beatley, Timothy (1994) *Habitat Conservation Planning*, Austin: University of Texas Press.

Beatley, Timothy (2000) *Green Urbanism: Learning from European Cities*, Washington, DC: Island Press.

Beatley, Timothy and Kristy Manning (1997) *The Ecology of Place: Planning for Environment, Economy, and Community*, Washington, DC: Island Press.

Benfield, F. Kaid, Matthew D. Raimi, and Donald D.T. Chen (1999) *Once There Were Greenfields*, Washington, DC: Natural Resources Defense Council.

Benfield, F. Kaid, Jutka Terris, and Nancy Vorsanger (2001) *Solving Sprawl: Models of Smart Growth in Communities Across America*, New York: Natural Resources Defense Council.

Blowers, Andrew (ed.) (1993) *Planning for a Sustainable Environment: A Report by the Town and County Planning Association*, London: Earthscan.

Braidotti, Rosi *et al.* (1994) *Women, the Environment and Sustainable Development: Towards a Theoretical Synthesis*, London: Zed Books.

Brown, Lester R. (1981) *Building a Sustainable Society*, New York: Norton.

Brown, Lester R., Christopher Flavin, and Sandra Postel (1991) *Saving the Planet: How to Shape an Environmentally Sustainable Global Economy*, New York: Norton.

Bullard, Robert D. (1990) *Dumping in Dixie: Race, Class, and Environmental Quality*, Boulder, CO: Westview Press.

Bullard, Robert D., Glenn S. Johnson, and Angel O. Torres (2000) *Sprawl City: Race, Politics, and Planning in Atlanta*, Washington, DC: Island Press.

Callenbach, Ernest (1975) *Ecotopia*, New York: Bantam Books.

Calthorpe, Peter (1993) *The Next American Metropolis: Ecology, Community and the American Dream*, New York: Princeton Architectural Press.

Calthorpe, Peter and William Fulton (2001) *The Regional City: Planning for the End of Sprawl*, Washington, DC: Island Press.

Cerdà, Ildefons (1999 [1863]) *The Five Bases of the General Theory of Urbanization*. Ed by Arturo Spriay Puig. Trans by Bernard Miller and Mary Fonsi Fleming, Barcelona: Electra.

Cervero, Robert (1998) *The Transit Metropolis: A Global Inquiry*, Washington, DC: Island Press.

Condon, Patrick and Stacy Moriarty (eds) (1999) *Second Nature: Adapting LA's Landscape for Sustainable Living*, Los Angeles, CA: Treepeople.

Corbett, Judy and Michael Corbett (2000) *Designing Sustainable Communities: Learning from Village Homes*, Washington, DC: Island Press.

Costanza, Robert (ed.) (1991) *Ecological Economics: The Science and Management of Sustainability*, New York: Columbia University Press.

Daly, Herman and John B. Cobb, Jr. (1989) *For the Common Good: Redirecting the Economy Toward Community, the Environment, and a Sustainable Future*, Boston, MA: Beacon Press.

Downs, Anthony (1994) *New Visions for Metropolitan America*, Washington, DC: Brookings Institution.

Duany, Andres, Elizabeth Plater-Zyberk, and Jeff Speck (2000) *Suburban Nation: The Rise of Sprawl and the Decline of the American Dream*, New York: North Point.

Elkin, Tim and Duncan McLaren with Mayer Hillman (1991) *Reviving the City: Towards Sustainable Urban Development*, London: Friends of the Earth.

Engwicht, David (1993) *Reclaiming Our Cities & Towns: Better Living with Less Traffic*, Philadelphia, PA: New Society Publishers.

Evans, Peter (ed.) (2002) *Livable Cities?: Urban Struggles for Livelihood and Sustainability*, Berkeley: University of California Press.

Furman, Richard T.T. and Michael Godran (1986) *Landscape Ecology*, New York: John Wiley & Sons.

Generalitat de Catalunya (2002) *Cerdà: Urbs i Territor: A Catalogue of the Exhibit Mostra Cerdà*; Barcelona.

Girardet, Herbert (1999) *Creating Sustainable Cities*, Devon, UK: Green Books.

Global Cities Project (1991) *Building Sustainable Communities: An Environmental Guide for Local Government*, San Francisco, CA: The Center for the Study of Law and Politics.

Goldsmith, Edward (1993) *The Way: An Ecological World View*, Boston, MA: Shambhala.

Greenbelt Alliance (1989) *Reviving the Sustainable Metropolis: Guiding Bay Area Conservation and Development into the 21st Century*, San Francisco, CA.

Hamm, Bernd and Pandurang K. Muttagi (eds) (1998) *Sustainable Development and the Future of Cities*, London: Centre for European Studies.

Hardoy, Jorge E., Diana Mitlin, and David Satterthwaite (1992) *Environmental Problems in Third World Cities*, London: Earthscan.

Harris, Jonathan M., Timothy A. Wise, Kevin P. Gallagher, and Neva R. Goodwin (2001) *A Survey of Sustainable Development: Social and Economic Dimensions*, Washington, DC: Island Press.

Hart, Maureen (1995) *Guide to Sustainable Community Indicators*, Ipswich, MA: QLF/Atlantic Center for the Environment.

Hawken, Paul (1993) *The Ecology of Commerce: A Declaration of Sustainability*, New York: HarperCollins.

Holmberg, Johan (ed.) (1992) *Making Development Sustainable: Redefining Institutions, Policy, and Economics*, Washington, DC: Island Press.

Jacobs, Allan (1985) *Looking at Cities*, Cambridge, MA: Harvard University Press.

Jacobs, Allan (1993) *Great Streets*, Cambridge, MA: MIT Press.

Jacobs, Jane (1961) *The Death and Life of Great American Cities*, New York: Random House.

Jacobs, Jane (1969) *The Economy of Cities*, New York: Random House.

Kelbaugh, Douglas (1997) *Common Place: Toward Neighborhood and Regional Design*, Seattle: University of Washington Press.

Kunstler, James Howard (1993) *The Geography of Nowhere: The Rise and Decline of America's Man-Made Landscape*, New York: Simon & Schuster.

Layard, Antonia, Simin Davoidi, and Susan Batty (eds) (2001) *Planning for a Sustainable Future*, New York: Spon.

Lyle, John Tillman (1994) *Regenerative Design for Sustainable Development*, New York: John Wiley & Sons.

Lyle, John Tillman (1999) *Design for Human Ecosystems: Landscape, Land Use, and Natural Resources*, Washington, DC: Island Press.

McHarg, Ian L. (1969) *Design With Nature*, Garden City, NY: The Natural History Press.

McHarg, Ian L. and Frederick Steiner (1998) *To Heal the Earth: Selected Writings of Ian L. McHarg*, Washington, DC: Island Press.

Mazmanian, Daniel A. and Michael E. Kraft (eds) (1999) *Toward Sustainable Communities: Transition and Transformations in Environmental Policy*, Cambridge, MA: MIT Press.

Meadows, Donella, Dennis L. Meadows, and Jörgen Randers (1992) *Beyond the Limits: Confronting Global Collapse, Envisioning a Sustainable Future*, Post Mills, VT: Chelsea Green.

Meadows, Donella, Dennis L. Meadows, Jörgen Randers, and William W. Behrens III (1972) *The Limits to Growth*, New York: Universe Books.

Merchant, Carolyn (1992) *Radical Ecology*, New York: Routledge.

Mitlin, Diana (1992) Sustainable Development: A Guide to the Literature. *Environment and Urbanization* 4(1):111–24.

Newman, Peter and Jeffrey Kenworthy (1989) *Cities and Automobile Dependence: An International Sourcebook*, Aldershot: Gower.

Newman, Peter and Jeffrey Kenworthy (1999) *Sustainability and Cities: Overcoming Automobile Dependence*, Washington, DC: Island Press.

Norgaard, Richard (1994) *Development Betrayed: The End of Progress and a Coevolutionary Revisioning of the Future*, New York: Routledge.

O'Connor, Martin (ed.) (1994) *Is Capitalism Sustainable? Political Economy and the Politics of Ecology*, New York: The Guilford Press.

Pearce, David and Edward B. Barbier (2000) *Blueprint for a Sustainable Economy*, London: Earthscan.

Pearce, David and Jeremy J. Warford (1993) *World Without End: Economics, Environment and Sustainable Development*, New York: Oxford University Press.

Pearce, David, Edward Barbier and Anil Markandya (1989) *Blueprint for a Green Economy*, London: Earthscan.

Pearce, David, Edward Barbier, and Anil Markandya (1990) *Sustainable Development: Economics and Environment in the Third World*, London: Edward Elgar.

Peck, Sheila (1998) *Planning for Biodiversity: Issues and Examples*, Washington, DC: Island Press.

Pratt, Rutherford H., Rowan A. Rountree and Pamela C. Muick (1994) *The Ecological City: Preserving and Restoring Urban Biodiversity*, Amherst: University of Massachusetts Press.

Redclift, Michael (1987) *Sustainable Development: Exploring the Contradictions*, London: Methuen.

Register, Richard (1987) *Ecocity Berkeley: Building Cities for a Healthy Future*, Berkeley, CA: North Atlantic Books.

Register, Richard (2002) *Ecocities: Building Cities in Balance with Nature*, Berkeley, CA: Berkeley Hills Books.

Riley, Ann L. (1998) *Restoring Streams in Cities*, Washington, DC: Island Press.

Roseland, Mark (1998) *Toward Sustainable Communities: Resources for Citizens and their Governments*, Stony Creek, CT: New Society Publishers.

Satterthwaite, David (ed.) (1999) *The Earthscan Reader in Sustainable Cities*, London: Earthscan.

Spirn, Ann Whiston (1984) *The Granite Garden: Urban Nature and Human Design*, New York: Basic Books.

Stivers, Robert L. (1976) *The Sustainable Society: Ethics and Economic Growth*, Philadelphia, PA: Westminster Press.

Stren, Richard, Rodney White, and B.R. Whitney (eds) (1992) *Sustainable Cities: Urbanization and the Environment in International Perspective*, Boulder, CO: Westview Press.

Thompson, J. William and Kim Sorvig (2000) *Sustainable Landscape Construction: A Guide to Green Building Outdoors*, Washington, DC: Island Press.

Todd, Nancy Jack and John Todd (1993) *From Eco-Cities to Living Machines*, Berkeley, CA: North Atlantic Books.

Urban Ecology, Inc. (1996) *Blueprint for a Sustainable Bay Area*, Oakland.

Van der Ryn, Sim and Peter Calthorpe (eds) (1984) *Sustainable Communities*, San Francisco, CA: Sierra Club Books.

Van der Ryn, Sim and Stuart Cowan (1996) *Ecological Design*, Washington, DC: Island Press.

World Commission on Environment and Development (1987) *Our Common Future*, New York: Oxford University Press.

ILLUSTRATION CREDITS

Every effort has been made to contact copyright holders for their permission to reprint plates and figures in this book. The publishers would be grateful to hear from any copyright holder who is not here acknowledged and will undertake to rectify any errors or omissions in future editions of this book. The following is copyright information for the figures and plates that appear in this book.

PART 2: DIMENSIONS OF URBAN SUSTAINABILITY

NEWMAN and KENWORTHY. Figure 1. Energy use per capita in private passenger travel versus urban density in global cities, 1990. Peter Newman and Jeffrey Kenworthy, *Sustainability and Cities: Overcoming Automobile Dependence* (Washington, DC: Island Press, 1999). Reprinted with the permission of Island Press.

HAYDEN. Plate 2. Inhospitable environments. *Redesigning the American Dream* (New York: Norton, 1984). Reprinted with the permission of Dolores Hayden.

EISENBERG and YOST. Figure 1. Ms. Truly Green's home. In "Sustainability and Building Codes," *Environmental Building News*, 10(9), September 2001. Reprinted with the permission of Bruce Coldham.

PART 4: SUSTAINABLE URBAN DEVELOPMENT INTERNATIONALLY

RABINOVITCH and LEITMAN. Figure 1. Location Map of Curitiba. *Scientific American*. March 1996. Reprinted by permission of Karl Gude.

RABINOVITCH and LEITMAN. Figure 2. Curitiba Transit System. *Scientific American*. March 1996. Reprinted by permission of Karl Gude.

PART 6: CASE STUDIES OF URBAN SUSTANABILITY

Plate CS2. Sky garden in Commerzbank, Frankfurt, Germany. Photo by Nigel Young. Permission from Norman Foster and Partners, London.

Plate CS3. Menara Mesiniaga, a bio-climatic skyscraper designed by architect Ken Yeang. Photo by K.L. Ng. General permission given by ArchNet, online slide library.

Plate CS4. Menara Mesiniaga, garden terraces. Photo by K.L. Ng. General permission given by ArchNet, online slide library.

Plate CS5. Artist's rendering of Christie Walk, Adelaide, Australia. Permission to reproduce given by Paul Downton, Urban Ecology Australia.

Plate CS6. Photo of Christie Walk, Adelaide, Australia. Photo by Paul Downton. Permission for use granted by Paul Downton, Urban Ecology Australia.

Plate CS19. Photo of traffic calming brunch, LA Eco-Village. Photo by Jesse Moorman. Permission for use given by Jesse Moorman, LA Eco-Village.

Plate CS20. Photo of tree planting, LA Eco-Village. Photo by Jesse Moorman. Permission for use given by Jesse Moorman, LA Eco-Village.

Plate CS21. Photo of solar house, Civano Community, Tucson, AZ. Permission for use given by GEO Advertising and Marketing.

Plate CS22. Photo of Civano streetscape, Tucson, AZ. Permission for use given by GEO Advertising and Marketing.

Plate CS23. Cleveland EcoVillage Conceptual Master Plan. Permission to reproduce given by David Beach, President, EcoCity Cleveland.

Plate CS24. Photo of Cleveland EcoVillage, green townhouses. Permission for use given by David Beach, President, EcoCity Cleveland.

Plate CS26. Map of Vancouver, BC, region, showing protected Green Zone. Permission to reproduce given by the Greater Vancouver Regional District (GVRD).

Plate CS27. Bogotá TransMilenio, transit system. Photo by TransMilenio S.A. Permission for use given by Enrique Peñalosa.

Plate CS28. Bogotá Ciclovia. Photo by Instituto Distrital de Recreación y Deporte de Bogotá. Permission for use granted by Enrique Peñalosa.

Plate CS29. Paolo Lugari sitting on one of Gaviotas' famous water pumps, Gaviotas, Colombia. Photo by ZERI Foundation. Permission for use given by Gunter Pauli, ZERI Foundation.

Plate CS30. Self-sufficient hospital, Gaviotas, Colombia. Photo from ZERI Foundation. Permission for use given by Gunter Pauli, ZERI Foundation.

Plate CS31. Image of Auroville Galaxy Plan, Auroville, India. Photo by Pino Marchese. Permission for use given by Auroville Planning Office.

Plate CS32. Photo of Matrimandir (temple) and Peace Area, Auroville, India. Photo by Pino Marcheses. Permission for use granted by Auroville Planning Office.

Plate CS38. Austin Smart Growth Map. Permission to reproduce given by City of Austin, Department of Growth Management.

Plate CS40. Map of Portland Metro 2040. Permission to reproduce given by the Portland Metropolitan Services District.

Plate CS41. Photo of the McNeil Generating Plant, Burlington, Vermont. Permission for use of photo given by the Burlington Electric Department.

Plate CS42. Photo of young and old together at the McClure Intergenerational Center, Burlington, Vermont. Photo by Nicole Craft. Permission given by the McClure Intergenerational Center.

Plate CS43. The Ohlone Chynoweth apartments, San José, California. Photo by and permission given by Jeffrey Peters.

Plate CS44. Map of urban growth boundary, San José, CA. Permission to reproduce granted by the City of San José.

Plate CS45. Photo of Colorado Court housing project, Santa Monica, California. Permission for use given by City of Santa Monica.

Plate CS46. Photo of Santa Monica Urban Runoff Recycling Facility, Santa Monica, California. Permission for use given by City of Santa Monica.

COPYRIGHT INFORMATION

Every effort has been made to contact copyright holders for their permission to reprint selections in this book. The publishers would be grateful to hear from any copyright holder who is not here acknowledged and will undertake to rectify any errors or omissions in future editions of this book. The following is copyright information.

1 ORIGINS OF THE SUSTAINABILITY CONCEPT

HOWARD, Ebenezer, "The Three Magnets" from *Garden Cities of To-morrow* (London: Faber and Faber, 1898 [1945]). Public domain.

MUMFORD, Lewis, "Cities and the Crisis of Civilization" from "Introduction" in *The Culture of Cities*. Copyright ©1938 by Harcourt, Inc. and renewed 1966. Reprinted by permission of the publisher.

LEOPOLD, Aldo, "The Land Ethic" from *A Sand County Almanac: And Sketches Here and There*. Copyright © 1949, 1977 by Oxford University Press, Inc. Used by permission of Oxford University Press, Inc.

JACOBS, Jane, "Orthodox Planning and The North End" from *The Death and Life of Great American Cities*. Copyright © 1961 by Jane Jacobs. Used by permission of Random House, Inc.

McHARG, Ian L., "Plight and Prospect" from *Design With Nature*. Copyright © 1969 by Ian L. McHarg. Used by permission of Wiley-Liss, Inc., a subsidiary of John Wiley & Sons, Inc.

FRANK, Andre Gunder, "The Development of Underdevelopment" from *Capitalism and Underdevelopment in Latin America* (New York: Monthly Review Press, 1969). Copyright © 1967, 1969 by Andre Gunder Frank. Used by permission of Monthly Review Press.

MEADOWS, Donella H., MEADOWS, Dennis L., RANDERS, Jörgen and BEHRENS III, William W. "Perspectives, Problems, and Models" from the "Introduction" to *The Limits to Growth* (New York: Universe Books, 1972). Copyright © 1972 by Dennis L. Meadows. Used by permission of Dennis L. Meadows.

DALY, Herman E., "The Steady-State Economy" from *Toward a Steady-State Economy* (San Francisco, CA: W.H. Freeman, 1973). Copyright © 1973 by Herman E. Daly. Used by permission of Herman E. Daly.

"Towards Sustainable Development" from *Our Common Future* by the World Commission on Environment and Development. Copyright © 1987 by the World Commission on Environment and Development. Reprinted by permission of Oxford University Press.

"The Rio Declaration on Environment and Development" from *Agenda 21* (New York: United Nations, 1992). Public domain.

2 DIMENSIONS OF URBAN SUSTAINABILITY

Index

Aalborg Charter 249–50
Abercrombie, Patrick 11
Abers, Rebecca 237
accidents 100–1, 107
Accumulation on a World Scale (Amin) 38
Adams, W.M. 235
adaptive cities 89, 91, 92
adaptive transit 89, 91, 92
Adelaide EcoVillage 300, Plate CS5, Plate CS6
"Advocacy and Pluralism in Planning" (Davidoff) 225–6
advocacy planning 225–6
After the Planners (Goodman) 31
Agenda 21 1, 58–63, 277
Agyeman, Julian 143
air pollution 101, 145–6
Aldo Leopold (Lorbiecke) 20
Alexander, Christopher 73
Always Coming Home (Le Guin) 277
American Dream 75–8
American Institute of Architects 188
American Planning Association 201
Amersfoort 304, Plate CS15, Plate CS16
Amin, Samir 38
Amory Park, Tucson 306
anarchism 288
Appleyard, Donald 97
appropriate technology 125
aquatic sewage treatment 138–40
Arcata, California 2, 139–40
architecture 3, 181–7, 188–92
Association of Pedestrian and Bicycle Professionals 104
Auroville 310–11, Plate CS31, Plate CS32
Austin, Texas 313, Plate CS38
automobiles: Calthorpe 77; car sharing 252; costs 106, 107–8; Jacobs 31; reducing use 89; traffic calming 97, 98–103

Bainbridge, David 193
Baldwin Hills Village 11
Baran, Paul 38, 39
Barbier, Edward B. 3, 159–61
Beatley, Timothy 2, 3, 116–19, 249–58
Beauregard, Robert 31
Beddington Zero Energy Development (BedZED), London 303, Plate CS11, Plate CS12
Behrens, William W., III 1, 7, 9, 42–6
Bell, Simon 203
Bellamy, Edward 277
Best Foot Forward 211
Beyond Economics (Boulding) 47
Beyond the Limits (Meadows *et al.*) 43
bicycling 2, 104–10, 253–4
bio-climatic skyscrapers 299
biodiversity 2, 116–18, 183
biological reassimilation 138–40
biotechnic era 16
Blueprint 2 (Pearce, Barbier and Markandya) 159
Blueprint for a Green Economy (Pearce, Barbier and Markandya) 159
Blueprint for a Sustainable Economy (Pearce and Barbier) 159
Bogotá 309–10, Plate CS27, Plate CS28
Bombay 235
Boston 30, 31, 32–3
Boulding, Kenneth 8, 38, 47, 50
Boulevard Book, The (Jacobs, Macdonald and Rofé) 220
Boyer, Christine 31
Braidotti, Rosi 150
Braungart, Michael 181, 184
Brazil 39–41
British Columbia 204–5, 207, 309, Plate CS25, Plate CS26
Brown, Lester 162
Brundtland, Gro Harlem 53
Brundtland Commission 1, 9, 53–7, 277, 321

Bryant, Dorothy 277
Buchanan, Colin 98–9
building codes 3, 193–8
Bullard, Robert 2, 143–9
Burlington, Vermont 314–15, Plate CS41, Plate CS42
bus transport 93–4; Curitiba 241, 242, 243, 246
Bush, George W. 226
"Bicycling Renaisissance" (Pucher, Komanoff and Shimek)

Cadbury, George 11
Cadoszo, Fernando 38
Cairo 129
Callenbach, Ernest 4, 277, 282–7, 322
Callicott, J. Baird 20
Calthorpe, Peter 2, 11, 73–80
Camacho, David E. 143
Canada 53
Canadian Institute of Planners 201
capitalism, natural 3, 162–70
Capitalism and Underdevelopment in Latin America (Frank) 39–41
Caporaso, James 38
carrying capacity 321
cars *see* automobiles
Carson, Rachel 35, 42
Case Against the Global Economy, The (Mander and Goldsmith) 171
Castells, Manuel 31
Cerdà, Ildefons 11
Cervero, Robert 2, 89–96
Chambers, Nicky 211
Charles, Prince 74
Charter of the New Urbanism 74, 277
Chicago 312–13, Plate CS37
children 153, 241, 244
Chomsky, Noam 38–9
CIAM 74
Cities and Automobile Dependence (Newman and Kenworthy) 97

Cities and Natural Process (Hough) 114

Cities and the Wealth of Nations (Jacobs) 31, 171

City Form and Natural Processes (Hough) 114

City Summit 9, 58, 63–5

Civano, Tucson 305–6, Plate CS21, Plate CS22

climate 107, 189

Clinton, Bill 226

Closing Circle, The (Commoner) 35

Club of Rome 42, 45

Co-op America 171

Cobb, John B, Jr. 48

cognitive mapping 320

Collaborative Planning (Healey) 225

"Collective Action" (Hsiao and Liu)

Colombia 309–10, Plate CS27, Plate CS28, Plate CS29, Plate CS30

combined heat and power (CHP) stations 130–1

Commerzbank Headquarters, Frankfurt 299, Plate CS1, Plate CS2

Commoner, Barry 35, 42

Communication and the Evolution of Society (Habermas) 225

communicative action 225

Communitas (Goodman and Goodman) 30, 277

commuter rail 95–6

compact cities 250–1

composting 138

Condé Nast building, New York 300, Plate CS7, Plate CS8

Conference on Environment and Development 1, 9, 53, 58–63, 186

Congrès Internationaux d'Architecture moderne 74

Congress for the New Urbanism (CNU) 74, 75, 278

conservation 2, 8–9, 23–5, 120

Copenhagen 81, 84

Corbett, Judy 304–5, Plate CS17, Plate CS18

Corbett, Michael 304–5, Plate CS17, Plate CS18

Costanza, Robert 159

Cowan, Stuart 181, 188

Cradle to Cradle (McDonough and Braungart) 181

Creating Sustainable Cities (Girardet) 125–32

creek mapping 120; exercises 329

crime 101–2

Culture of Cities, The (Mumford) 15, 16–19

Culture of Terrorism, The (Chomsky) 38

Curitiba 3, 237–48

cycling 2, 104–10, 253–4

Daly, Herman E. 1, 8, 47–52, 159, 164

Davidoff, Paul 225–6

Davis, California 304–5, Plate CS17, Plate CS18

Death and Life of Great American Cities, The (Jacobs) 30, 31–4

dependency theory 38

Design for Human Ecosystems (Lyle) 120

Design with Nature (McHarg) 35–7

Deterring Democracy (Chomsky) 38–9

developing countries, exercises 332–4

Development as Freedom (Sen) 39

Dickens, Charles 7

Dijkstra, Lewis 104

discursive democracy 225

Downs, Anthony 89

Doxiotis, Constantine 73

Dreaming the Rational City (Boyer) 31

Dryzek, John 225

Duany, Andres 74, 75

Dubos, Rene 42

Dumping in Dixie (Bullard) 2, 143–9

Dwelling, Seeing, and Designing (Seamons) 220

Earth in Mind (Orr) 182

Earth Rising (Shabecoff) 20

Earth Summit 1, 9, 53, 58–63, 186

Earthscan Reader in Sustainable Cities, The (Satterthwaite) 236

EcoCity Cleveland and EcoVillage 306, Plate CS23, Plate CS24

Eco-Economy (Brown) 162

Ecological City, The (Platt, Rowntree and Muick) 120

ecological design 3, 181–7; building codes 193–8; principles 188–92

Ecological Design (Van der Ryn and Cowan) 181

Ecological Design Association 188

Ecological Design Principles 188

ecological economics 159

Ecological Economics 159

ecological footprint analysis 3, 201, 211–19; exercises 324–5

Ecological Literacy (Orr) 182

ecological restoration 102–3

ecological site plans 331–2

ecology 116, *see also* urban ecology

Ecology of Place, The (Beatley) 116, 249

Economic Commission on Latin America 38

economic development 2–3, 8; environmental economics 159–61; exercises 327–9; Frank 38–41; import replacement 171–8; Jacob 30–1; natural capitalism 162–70; and traffic calming 102

economic growth 53, 164–5

economy 143; Daly 47–52; exercises 322–4

Economy of Cities, The (Jacobs) 30–1

Ecotopia (Callenbach) 4, 277, 282–7

Ecotopia Emerging (Callenbach) 282

Ehrlich, Paul 42, 43

Eisenberg, David 3, 193–8

energy 2, 125; conservation 189; Girardet 129–32

Engels, Frederick 7

Engwicht, David 89, 97

environment 8, 12, 143; Brundtland Commission 53–7; conservation and preservation 8–9; exercises 322–4; *Limits to Growth, The* 42–6; McHarg 35–7; restoration 120–2; Rio Declaration 58–65

Environmental Building News 193–8

environmental design 81

environmental economics 159–61; natural capitalism 162–70

Environmental Economics (Field and Field) 159

Environmental Indicators (World Resources Institute) 203

Environmental Injustice, Political Struggles (Camacho) 143

environmental justice 2, 143–9

Environmental Problems in Third World Cities (Hardoy *et al.*) 235

equity 2, 8, 38, 143, 269–70; environmental justice 143–9; exercises 322–4; people of color 143–9; women 150–6

equity planning 226

Ethical Land Use (Beatley) 116

ethics 21; ecological conscience 23–4, *see also* land ethic

Europe 3, 249–58

European Housing Ecology Network 188

European Sustainable Cities and Towns Campaign 235

Evaluative Image of the City, The (Nazar) 220

Evans, Bob 143

Evans, Peter 236

Experience of Place, The (Hiss) 220

Fair Trade 171
Fair Trade Federation 171
Fair Trade Foundation 171
farming 174–7
Farrar, Dean 12
Farsta 11
feng shui 181
Field, Barry C. 159
Field, Martha K. 159
Fierce Green Fire, A (Shabecoff) 20
Flader, Susan L. 20
For the Common Good (Daly and Cobb) 48
Forcing the Spring (Gottlieb) 20
Forester, Jay 42
Forester, John 225
forestry 8
Forman, Richard 116
Forward-looking indicators 206–7
Foster, Norman 299, Plate CS1, Plate CS2, Plate CS35, Plate CS36
Fourier, Charles 11
Francis, Carolyn 81
Frank, Andre Gunder 1, 38–41
Frankfurt 299, Plate CS1, Plate CS2
Friedmann, John 226
From Eco-Cities to Living Machines (Todd and Todd) 181
fuel cells 132
Fulton, William 75

Gaia Atlas of Cities, The (Girardet) 125
garden cities 4, 7, 8, 11–14, 277, 278, 279–81
Garden Cities of To-morrow (Howard) 11–14, 277, 279–81
Gaviotas 310, Plate CS29, Plate CS30
Geddes, Patrick 7, 15
Gehl, Jan 2, 81–5
Genuine Progress Indicator 48
Germany 8, 11
Giddens, Anthony 225
Gilman, Charlotte Perkins 277
Girardet, Herbert 2, 125–32
Global 2000 Report 53
Global Rift (Stavrianos) 38
Going Local (Shuman) 171–8
Goldman, Emma 288
Goldsmith, Edward 171
Good City Form (Lynch) 15, 73, 220, 277–8
Goodman, Paul 30, 277, 288
Goodman, Percival 30, 277
Goodman, Robert 31
Gorst, Sir John 12, 14
Gottlieb, Robert 20
Granite Garden, The (Spirn) 2, 113–15, 277
Great Streets (Jacobs) 220

green architecture 3; McDonough 181–7; principles 188–92
Green Architecture (Vale and Vale) 181, 188–92
green building 3, 181; codes 193–8
Green Building Council 181, 188
Green Development (Adams) 235
Green Urbanism (Beatley) 249
Greenbelt, Maryland 11
Greendale, Wisconsin 11
Greenhills, Ohio 11
Greenwich Millennium Village, London 304, Plate CS13, Plate CS14
growth 47–9, 52, 322
growth management 74, 97
Gruen, Victor 73
Guide to Sustainability Community Indicators (Hart) 203

Habermas, Jürgen 225
Habitat Conservation Planning (Beatley) 2, 116–19
habitat conservation plans (HCPs) 2, 116, 118–19
Habitat II "City Summit" 9, 58, 63–5
Hamilton-Wentworth, Ontario 204
Hannover 303, Plate CS9, Plate CS10
Hannover Principles 181, 188, 277
Hardoy, Jorge E. 235
Hart, Maureen 203
Harvey, David 31
Hawken, Paul 3, 162–70
Hayden, Dolores 2, 150–6
Healey, Patsy 225
health 109
heavy rail systems 94–5
Henderson, Hazel 47
Herland (Gilman) 277
Hiss, Tony 220
holism 192
Holmberg, Johan 235
Hough, Michael 114
Housing as if People Mattered (Marcus and Sarkissian) 81, 150
Houten 11
Howard, Ebenezer 1, 4, 7, 8, 11–14, 15, 74, 277, 278, 279–81
Hsiao, Hsin-Huang Michael 3, 259–74
Human Development Index 203, 206, 209
hybrids 92

IBA Emscher Park 311–12, Plate CS33, Plate CS34
If Women Counted (Warring) 211
Image of the City, The (Lynch) 81, 220

import replacement 3, 31, 171–8
Index of Sustainable Economic Welfare 48
India 310–11, Plate CS31, Plate CS32
inequity *see* equity
Innes, Judith 225
Institute for Local Self-Reliance 171
integrating indicators 206
International City/County Management Association (ICMA) 201
International Council on Local Environmental Initiatives (ICLEI) 58, 235
International Institute for Sustainable Development 235
International Sustainability Indicators Network (ISIN) 203
International Union for the Conservation of Nature 322
Introduction to Ecological Economics, An (Costanza *et al.*) 159
Inventing Local Democracy (Abers) 237
Is Capitalism Sustainable? (O'Connor) 162
Istanbul Declaration on Human Settlements 63–5

Jacobs, Allan B. 3, 201, 220–4, 325
Jacobs, Jane 1, 11, 30–4, 35, 39, 69, 73, 81, 171
jitneys 92–3
Johnson, Glenn S. 143
Journal of the American Planning Association 3, 201, 203–10
Just Sustainabilities (Agyeman, Bullard and Evans) 143

Kentlands 74
Kenworthy, Jeffrey 2, 89, 97–103, 250
Kin of Ata Are Waiting For You, The (Bryant) 277
Komanoff, Charles 2, 104–10
Krier, Leon 74
Krier, Rob 74
Kronsberg Ecological District, Hannover 303, Plate CS9, Plate CS10
Kropotkin, Peter 11, 288
Krumholtz, Norman 226
Kuala Lampur 299, Plate CS3, Plate CS4
Kuhn, Thomas 43, 49

Lagos 235
Laguna West 74
Lambert, Jaques 39

land ethic 21–2, 24–5, 27–8
land pyramid 25–7
land use 69, 74, 97
landscape architecture 8
landscape ecology 116
Landscape Park, Duisburg North
    311–12, Plate CS33, Plate CS34
Lawrence, D.H. 7
Lawrence, Kansas 181, 185–6
LEED (Leadership for Energy and
    Environmental Design) 181
Le Guin, Ursula K. 4, 277, 288–91
Leibniz, Gottfried Wilhelm 52
Leitman, Josef 3, 237–48
Leopold, Aldo 1, 9, 20–9, 47
Lerner, Jaime 237
Lerner, Michael 3, 201, 225–31
Letchworth 11, 279
Lever, William 11
Lewis Mumford (Miller) 16
Lewis Mumford and the Ecological
    Region (Luccarelli) 16
Life Between Buildings (Gehl) 81–5
light rail transit (LRT) 94
Limits to Growth, The (Meadows et al.)
    1, 7, 9, 42–6
Litman, Todd 104
Liu, Hwa-Jen 3, 259–74
Livable Cities? (Evans) 236, 259–74
Livable Streets (Appleyard) 97
Living Planet Report (World Wildlife
    Federation) 211
Local Agenda 21 58
local exchange trading systems
    (LETS) 171
Local Government Commission 75
Local Politics of Global Sustainability,
    The (Prugh, Costanza and Daly)
    159
London 7, 312, Plate CS35, Plate
    CS36; BedZED 303, Plate CS11,
    Plate CS12; Greenwich
    Millennium Village 304, Plate
    CS13, Plate CS14; metabolism 2,
    126, 128, 129, 130
Looking at Cities (Jacobs) 3, 201,
    220–4, 325
Looking Backwards (Bellamy) 277
Lorbiecke, Marybeth 20
Los Angeles 74, 305, Plate CS19,
    Plate CS20
Lovins, Amory 125, 166
Luccarelli, Mark 16
Lyle, John Tillman 2, 120, 133–40,
    181, 188
Lynch, Kevin 15, 73, 81, 220, 277–8,
    320

McDonough, William 3, 181–7, 188,
    277

McHarg, Ian L. 1, 11, 35–7, 39, 42,
    69, 73, 113, 114
Maclaren, Virginia W. 3, 201,
    203–10
Macdonald, Elizabeth 220
magnets 11–14
Making Development Sustainable
    (Holmberg) 235
"Making Walking and Cycling Safer"
    (Pucher and Dijkstra) 104
Maldevelopment (Amin) 38
Malthus, Thomas 42
Mander, Jerry 171
Marcus, Clare Cooper 81, 150
Markandya, Anil 159
Marxism 38, 39
materials use 2; Girardet 125–9; Lyle
    133–40
Meadows, Dennis L. 1, 7, 9, 42–6
Meadows, Donella H. 1, 7, 9, 42–6,
    47
Melbourne 84
Menara Mesiniaga tower 299, Plate
    CS3, Plate CS4
Merchant, Carolyn 20
metros 94–5
Mexico City 235
Mill, John Stuart 8, 47, 49, 50, 51–2
Miller, Danold L. 16
minibuses 92–3
Modern Urban Landscape, The (Relph)
    220
Mollison, Bill 120
moral growth 52
Morris, David 171
Morse, Stephen 203
Moses, Robert 30
Mother Jones 162–70
Moule, Elizabeth 74
Muick, Pamela C. 120
Muir, John 9, 120
Mumford, Lewis 1, 7, 11, 15–19, 30,
    69, 74, 277
Myrdahl, Gunnar 38

National Roundtable on the
    Environment and Energy
    (Canada) 53
"Natural Capitalism" (Hawken) 3,
    162–70, 321
Natural House Book, The (Pearson)
    188
nature 8; Curitiba 238–40, 242;
    McHarg 35–7; and restoration
    122; Spirn 2, 113–15
Nature of Design, The (Orr) 182
Nazar, Jack 220
neighborhood planning 30–4;
    exercises 330–1
neotechnic era 16

Netherlands, The: ecological
    footprint 217, 218, 219; new
    towns 11
New Autonomous House, The (Vale
    and Vale) 181–2
New Natural House Book, The
    (Pearson) 188
New Organic Architecture (Pearson)
    188
New Urbanism 2, 30, 73–80, 91
New York City 30, 113, 300, Plate
    CS7, Plate CS8
Newman, Peter 2, 89, 97–103, 250
Next American Metropolis, The
    (Calthorpe) 73, 74–80
Nieuwland, Amersfoort 304, Plate
    CS15, Plate CS16
noise 101
Nolen, John 11
Norgaard, Richard 322
North America: bicycling 104–10,
    see also Canada; United States

observation 3, 201, 220–4; exercises
    325–6
O'Connor, Martin 162
open spaces 255
Oregon 206–7
Orr, David 182
Our Common Future (Brundtland
    Commission) 53, 54–7
Our Ecological Footprint
    (Wackernagel and Rees) 3,
    211–19
outdoor spaces 81–4
Owen, Robert 11

paleotechnic era 16
paratransit 92–3
Parker, Barry 15
Pattern Language, A (Alexander et al.)
    73
Pearce, David 3, 159–61, 321
Pearson, David 188
Peccei, Aurelio 42
Pedestrian and Bicycle Planning
    (Litman et al.) 104
pedestrians 101, 254
people-of-color 143–9
People Places (Marcus and Francis)
    81
permaculture 120
Permaculture (Mollison) 120
Persky, Joseph 173–4
phenomenology 220
Piercy, Marge 277
Pinchot, Gifford 8, 120
planning see urban planning
Planning in the Face of Power
    (Forester) 225

Planning in the Public Domain (Friedmann) 226
"Planning Theory's Emerging Paradigm" (Innes) 225
Plater-Zyberk, Elizabeth 74, 75
Platt, Rutherford H. 120
Political Economy of Growth, The (Baran) 38
politics of meaning 3, 201, 225–31
Politics of Meaning, The (Lerner) 225–31
Polyzoides, Stefanos 74
Population Bomb, The (Ehrlich) 42
Portland, Oregon 74, 313–14, Plate CS39, Plate CS40
Porto Alegre 237
Power, Thomas Michael 174
Practice of Local Government Planning, The (International City/County Management Association) 201
Prebisch, Raul 38
preservation 9, 120
Prince of Wales Institute 74
Proudhon, Pierre-Joseph 11, 288
Prugh, Tom 159
public space, and women 150–6
public transport 89, 90–6; Curitiba 241, 242, 243–4, 246; Europe 251–2; women 154
Pucher, John 2, 104–10
Pullman, George 11

Rabinovitch, Jonas 3, 237–48
Radburn, New Jersey 11
Radical Ecology (Merchant) 20
rail transport 94–6
Randers, Jörgen 1, 7, 9, 42–6
Ranney, David 173–4
rapid rail transit 94–5
Reclaiming Our Cities and Towns (Engwicht) 89, 97
recycling 128–9, 137–8, 190, 247
Redefining Progress 48, 211
Redesigning the American Dream (Hayden) 2, 150–6
Rees, William 3, 201, 211–19, 322
Regenerative Design for Sustainable Development (Lyle) 133–40, 181
Regenerative Design Principles 188
Regional City, The (Calthorpe and Fulton) 75
regional vision 326–7
Relph, Edward 220
renewable energy 125, 131–2, 255–6
resources 2, 125, 165–6; Girardet 125–9; Lyle 133–40; minimizing use 189–90; politics 167–9; productivity 166–7
respect 190–2
Reston, Virginia 11

restoration 2, 120–2
Restoring Streams in Cities (Riley) 120–2
Riley, Ann L. 2, 120–2
Rio Declaration on Environment and Development 58–63
Rio Summit 1, 9, 53, 58–63, 186
River of the Mother, The (Flader and Callicott) 20
Roberts, John 98
Rofé, Yodan 220
Rosebery, Lord 12
Rowntree, Rowan A. 120
Royal Town Planning Institute 201
Rudofsky, Bernard 97
Russell, Bertrand 288

Saarbrücken 131
Safe Cities (Wekerle and Whitzman) 150
San José, California 315–16, Plate CS43, Plate CS44
Sand County Almanac, A (Leopold) 20–9
Santa Monica, California 316, Plate CS45, Plate CS46
Sarkissian, Wendy 81, 150
Satterthwaite, David 236
Schumacher, E.F. 8, 47, 125
science 43
Scientific American 3, 237–48
Seamons, David 220
Seaside 74
Seattle 203, 204, 206, 208
Self-Reliant Cities (Morris) 171
Sen, Amartya 39
sewage 2, 126–7, 138–40
Shabecoff, Philip 20
Sharing Nature's Interest (Chambers, Simmons and Wackernagel) 211
Shimek, Paul 2, 104–10
Shuman, Michael 3, 31, 171–8
Silent Spring (Carson) 35
Simmons, Craig 211
Simon, Julian 43
Sitte, Camillo 15
sky gardens 299, Plate CS2
Small is Beautiful (Schumacher) 47, 125
smart growth 2, 74
Smart Growth Network 75
Social Life of Small Urban Spaces, The (Whyte) 81, 84
Soemarwoto, Otto 322
Soft Energy Paths (Lovins) 125
Soja, Edward 73
solar energy 131–2, 181, 183, 304, Plate CS15, Plate CS16
solid waste 127–9
Sorvig, Kim 120

space 73
Speck, Jeff 75
Spirn, Anne Whiston 2, 113–15, 277
sprawl 74
Sprawl City (Bullard, Johnson and Torres) 143
State of the World (Worldwatch Institute) 53
Stavrianos, L.S. 38
steady-state economy 47, 49–50; distribution 51–2; resources 50–1
Steen, Athena Swentwell 193
Steen, Bill 193
Stimpson, Catherine R. 150
Story of Utopias, The (Mumford) 277
Straw Bale House, The (Eisenberg et al.) 193
Street Reclaiming (Engwicht) 97
Streets for People (Rudofsky) 97
Stren, Richard 235
Strong-core cities 91–2
Structure of Scientific Revolutions, The (Kuhn) 43, 49
Stuck in Traffic (Downs) 89
Suburban Nation (Duany, Plater-Zyberk and Speck) 75
suburban rail 95–6
suburbanization 8, 35, 74, 90
Surface Transportation Policy Project 89
Sustainability and Cities (Newman and Kenworthy) 89, 97–103
sustainability indicators 3, 201, 203–10; exercises 324
Sustainability Indicators (Bell and Morse) 203
Sustainable Cities (Stren et al.) 235
Sustainable Communities (Calthorpe and Van der Ryn) 74
sustainable development 7; definitions 53, 56, 321–2
Sustainable Homes 188
Sustainable Landscape Construction (Thompson and Sorvig) 120
Sustainable Seattle 203, 204, 206, 208
Sweden 11
Sweezy, Paul 38, 39
Swiss Reinsurance building Plate CS36
systems dynamics 42

Taipei 3, 259–74
Thant, U. 43, 44
Third World 235
Thompson, J. William 120
Thoreau, Henry David 7
Todd, John 181, 183
Todd, Nancy Jack 181
Torres, Angel O. 143

Touch-this-earth-lightly 191–2
Toward a Steady-State Economy (Daly) 47, 48–52
Towards Sustainable Development (Bell and Morse) 203
Toynbee, Arnold 39
trade 3, 171
traditional neighbourhood design 74
traffic calming 2, 97, 98–103
trams 94
Transit Metropolis, The (Cervero) 89, 90–6
transit metropolises 89, 91–2
transit-oriented development (TOD) 75, 76, 78–80, 91
transportation 2, 69; bicycling 104–10; Cervero 89–96; Curitiba 241, 242, 243–4, 246; Europe 251–2; traffic calming 97–103; women 154
Transportation Research, A 104–10
Tucson, Arizona 305–6, Plate CS21, Plate CS22

Ultimate Resource (Simon) 43
United Kingdom 11, 58
United Nations 1, 237; Brundtland Commission 1, 9, 53–7, 277, 321; Commission on Sustainable Development 58; Conference on Environment and Development 1, 9, 53, 58–63, 186; Habitat II "City Summit" 9, 58, 63–5; Human Development Index 203, 206, 209; World Summit on Sustainable Development 58
United Nations Division for Sustainable Development 58, 235
United Nations Human Settlements Programme 235

United States: garden cities 11; New Urbanism 73–80; World Summit on Sustainable Development 58
Unwin, Raymond 11, 15
urban design 1–2; Calthorpe 73–80; Gehl 81–5
urban ecology 2; biodiversity and habitat conservation plans 116–19; city and nature 113–15
urban planning 3, 8, 69–70, 73, 201, 277, 278; ecological footprint analysis 211–19; exercises 322–4; Jacobs 31–4; observation 220–4; politics of meaning 225–31; sustainability indicators 203–10
urban sustainability 205
utopias *see* visions

Vale, Brenda 3, 181–2, 188–92
Vale, Robert 3, 181–2, 188–92
Vällingby 11
Van der Ryn, Sim 74, 181, 188, 322
Vancouver 309, Plate CS25, Plate CS26
Victoria Transportation Policy Institute 89, 104
Village Homes, Davis 304–5, Plate CS17, Plate CS18
visions 3–4, 277–8; Callenbach 282–7; exercises 320–1; Howard 279–81; Le Guin 288–91

Wachs, Martin 109
Wackernagel, Mathis 3, 201, 211–19
Wal Mart 181, 185–6
Warford, Jeremy J. 159
Warring, Madeline 211
waste 2; Curitiba 245, 247; Girardet 127–9; Lyle 133–40; McDonough 183, 184–5

water 126–7, 303, Plate CS9, Plate CS10
Wekerle, Gerda R. 150
Welwyn 11, 279
Whitzman, Carolyn 150
Whyte, William H. 30, 81, 84
Wiewel, Wim 173–4
wilderness 28–9
Woman on the Edge of Time (Piercy) 277
women 2, 150–6
Women and the American City (Stimpson) 150
Women, the Environment and Sustainable Development (Braidotti) 150
World3 42, 45–6
World Bank 47
World Business Council for Sustainable Development 235
World Commission on Environment and Development 1, 9, 53–7, 277, 321
World Conservation Strategy (World Conservation Union) 53
World Conservation Union 53, 321
World Resources Institute 203
World Summit on Sustainable Development (Johannesburg) 58
World Trade Organization 59
World Wide Web Library on Sustainable Development 235
World Wildlife Federation 211
World Without End (Pearce and Warford) 159
Worldwatch Institute 53

Yeang, Ken 299, Plate CS3, Plate CS4
Yost, Peter 3, 193–8